Springer Proceedings in Complexity

Springer Proceedings in Complexity publishes proceedings from scholarly meetings on all topics relating to the interdisciplinary studies of complex systems science. Springer welcomes book ideas from authors. The series is indexed in Scopus.

Proposals must include the following:

- name, place and date of the scientific meeting
- a link to the committees (local organization, international advisors etc.)
- scientific description of the meeting
- list of invited/plenary speakers
- an estimate of the planned proceedings book parameters (number of pages/articles, requested number of bulk copies, submission deadline)

Submit your proposals to: Hisako.Niko@springer.com

Anna Visvizi · Mara Grimaldi ·
Mady A. A. Mohamed
Editors

Research and Innovation Forum 2025 Vol2

Competitiveness, Growth and Well-Being Through Technology, Innovation and Education

Springer

Editors
Anna Visvizi
SGH Warsaw School of Economics
Warsaw, Poland

Effat University
Jeddah, Saudi Arabia

Mady A. A. Mohamed
Effat University
Jeddah, Saudi Arabia

Mara Grimaldi
University of Salerno
Fisciano, Italy

ISSN 2213-8684 ISSN 2213-8692 (electronic)
Springer Proceedings in Complexity
ISBN 978-3-032-09061-4 ISBN 978-3-032-09062-1 (eBook)
https://doi.org/10.1007/978-3-032-09062-1

This Springer imprint is published by the registered company Springer Nature Switzerland AG
The registered company address is: Gewerbestrasse 11, 6330 Cham, Switzerland

Organization

Rii25 Team and People

The Chairs

Anna Visvizi, Ph.D. (dr hab.), Professor, SGH Warsaw School of Economics, Poland & Effat University, Jeddah, Saudi Arabia
Orlando Troisi, Ph.D., Professor, University of Salerno, Italy
Enric Serradell-López, Ph.D., Professor, Universitat Oberta de Catalunya, Spain

The Steering Committee

Xavier Baraza-Sanchez, Ph.D., Professor, Dean of the Faculty of Business and Economics, Universitat Oberta de Catalunya, Spain
Vincenzo Corvello, Ph.D., Professor, University of Calabria, Italy
Mara Grimaldi, Ph.D., Research Fellow, University of Salerno, Fisciano, Italy
Giuseppe Ioppolo, Ph.D., Professor, University of Messina, Italy
Jolanta Klemens, M.D., PSYCHOMedical, Bielsko-Biała, Poland
Higinio Mora, Ph.D., Professor, University of Alicante, Spain
Maria Jesús Martínez-Argüelles, Ph.D., Vice-Rector of Governance and Academic Policy, Universitat Oberta de Catalunya, Spain
Francesco Polese, Ph.D., Professor, Director of SIMAS [Centre for Innovation Systems and Health Management], University of Salerno, Fisciano, Italy
Jordi Sanchez-Navarro, Ph.D., Professor, Dean of the Faculty of Information and Communication Sciences, Universitat Oberta de Catalunya, Spain
Orlando Troisi, Ph.D., Professor, University of Salerno, Fisciano, Italy
Anna Visvizi, Ph.D. (dr hab.), Professor, SGH Warsaw School of Economics, Warsaw, Poland
Yves Wautelet, Ph.D., Professor, KU Leuven, Brussels Campus, Belgium

The Academic Committee

Rii Forum Team

Preface

This volume is the outcome of the vibrant intellectual exchange that took place during the 7th Research and Innovation Forum (Rii Forum 2025), held in a hybrid format on April 24–25, 2025, in Barcelona, Spain. Convened under the overarching theme "Competitiveness, Growth and Well-Being Through Technology, Innovation and Education", the Rii Forum reaffirmed its standing as a high-impact platform fostering interdisciplinary collaboration among academia, industry, and civil society. Hosted in partnership with the Universitat Oberta de Catalunya (UOC), a global leader in digital education and applied research, also this edition of Rii Forum brought together scholars and practitioners committed to advancing rigorous, evidence-based dialogue on the most pressing technological, economic, and societal challenges of our time. The opening remarks of Prof. Àngels Fitó Bertran, President of the Universitat Oberta de Catalunya (UOC), confirmed the value of Rii Forum and the role and impact research excellence and collaboration have in advancing science, education, and innovation.

The Rii Forum 2025 Chairs

Rii Forum 2025 was co-chaired by Prof. Anna Visvizi (SGH Warsaw School of Economics, Poland), Prof. Orlando Troisi (University of Salerno, Italy), and Prof. Enric Serradell-López (Universitat Oberta de Catalunya, Spain), all experienced and well-respected researchers, experienced academics, and leaders. The program was organized around six core thematic tracks, alongside a dedicated Ph.D. workshop:

1. The individual, the state, and technology—exploring cities, urbanism, democracy, and resilience
2. Innovation, quality, and technology management exploring their local, regional, and global contexts
3. Education in the GenAI era addressing pedagogical, content-related, and managerial challenges

4. Design thinking and agile methodologies highlighting foresight, future studies, and strategic planning
5. Regulatory frameworks and policy-making dwelling on issues pertaining to competitiveness, growth, and well-being
6. Ph.D. Track focused on research impact, publication strategy, and supervisory collaboration, served also as an opportunity to empower early career researchers.

Key Events and Highlights

Keynote Address

The Rii Forum 2025 opened with a spot-on keynote by Prof. Johan Magnusson, Director of the Swedish Center for Digital Innovation (www.scdi.se) at the University of Gothenburg. Professor Magnusson's research covers issues pertaining to how governance needs to be designed to facilitate transformation. In his roles as researcher, lecturer, educator, and thought-leader, Prof. Magnusson is highly visible in the national, Swedish, and international debates concerning digital and green transformation. He is ranked as one of the most influential individuals in tech. His keynote address offered strategic insights into digital governance, public sector transformation, and the interplay between regulatory frameworks and innovation ecosystems.

Round Table Debate

The high-profile Round Table Debate, titled "*The Great (?) Leap (?): Tech, Talent, and Transformation in the 21st Century*", was moderated by Prof. Anna Visvizi and featured:

- Prof. Johan Magnusson (University of Gothenburg, Sweden)
- Prof. Enric Serradell-López (UOC, Spain)
- Prof. Orlando Troisi (University of Salerno, Italy)
- Prof. Igor Calzada (University of the Basque Country and Cardiff University)
- Prof. Giuseppe Ioppolo (University of Messina, Italy).

The debate addressed the strategic risks and opportunities posed by accelerated technological change, talent mobility, and geopolitical fragmentation.

Online Panel Debate

The Rii Forum online panel debate "*From Community Engagement to Labor Market Participation: Rethinking Talent in the Web 3.0 Era*", moderated by Prof. Radosław Malik (SGH Warsaw School of Economics), explored the issue of how digital inclusion, migration, and mobility are reshaping the very notion of "talent" today. Regional case studies from Greece, Brazil, and Lithuania provided practical, grounded perspectives. The speakers included:

- Anis Barnat (CAN Network and El Sistema Greece)
- Prof. Regina Negri-Pagani (Federal University of Technology Paraná, Brazil)
- Prof. Mangirdas Morkūnas (Vilnius University, Lithuania).

The Rii Forum Industry–Academia Dialogue

A highlight of the Forum, the Rii Forum Industry–Academia Dialogue explored how cross-sector collaboration can catalyze innovation and social impact. Moderated by Prof. Lloyd Waller (University of the West Indies, Jamaica), the session brought together voices from business, design, entrepreneurship, and academia:

- Petros Georgiakakis, MBA (Senior Clinical Operations Manager, ICON Pharma Development Solutions)
- Dr Francesc Font-Cot (Co-Founder and Co-CEO, Incapto Coffee)
- Dr Alessandro Manetti (Vice President Europe, LCI Education; Barcelona Design Week Curator)
- Prof. Katarzyna Szopik-Depczyńska (University of Szczecin, Poland).

The panel discussed how academia can meaningfully contribute to industry through talent development, strategic foresight, knowledge co-creation, and technology transfer, emphasizing the value of long-term, sustainable partnerships.

Academic Diplomacy

A special moment was marked by the remarks of Her Excellency Ilona Kałdońska, Consul General of the Republic of Poland in Barcelona, who emphasized the role of research and innovation in shaping EU-wide policy, particularly in the context of Poland's Presidency of the Council of the European Union (January–June 2025). The Rii Forum 2025 delegates had the opportunity to meet also His Excellency Łukasz Knurowski, Consul of the Republic of Poland in Barcelona. During the networking meeting that followed the Rii Forum Industry–Academia Dialogue, the scholars, academics, administrators and industry representatives had the opportunity to explore the prospects of collaboration. In this way, the Rii Forum 2025 aligned itself with goals and objectives of the Polish Presidency of the Council of the European Union.

Ph.D. Workshop

Convened by Dr Mara Grimaldi (University of Salerno), the Ph.D. Workshop focused on the thesis-by-compendium model, offering doctoral candidates actionable strategies for writing, publishing, engaging with supervisors, and achieving research impact.

Social Impact: The Power of Music, Inclusion, and Education

At the heart of the Rii Forum lies a deep commitment to social impact. This year, the Forum proudly reaffirmed its collaboration with El Sistema Greece, a pioneering initiative inspired by El Sistema Venezuela, by Anis Barnat. Since 2016, El Sistema Greece, led by Anis Barnat, experienced manager, leader, and art manager, has provided music education to over 3000 children and youth from more than 40 countries, including refugee and migrant communities. Through structured instrumental, choral, and orchestral training, El Sistema Greece promotes self-esteem, teamwork, emotional resilience, and cultural inclusion. Programs culminate in public performances, such as the Annual Christmas Gala Concert, bringing together families and communities through the unifying power of music. El Sistema Greece exemplifies how education, creativity, and compassion intersect to create tangible social change. The Rii Forum community is proud to support its mission.

Acknowledgments and Gratitude

We extend our deepest thanks to all authors and contributors to this volume. Your work continues to advance the conversation at the intersection of technology, innovation, education and their social impact, policy-making considerations including. We thank our peer reviewers for their diligence and commitment to academic excellence.

Special appreciation goes to the Rii Forum 2025 organizing team, including:

- Dr Jolanta Klemens, M.D., for her unconditional support, and for enabling and facilitating collaboration with Rii Forum strategic partners.
- Anna Moni, Krzysztof Kozłowski, and Bilal Afzal, for their leadership in managing the on-site component of the conference. A simple "thank you" is not enough to express the value of your input to Rii Forum 2025.

- The UOC local team, whose hospitality and professionalism made the event in Barcelona a truly memorable experience.

We would also like to thank Anna Guerrero i Olivella (UOC) who generously offered her time and expertise to enable the Rii Forum team to run the conference.

Strategic Partners

We are grateful to our strategic partners, who continue to support the Rii Forum's values of research excellence, innovation, and social impact:

- Universitat Oberta de Catalunya (UOC)—Host institution
- EkaMedica (Poland)—advocates for natural health and wellness
- Polskie Piekarnie (Poland)—traditional and innovative bakery products
- PSYCHOMedical (Poland)—holistic healthcare and mental wellness
- INTER-TECH (Poland)—advanced solutions for industrial automation and machine relocation
- Polmotors (Poland)—leader in lightweight, eco-friendly automotive components for global OEMs.

We also express our sincere gratitude to Springer Nature, our publishing partner, for their trust, professionalism, and continuous support.

The Rii Forum Community

The Research and Innovation Forum is more than just a conference. Over the years it has evolved into a community of researchers and thought leaders committed to Research excellence and ethical integrity; interdisciplinary, thought-provoking dialogue; constructive collaboration across academia, industry, and policy-making; supporting early-career scholars; promoting research that is rigorous, relevant, and socially responsible. By participating in the Rii Forum, scholars gain exposure to critical debates, receive constructive feedback, and connect with a global network of like-minded professionals. This book is a testament to that ethos, and an invitation to carry the conversation forward. We look forward to reading your work, collaborating throughout the year, and welcoming you to the 8th edition of the Research and Innovation Forum, scheduled to take place in Berlin in Spring 2026.

Saudi Arabia, Poland — Anna Visvizi
Fisciano, Italy — Mara Grimaldi
Jeddah, Saudi Arabia — Mady A. A. Mohamed

Contents

Regulatory Frameworks and Policy Making for Competitiveness, Growth and Well-Being

Education and Technology in the AI Era

Contributors

Martti Aho HAMK Häme University of Applied Science, Hämeenlinna, Finland

Néstor Aldana-Rodríguez Universidad Centroamericana José Simeón Cañas, La Libertad, El Salvador

Jaime Almeida-Cepeda Career of Social Work, Universidad Central del Ecuador, Quito, Ecuador

Domenica Barile Department of Engineering, LUM University Giuseppe Degennaro, Casamassima (Bari), Italy

Anass Bendarkawi Research Laboratory in Strategy and Organizational Management, National School of Business and Management, Hassan First University, Settat, Morocco

Alberto Bongiorno Department of Economics, University of Messina, Messina, Italy

Grazia Calabrò Università Degli Studi Di Messina, Messina, Italy

Giuseppe Caristi Università Degli Studi Di Messina, Messina, Italy

Vilma Çekani Epoka University, Tirana, Albania

Jawad Chahine College of Business, Australian University, Kuwait City, Kuwait

Valentina Cicalese Department of Management and Innovation Systems, University of Salerno, Fisciano, Italy

Maria Cristina Cinici Department of Economics, University of Messina, Messina, Italy

Nicola Cobelli University of Verona, Verona, Italy

Carmela Cornacchia CNR-IMAA (Italy), Tito, Italy

Carlotta D'Alessandro Department of Economics, University of Messina, Messina, Italy

Corrado De Cristofaro University of Salerno, Salerno, Italy

Roberto De Luca Department of Management and Innovation Systems, University of Salerno, Fisciano, Italy

Pasquale Del Vecchio Department of Engineering, LUM University Giuseppe Degennaro, Casamassima (Bari), Italy

Marcella De Martino CNR-IMAA (Italy), Tito, Italy

João Batista Furlan Duarte Graduate Program in Applied Informatics, University of Fortaleza (UNIFOR), Fortaleza, Brazil

Nicole Dujany University of Turin, Turin, Italy

Aldrin Espín-León Career of Social Work, Universidad Central del Ecuador, Quito, Ecuador;
Research Department, Instituto Tecnológico Superior Sudamericano Quito, Quito, Ecuador

Alberto Ferraris University of Turin, Turin, Italy;
University of Nicosia, Nicosia, Cyprus

Kata Fredheim Stockholm School of Economics in Riga, Riga, Latvia

Simone Gagliardi CNR-IMAA (Italy), Tito, Italy

Giuseppe Gargano CNR-IMAA (Italy), Tito, Italy

Giuliano Gaveglia University of Salerno, Fisciano, Salerno, Italy

Giuseppe Giordano University of Salerno (Italy), Fisciano (Salerno), Italy

Mara Grimaldi University of Salerno, Salerno, Italy

Telle Hailikari HAMK Häme University of Applied Science, Hämeenlinna, Finland

Noémi Hajdú University of Miskolc, Miskolc, Hungary

Ana Beatriz Hernández-Lara Department of Business Management, Universitat Rovira i Virgili, Reus, Spain

Laura-Maija Hero HAMK Häme University of Applied Science, Hämeenlinna, Finland

Asmaa Ibrahim Urban Design Lab, College of Architecture and Design, Effat University, Jeddah, Saudi Arabia

Giuseppe Ioppolo Università Degli Studi Di Messina, Messina, Italy

Anne-Mari Järvenpää Häme University of Applied Sciences (HAMK), Hämeenlinna, Finland

Carlos Juárez-Ramos Universidad Centroamericana José Simeón Cañas, La Libertad, El Salvador

Jari Jussila Häme University of Applied Sciences (HAMK), Hämeenlinna, Finland

Krzysztof Karwowski SGH Warsaw School of Economics, Warsaw, Poland

Ulla-Maija Knuutti HAMK Häme University of Applied Science, Hämeenlinna, Finland

Iivari Kunttu Häme University of Applied Sciences, Hämeenlinna, Finland

Leena Kunttu Finnish Environment Institute, Helsinki, Finland

Ana Beatriz Hernández Lara Department of Business Management, Universitat Rovira I Virgili, Reus, Spain

Chenxiang Long College of Biology and Engineering, Guizhou Medical University, Guiyang, China

Zenah Mahmoud Effat College of Architecture and Design, Effat University, Jeddah, Saudi Arabia

Gennaro Maione San Raffaele Roma University, Roma, Italy

Ornella Malandrino Department of Management and Innovation Systems, University of Salerno, Fisciano, Italy

Antonieta Martínez-Velasco Facultad de Ingeniería, Universidad Panamericana, Ciudad de México, Mexico

Mady A. A. Mohamed Architecture Department, Effat University, Jeddah, Saudi Arabia;
Architecture Department, Faculty of Engineering, Zagazig University, Zagazig, Egypt

Anna Moni UOC (Universitat Oberta de Catalunya) & Deree College–The American College of Greece, Athens, Greece

Ali Mohamad Mouazen Business Management Department, Lebanese International University, Beirut, Lebanon

Malandrino Ornella Department of Management and Innovation Systems, University of Salerno, Salerno, Italy

Kristiina Paju Forssa Vocational Institute, Forssa, Finland

Rosa Maria Petracca Altieri CNR-IMAA (Italy), Tito, Italy

Vincenzo Piluso Department of Management and Innovation Systems, University of Salerno, Fisciano, Italy

Plácido Rogério Pinheiro Graduate Program in Applied Informatics, University of Fortaleza (UNIFOR), Fortaleza, Brazil

Mauricio Pohl-Alfaro Universidad Centroamericana José Simeón Cañas, La Libertad, El Salvador

Herleson Paiva Pontes Federal Institute of Ceará (IFCE), Jaguaribe, Brazil

De Luca Roberto Department of Management and Innovation Systems, University of Salerno, Salerno, Italy

Lucie Rohlíková Faculty of Education, University of West Bohemia, Plzeň, Czech Republic

Sessa Maria Rosaria Department of Management and Innovation Systems, University of Salerno, Salerno, Italy

Giulia Saponaro ACTRIS-ERIC, Helsinki, Finland

Péter Sasvári University of Miskolc, Miskolc, Hungary; Ludovika University of Public Service, Budapest, Hungary

Giustina Secundo Department of Engineering, LUM University Giuseppe Degennaro, Casamassima (Bari), Italy

S. M. Siintoharju Häme University of Applied Sciences (HAMK), Hämeenlinna, Finland

Jana Šindlerová Faculty of Arts, Charles University, Prague, Czech Republic

Katarzyna Szopik-Depczyńska Department of Business Management, Institute of Management, University of Szczecin, Szczecin, Poland

Antonia Terán-Bustamante Facultad de Ciencias Económicas y Empresariales, Universidad Panamericana, Ciudad de México, México

Yao Tong Huaqiao University, Quanzhou, Fujian, China

Edison Torres-Parra Research Department, Instituto Tecnológico Superior Sudamericano Quito, Quito, Ecuador

Marta Maria Ugolini University of Verona, Verona, Italy

Cicalese Valentina Department of Management and Innovation Systems, University of Salerno, Salerno, Italy

Erick Varela-Guzmán Kriete Institute of Engineering and Science, La Libertad, El Salvador

Zane Varpina Stockholm School of Economics in Riga, Riga, Latvia

Jana Vejvodová Faculty of Education, University of West Bohemia, Plzeň, Czech Republic

Anna Visvizi SGH Warsaw School of Economics, Warsaw, Poland; Effat University, Jeddah, Saudi Arabia

Wei Wang Huaqiao University, Quanzhou, Fujian, China

Yenchun Jim Wu Graduate Institute of Global Business and Strategy, National Taiwan Normal University, Taipei, Taiwan

Simian Zhu College of Biology and Engineering, Guizhou Medical University, Guiyang, China

Artificial Intelligence (AI) and Its Application in Education

Education Management in the GenAI Era: Exploring the Impact on Course Design and Institutional Strategy

Nicola Cobelli, Mara Grimaldi, Marta Maria Ugolini, Gennaro Maione, and Corrado De Cristofaro

Abstract Over the past few years, Large Language Models (LLMs) have attracted increasing attention from the scientific community for their transformative potential in numerous sectors, including education (Tayan et al. in Mach. Learn. Appl., 2024 [1]). In the educational context, the adoption of generative artificial intelligence (GenAI) tools is experiencing rapid acceleration, fostered by their ease of use and increasing integration into learning processes, among other factors. A growing body of empirical evidence has highlighted the positive effects of using GenAI on key variables such as academic performance, perception of self-efficacy, and student satisfaction (Al-Abdullatif in Educ. Sci. 13:1151, 2023 [2]; Labadze et al. in Int. J. Educ. Technol. High. Educ. 20:56, 2023 [3]; Deng and Yu in Sustainability 15:2940, 2023 [4]). However, the existing literature has only recently begun to consider the analysis of the organizational and educational transformations that derive from adopting this technology. The aim of this work is to foster the understanding of the organizational changes induced using GenAI in the education sector. To explain organizational change, we propose using the theoretical framework Diffusion of Innovation Theory (DIT) (Rogers in Diffusion of Innovations, Free Press, Tampa, FL, 2003 [5]).

Keywords Education management · Artificial intelligence (AI) · Course design · SDG 4

N. Cobelli · M. M. Ugolini
University of Verona, Verona, Italy

M. Grimaldi (✉) · C. De Cristofaro
University of Salerno, Salerno, Italy
e-mail: margrimaldi@unisa.it

G. Maione
San Raffaele Roma University, Roma, Italy

A. Visvizi et al. (eds.), *Research and Innovation Forum 2025 Vol2*, Springer Proceedings in Complexity, https://doi.org/10.1007/978-3-032-09062-1_1

1 Introduction

In recent years, generative artificial intelligence (GenAI) tools have attracted significant attention from the scientific community for their transformative potential in numerous sectors, including education [1, 6–10]. The rapid acceleration in the adoption of GenAI tools in the educational context is due to the multiple benefits that derive from its use, both for educators who integrate it into teaching processes and for students who use GenAI to streamline their learning processes [11].

In particular, according to Pereira et al. [12] the integration of GenAI into teaching processes allows educators to increase the degree of personalization of student learning experiences, an element that positively affects academic performance. According to Al-Dokhny et al. [13], Kasneci et al. [14], Bewersdorff et al. [15], and Perkins [16], GenAI enables educators to generate creative and more impactful educational content for students. Deri et al. [17] highlight that the use of GenAI in the educational context fosters the formation of a more inclusive and engaging educational environment.

The analysis of the impact of GenAI tools on student learning outcomes is widely explored in the literature, several studies show that the use of GenAI positively affects academic performance, student self-efficacy, and overall satisfaction [2, 3]. According to Deng and Yu [4], using GenAI in education improves student engagement, stimulates critical thinking, and promotes creativity. Jia et al. [18] note that using GenAI promotes peer learning. Other scholars have shown that the use of GenAI positively affects the learning process of students by reducing learning anxiety [19–22].

Despite the attention paid to the topic under analysis, it should be noted that only recently has the existing literature begun to consider the analysis of the organizational and didactic transformations resulting from the use of GenAI. This study pays particular attention to the effects produced by the implementation of GenAI on the redesign of teaching and the forms of collaboration and coordination between teacher and student, exploring: (1) the impact of the adaptation between GenAI, students, and institutions on educational redesign; (2) the enabling factors and barriers to the effective integration of GenAI in university courses.

The results of this study allow us to build an explanatory model of the conditions for the successful integration of GenAI in the educational context and allow us to define institutional policies and educational design strategies in the context of GenAI. Therefore, these findings are important for educational organizations seeking to innovate instructional strategies and promote the formation of an engaging learning environment via the integration of GenAI into teaching processes.

The rest of the document is structured as follows. Section 2 provides the theoretical background of the Work. Section 3 provides a detail of the theoretical reference framework. Section 4 describes the research methodology. Section 5 shows research results and contains a discussion of them. Section 6 concludes the paper by highlighting some managerial implications and providing some ideas for future research.

2 The New Era of the Education Industry

In recent years, there has been a rapid evolution of technology and a strong expansion of its potential applications [23, 24]. The education sector has not remained indifferent to this evolution, and technology has increasingly become a fundamental input in the provision and growth of educational services [25]. In particular, the last major boost to the application of technologies to the education sector was induced by the Covid-19 pandemic that prompted governments and international organizations to require education technology (EdTech) solutions to minimize student learning interruptions.

The strong diffusion and application of EdTech solutions have led educators to question themselves on multiple aspects related to their integration into the educational context; in particular, the question of how to correctly use EdTech solutions, when to integrate them within the teaching process [26] and how to integrate them in such a way as to make the students' learning process more efficient and stimulate critical reasoning [27]. These concerns have been heightened following the dissemination and subsequent application of large language models (LLMs) to the education sector [28]. LLMs are advanced artificial intelligence (AI) tools designed in such a way that they can mimic human understanding and responses [29].

3 Rethinking Education in the Context of AI

Despite concerns about its use, the integration of LLMs into the education sector is seen in many cases as an opportunity to improve students' learning processes [11]. Many scholars have found that the use of GenAI positively affects students' academic performance, self-efficacy, and overall satisfaction [3, 13]. According to Kasneci et al. [14], AI tools can help students in language learning, writing, research activities, and the development of critical thinking; In addition, they note that AI tools can be used to facilitate the didactic learning of students with disabilities.

These benefits are connected to a series of opportunities that derive from the use of GenAI in teaching processes. The integration of GenAI into teaching practices allows educators to personalize learning experiences and adapt teaching content and strategies to students' needs [12, 30]. In this sense, the integration of AI into teaching practices can make lessons more effective by transforming traditional passive learning and promoting project- and activity-based learning that can stimulate problem-solving skills [31, 32], the teacher, therefore, assumes a role of guide and not of mere transmitter of knowledge [33].

Ilayaraja et al. [34] acknowledge that the benefits of applying GenAI to education do not necessarily derive from a pedagogical renewal. GenAI can also be integrated into teaching processes in order to automate activities such as the evaluation of papers in order to achieve greater productivity, efficiency, and accessibility for students and teachers [14, 35–38].

Regardless of the intensity of the pedagogical renewal related to the adoption of GenAI in education, an analysis of the limits and threats associated with its use is essential. Pavlik [39] recognizes that AI generated information could have a negative impact on students' critical thinking, and for this reason, he believes it is appropriate for GenAI to support learning without replacing the role of the teacher or other authoritative sources.

Redecker et al. [40] believe that another major threat to the proper use of GenAI for educational purposes is the absence of adequate knowledge of the use of such tools by educators. Cotton et al. [41] and Duc et al. [42] place an emphasis on academic integrity, students may be tempted to cheat if they do not fully understand the value of their education. Such evidence points to the need for a review of academic integrity policies so that the use of GenAI tools can be effectively addressed [43].

The many challenges and opportunities related to the integration of GenAI into the education sector mean that institutional policies become a fundamental guide for the integration of GenAI within the education sector [44–46]. Such policies foster the proper and widespread application of GenAI and the maintenance of academic integrity [47, 48].

It follows that in order to fully reap the benefits associated with the use of AI, there is a need to shift from traditional teaching methods to a more technology-oriented educational paradigm.

Given the ability of GenAI to radically change learning and teaching processes, educational redesign becomes a crucial challenge to be addressed for an effective integration of GenAI into teaching practices. However, only recently has the existing literature begun to consider the analysis of organizational and didactic transformations resulting from the adoption of this technology, little space has been dedicated to the analysis of didactic redesign due to the integration of GenAI into teaching practices and to the analysis of the barriers that hinder a profound revolution in pedagogical processes. This gap led to the formulation of the following two research questions:

RQ 1: How does the interaction between students, teachers, and GenAI tools redefine the didactic design of courses?

RQ 2: What are the organizational, cultural, and individual factors that facilitate or hinder the effective integration of GenAI in university courses?

4 Research Design

To achieve the research objective, a qualitative multiple-case study design was adopted. The multiple-case study allowed us to select multiple cases to investigate in depth the adoption of GenAI in university teaching and consequently allowed us to accurately explain the elements of distinction between the various educational contexts and their determinants [49, 50].

4.1 Framework Contextualization

The present study, based on the theoretical references above, proposes an integrated theoretical framework that combines the Diffusion of Innovation Theory (DIT) [5] with the Organizational Learning and Change framework [51, 52].

In particular, the DIT provides a solid framework for analyzing how innovations such as GenAI spread across social systems, including universities. This model will help explain why some teachers or departments become early adopters while others delay or reject adoption. Organizational Learning and Change allows us to detect how individuals adapt their mental models and routines in response to change. In particular, Argyris and Schön believe that two types of learning can be distinguished: single-cycle learning, which involves correcting actions without questioning the underlying objectives or assumptions, and double-cycle learning, which, in contrast, involves a critical re-evaluation of underlying beliefs and consequently leads to transformative changes in objectives [51], in roles and strategies. March's distinction between exploration (innovation, experimentation, risk-taking) and exploitation (efficiency, refinement, incremental improvement) further informs our framework [52].

4.2 Sampling, Data Collection, and Analysis

This work involved university professors from different disciplinary fields and institutions. In particular, convenience sampling allowed us to select professors already engaged in adopting GenAI in the definition of their data programs, easily available and interested in participating in this study.

The data were collected through semi-structured interviews involving 18 university professors not involved in management or strategic planning roles at the university, which allowed us to collect both reflections on the use of GenAI and identify educational redesign processes. The semi-structured interviews were associated with the documentary analysis of the teaching material and the evaluation criteria adopted, this allowed us to investigate the presence of any changes in the evaluation methods and in the definition of learning objectives. Finally, field observation allowed an analysis of the implications of the relationship between students and teachers. The data collected were analyzed through thematic analysis.

5 Results and Discussion

In examining the integration of (GenAI) within a series of university courses, researchers observed varied patterns of instructional modification that revealed an intricate interplay among individual predispositions, institutional guidelines, and cultural traditions shaping teaching activities and learning objectives. Early adopters, labeled as transformative pioneers, introduced these tools with an emphasis on far-reaching adjustments to pedagogical aims, conceptual frameworks, and evaluative procedures, prompting a substantial departure from conventional approaches to syllabus design [33, 53].

Researchers noticed that these educators encouraged a process of questioning entrenched assumptions, shifting from a mere augmentation of existing coursework to a deeper realignment of professional goals and student-centered learning objectives. One professor reported rethinking fundamental course objectives, highlighting how GenAI compels learners to focus on interpretive skills and the critical use of automated outputs, thereby reframing the educators' position from content transmitters to guide across diverse analytical modalities. This mirrors the findings of Zhai [32], who revealed that GenAI has led teachers to reorient pedagogical models to align them with reflective practices. Investigations also revealed that incremental adopters displayed a more measured outlook, choosing to incorporate the AI platform for isolated functions such as automated grading assistance, while largely preserving traditional objectives and assessment norms. This is in line with Qureshi [43] who observed similar trends in courses using GenAI primarily for operational efficiency without altering pedagogical frameworks.

Those instructors regarded the generative system as a resource for boosting existing practices rather than initiating major transformations, illustrating a tendency toward single-cycle learning focused on operational refinement rather than a significant reshaping of pedagogical principles [51]. An example was a course that integrated chat-based text generation primarily to streamline instructor workload, with no explicit adaptation of the overarching curriculum, thereby retaining conventional aims and grading procedures despite the presence of advanced automation. This evidence is in line with the studies of Ilayaraja et al. [34], Sergeeva et al. [35], and Bozkurt [36], which recognize the ability of AI to make educational processes more efficient. Investigators also identified a unique subset labeled marginal explorers, who applied generative solutions intermittently within classroom tasks or evaluations, yet did not position these attempts within a systematic strategy or strong institutional framework.

Questions of strategic direction and long-term planning remained unaddressed in that subset, as random experiments with AI-driven prompts reflected personal curiosity more than a collaborative push for pedagogical advancement or structured integration. One participant mentioned trialing automated question generation for a single workshop but hesitated to repeat it, given the limited institutional endorsement and the absence of peer consultation regarding didactic benefits or possible risks. This echoes the statements of Tsai and Gasevic [44], Tsai et al. [45], and Cheng and Yim [46], who recognize that the lack of policy clarity results in cautious and low-impact experimentation.

A final group, labeled reflective resisters, declined to incorporate any form of AI, not because of a deficit in technical skills or unwillingness to consider new options, but rather based on philosophical standpoints prioritizing human-centered cognitive endeavors. An interviewee highlighted that mechanical generation of content might reduce the integral uncertainty and nuanced reasoning required for true scholastic maturation, indicating a conscious objection to embedding automated systems within fundamental course structure or evaluative moments. Such a position is in tune with Pavlik [39] who conceptualizes non-adoption as a form of reflective action rather than as resistance rooted in inertia. Interpreting such deliberate non-usage as reflective agency allows observers to grasp the spectrum of responses in which advanced technology can either spark questioning of entrenched norms or prompt unchanging reliance on prior routines.

Collected data shows that different rationales, institutional frameworks, and personal stances shape whether generative artificial intelligence will be harnessed for incremental optimization, ambitious reconfiguration, scattered experimentation, or outright refusal, revealing a nuanced set of outcomes in educational adoption. Observational evidence suggests that diverse outcomes originate from a confluence of perceived benefits, context-specific deterrents, and professional beliefs about the importance of human insight, shaping each course's stance on how far digital methods should extend into the learning process [54–56]. Synthesis of these findings indicates that systematic transformation, incremental improvement, sporadic trial, and philosophical disavowal represent four distinct trajectories, each of which exemplifies a specific blend of acceptance, revision, or refusal, underscoring the multifaceted influence of generative artificial intelligence on academic environments.

Analysis of the cross-case data, grounded in the integrated diffusion model and learning theory, showcased how the depth of pedagogical revision often aligned with perceptions of advantage, trialability, and reduced intricacy associated with generative methods in higher education.

Transformative pioneers exhibited a high willingness to experiment and reorient lessons, influenced by the substantial potential they detected in tailoring tasks and evaluations that prompt analytical reflection rather than mechanical memorization [30, 32]. Such an approach triggered a shift that involved the entire conceptual structure of course materials, inviting faculty to craft learning experiences where the generative system becomes a channel for fostering critical appraisal and creative problem-solving skills.

Incremental adopters, on the other hand, generally perceived the novel technology as an opportunity to fine-tune ongoing routines without confronting institutional norms or refashioning the intellectual framework of their syllabi. Individuals in that cluster introduced generative aids as optional features integrated.

Marginal explorers illustrated partial introductions of the technology through episodic pilots, missing coordinated planning, or strong collegial discourse, which led to isolated trials that occasionally sparked curiosity but lacked consistent follow-through or institutional support. A few participants confessed that they initiated small-scale applications, such as auto-generated quizzes or lecture enhancements, though absent a deliberate structure for analyzing the outcomes or sharing findings with peers, the effect remained limited. Within that context, sporadic engagement with generative platforms seldom progressed into a more formal recalibration of course philosophies and instead demonstrated the inclination of certain educators to remain cautious until institutional frameworks became more transparent.

Reflective resisters epitomized a distinct attitude characterized by a deliberate rejection of automated mechanisms, hinging on deeply held convictions regarding the importance of authentic human cognition, interactive dialogue, and the intangible attributes of scholarly growth. Some faculty within that profile insisted that unmediated intellectual struggle and exploration of messy cognitive challenges are indispensable for genuine mastery, leading them to disallow generative interventions that might shortcut the reasoning process or replace rigorous human deliberation.

Their stance signified not an absence of technological competence, but a purposeful decision grounded in the conviction that unassisted human intricacy fosters imaginative leaps and deeper student ownership of scholarly progress.

The analysis of the collected data suggests that the adoption of GenAI in the education sector entails a significant strategic impact for academic institutions, requiring a revision of policies and assessment models, the enhancement of targeted investments in faculty professional development, and the establishment of regulatory frameworks capable of addressing emerging ethical challenges. Institutional strategies are therefore called to evolve toward a flexible governance model oriented toward informed experimentation.

Ultimately, these findings demonstrate that the presence of GenAI does not automatically yield profound realignments in course design, but rather that a confluence of personal motivations, institutional frameworks, and disciplinary norms drives each educator's stance, shaping whether they pursue radical reinvention, incremental updates, scattered experiments, or reflective refusal.

A noteworthy observation from this cross-case review is the decisive influence of collaborative learning, since educators who participated in practice-oriented gatherings often reported richer interpersonal exchanges, recurrent feedback loops, and deeper readiness to question entrenched assumptions, culminating in more nuanced assimilation of generative approaches within their curriculum planning.

6 Implications and Conclusion

Perspectives gleaned from these diverse course experiences illuminate how GenAI can either reinforce existing pedagogical norms or prompt new vantage points, depending on instructors' readiness for re-examination of practices and alignment with institutional priorities. Investigators recognized that such technology does not operate in isolation, since dynamic collaboration among faculty members and administrators can shape policies guiding usage protocols, resource allocation, and capacity-building efforts that affect whether deeper instructional shifts gain traction or remain tentative. Some programs demonstrated that, with appropriate institutional reinforcement and robust dialogue among peers, generative technologies might accelerate meaningful transformations in learning objectives, rubrics, and student engagement strategies, creating momentum for reevaluation of longstanding academic conventions.

Researchers also noted an emphasis on professional development as a mechanism to broaden instructors' familiarity with practical applications, while at the same time fostering reflection about whether traditional assessment structures remain suitable in contexts enriched by digital tools. Instructors who embraced ongoing exchange of experiences and collective troubleshooting found fresh avenues for personal growth, revealing how reflective dialogue can dismantle prescriptive routines and stimulate a more dynamic attitude toward redesign, even among educators initially hesitant to challenge their longstanding methods.

More cautious instructors, conversely, sought incremental benefits by implementing generative solutions exclusively for logistical tasks, a choice that underscores how personal convictions and perceived institutional signals intersect to determine whether these platforms shape the deeper core of learning trajectories. Implications for organizational leaders include recognizing that technology acceptance is highly contextual, driven by disciplinary traditions, existing pedagogical philosophies, and the capacity of educators to modify entrenched routines when presented with alternatives that promise pedagogical value and administrative feasibility. Observation of marginal exploration patterns suggested that sporadic testing of generative systems, unaccompanied by reflection or collegial discussion, tends to result in subdued pedagogical impact that fails to leverage cross-functional insights and collective reconsideration of norms. Programs might discover stronger outcomes if faculty cooperation is institutionalized through workshops and shared pilot initiatives, catalyzing alignment among didactic strategies, resource distribution, and explicit guidelines for measuring the educational outcomes of intelligent automation.

Future policy proposals could address the tension between top-down directives, which often mandate uniform standards for new technology usage, and the need to empower instructors to exercise judgment regarding curriculum adaptation and alignment with their disciplinary ethos. Investigators caution that uniform acceptance strategies overlook contextual differences, an approach that risks marginalizing certain departments or educators who might be unprepared for major pedagogical reconfigurations or possess reservations due to philosophical objections. Meanwhile, reflective resisters spotlight the pivotal question of whether the mechanization of textual creation diminishes the instructional emphasis on genuine cognitive inquiry, nudging administrators to consider how diverse philosophies of teaching can cohabit under one institutional framework.

Practical suggestions stemming from these findings encourage designing inclusive decision-making mechanisms that accommodate varied stances, including those who regard automation as an intellectual catalyst and those who preserve a more traditional teacher-centered model championing deep human-authored analysis.

Academic policymakers who seek to harness artificial intelligence for institutional progress may realize that distributing resources for faculty training, forming cross-departmental partnerships, and incentivizing critical reflection can increase the likelihood of more profound pedagogical change. Throughout these multiple-case observations, generative platforms materialized as catalysts that might prompt a redefinition of academic norms or fortify existing structures, contingent on how faculty members weigh new approaches against their convictions and navigate organizational parameters. Ultimately, the insights summarized here underline the need to conceive strategies that respect distinctive pedagogical identities, encourage reflective practice, and foster flexible yet principled uses of GenAI within the shared realm of university teaching.

Conclusion emerges from the assertion that generative technologies are neither a panacea for educational hurdles nor a trivial add-on, but instead constitute a living instrument that can align with or reorient academic routines based on institutional culture and individual dedication. Investigators conclude that a purely mechanistic implementation, limited to automating processes without rethinking program aims or teacher roles, can perpetuate superficial expansions of existing routines instead of contributing to a more far-reaching upgrade of educational endeavors. On the contrary, attempts to integrate AI into classroom activities with a reflective mindset reveal how even incremental modifications, guided by critical dialogue among peers, can spark new trajectories in the arrangement of content, tasks, and evaluation protocols. Concrete gains that emerged include a heightened sensitivity to the instructor's position as curator of technologically supported experiences, with a sharper inclination to question conventional rubrics and learning outcomes in ways that tie knowledge creation to the strategic use of intelligent automation.

Others who pursued partial engagement highlight the reality that sporadic application of generative platforms may yield fleeting pedagogical advantages yet tend to remain siloed within individual endeavors unless institutional frameworks encourage collaborative reflection or formal evaluation. Refusers, driven by a principled commitment to ensuring that students confront uncertainties and hone interpretive abilities without algorithmic shortcuts, underscore the value of preserving distinct teaching philosophies that place human creativity at the core of the educational process. Such an attitude reminds decision-makers that not every introduction of advanced tools harmonizes fluidly with the core tenets of certain disciplinary perspectives, especially those emphasizing the experiential dimension of grappling with complex problems.

Future agendas might revolve around designing frameworks that incorporate both the potential of digital methodologies and the wisdom of deeply humanistic approaches, fostering room for productive interplay while respecting critical boundaries that preserve the essence of personal inquiry. Policymakers seeking coherence among diverse pedagogical stances could promote constructive dialogue sessions, incentivize knowledge-sharing platforms, and designate adequate resources that enable instructors at varied stages of readiness to feel supported when testing or scaling machine-assisted coursework.

Research that extends beyond immediate classroom outcomes and investigates longer-term professional evolution can clarify whether teachers who engage purposefully with these technologies adapt their instructional philosophies over multiple semesters or revert to status quo methods after initial enthusiasm subsides. Those inquiries may also illuminate how organizational cultures embed expectations, incentives, and peer dynamics that either bolster or undermine the impetus to incorporate generative platforms in a manner that transcends streamlined operations, channeling them toward creative reinventions.

References

1. Tayan, O., Hassan, A., Khankan, K., Askool, S.: Considerations for adapting higher education technology courses for AI large language models: a critical review of the impact of ChatGPT. Mach. Learn. Appl. (2024). https://doi.org/10.1016/j.mlwa.2023.100513
2. Al-Abdullatif, A.M.: Modeling students' perceptions of chatbots in learning: integrating technology acceptance with the value-based adoption model. Educ. Sci. **13**(11), 1151 (2023). https://doi.org/10.3390/educsci13111151
3. Labadze, L., Grigolia, M., Machaidze, L.: Role of AI chatbots in education: systematic literature review. Int. J. Educ. Technol. High. Educ. **20**(1), 56 (2023). https://doi.org/10.1186/s41239-023-00426-1
4. Deng, X., Yu, Z.: A meta-analysis and systematic review of the effect of chatbot technology use in sustainable education. Sustainability **15**(4), 2940 (2023). https://doi.org/10.3390/su15042940
5. Rogers, E.M.: Diffusion of Innovations, 5th edn. Free Press, Tampa, FL (2003)
6. Troisi, O., Grimaldi, M., Monda, A.: Managing smart service ecosystems through technology: how ICTs enable value cocreation. Tour. Anal. **24**(3), 377–393 (2019). https://doi.org/10.3727/108354219X15511865533103
7. Troisi, O., Grimaldi, M., Loia, F., Maione, G.: Big data and sentiment analysis to highlight decision behaviours: a case study for student population. Behav. Inf. Technol. **37**(10–11), 1111–1128 (2018). https://doi.org/10.1080/0144929X.2018.1502355
8. Maione, G., Leoni, G.: Artificial intelligence and the public sector: the case of accounting. In: Artificial Intelligence and Its Contexts: Security, Business and Governance, pp. 131–143. Springer, Cham (2021). https://doi.org/10.1007/978-3-030-88972-2_9
9. Maione, G.: Convergence of artificial intelligence and sustainable innovation reporting. In: Sustainable Innovation Reporting and Emerging Technologies, pp. 29–48. Emerald Publishing, Bingley (2024). https://doi.org/10.1108/978-1-83797-739-020241003
10. Visvizi, A., Troisi, O., Grimaldi, M.: Exploring the drivers for digital transformation in smart education: An ecosystems approach. ITM Web Conf. **51**, 01003 (2023). https://doi.org/10.1051/itmconf/20235101003
11. Huang, H., Zheng, O., Wang, D., Yin, J., Wang, Z., Ding, S., Yin, H., Xu, C., Yang, R., Zheng, Q., et al.: ChatGPT to shape the future of dentistry: the potential of the large multimodal model. Int. J. Sci. Oral. **15**, 29 (2023). https://doi.org/10.1038/s41368-023-00239-y
12. Pereira, E., Nsair, S., Pereira, L.R., Grant, K.: Constructive alignment in a graduate-level project management course: an innovative framework using large language models. Int. J. Educ. Technol. High. Educ. **21**(1), 25 (2024). https://doi.org/10.1186/s41239-024-00457-2
13. Al-Dokhny, A., Alismaiel, O., Youssif, S., Nasr, N., Drwish, A., Samir, A.: Can multimodal large language models enhance performance benefits among higher education students? An investigation based on the task-technology fit theory and the artificial intelligence device use acceptance model. Sustainability **16**(23), 10780 (2024). https://doi.org/10.3390/su162310780
14. Kasneci, E., Seßler, K., Küchemann, S., Bannert, M., Dementieva, D., Fischer, F., Kasneci, G.: ChatGPT for good? On the opportunities and challenges of the great linguistic models for education. Learn. Individ. Differ. **103**, 102274 (2023). https://doi.org/10.1016/j.lindif.2023.102274
15. Bewersdorff, A., Hartmann, C., Hornberger, M., Seßler, K., Bannert, M., Kasneci, E., Kasneci, G., Zhai, X., Nerdel, C.: Taking the next step with generative artificial intelligence: the transformative role of large multimodal language models in science education. arXiv preprint. arXiv:2401.00832 (2024). https://doi.org/10.1016/j.lindif.2024.102601

16. Perkins, M.: Considerations on the academic integrity of large AI language models in the post-pandemic era: ChatGPT and beyond. J. Univ. Insegna. Impara. Pratica. **20** (2023)
17. Deri, M.N., Singh, A., Zaazie, P., Anandene, D.: Leveraging artificial intelligence in higher educational institutions: a comprehensive overview. Rev. Educ. Derecho **30** (2024). https://doi.org/10.4314/ngjsd.v16i1.12
18. Jia, Q., Cui, J., Xiao, Y., Liu, C., Rashid, P., Gehringer, E.F.: ALL-IN-ONE: multi-task learning BERT models for evaluating peer assessments. arXiv preprint. arXiv:2110.03895 (2021). https://doi.org/10.48550/arXiv.2110.03895
19. Ji, H., Han, I., Ko, Y.: A systematic review of conversational AI in language education: focusing on the collaboration with human teachers. J. Res. Technol. Educ. **55**(1), 48–63 (2022). https://doi.org/10.1080/15391523.2022.2142873
20. El Shazly, R.: Effects of artificial intelligence on English speaking anxiety and speaking performance: a case study. Expert. Syst. **38**(3), e12667 (2021). https://doi.org/10.1111/exsy.12667
21. Jeon, J.: Chatbot-assisted dynamic assessment (ca-da) for L2 vocabulary learning and diagnosis. Comput. Assist. Lang. Learn. **36**(7), 1338–1364 (2021). https://doi.org/10.1080/09588221.2021.1987272
22. Lin, C.-J., Mubarok, H.: Learning analytics for investigating the mind map-guided AI Chatbot approach in an EFL flipped speaking classroom. Educ. Technol. Soc. **24**(4), 16–35 (2021). https://www.jstor.org/stable/48629242
23. Troisi, O., Maione, G., Grimaldi, M., Loia, F.: Growth hacking: insights on data-driven decision-making from three firms. Ind. Mark. Manag. **90**, 538–557 (2020). https://doi.org/10.1016/j.indmarman.2019.08.005
24. Visvizi, A., Troisi, O., Grimaldi, M., Loia, F.: Think human, act digital: activating data-driven orientation in innovative start-ups. Eur. J. Innov. Manag. **25**(6), 452–478 (2022). https://doi.org/10.1108/EJIM-04-2021-0206
25. Rodriguez-Segura, D.: EdTech in developing countries: a review of the evidence. World Bank Res. Obs. **37**(2), 171–203 (2022). https://doi.org/10.1093/wbro/lkab011
26. Caukin, N., (20189): Edtech—where do I start? Int. J. Whole Child **3**(2), 36–39 (2018)
27. Thomas, C.: Digital distractions in the classroom: 4 ways instructional technology can help. Emerging EdTech (2019). https://emergingedtech.com/.
28. Concannon, F., Costello, E., Farrell, O., Farrelly, T., Wolf, L.G.: There's an AI for that: rhetoric, reality, and reflections on EdTech in the dawn of GenAI. Ir. J. Technol. Enhanc. Learn. **7**(1) (2023). https://doi.org/10.22554/ijtel.v7i1.116
29. Hadi, M.U., Al Tashi, Q., Qureshi, R., Shah, A., Muneer, A., Irfan, M., Mirjalili, S.: Great language models: a comprehensive investigation of its applications, challenges, limitations and future prospects (2023) https://doi.org/10.36227/techrxiv.23589741.v4
30. Meli, K., Taouki, J., Pantazatos, D.: Empowering educators with generative AI: the GenAI education frontier initiative. In: EDULEARN24 Proceedings, pp. 4289–4299. IATED, Valencia (2024). https://doi.org/10.21125/edulearn.2024.1077
31. Zhai, X.: Chatgpt and AI: the game changer for education (2023)
32. Zhai, X.: ChatGPT: reforming education on five aspects. Shanghai Educ. 16–17 (2023)
33. Mollick, E.R., Mollick, L.: Using AI to implement effective teaching strategies in classrooms: five strategies, including prompts. The Wharton School Research Paper (2023). https://doi.org/10.2139/ssrn.4391243
34. Ilayaraja, L., Yukthika, S.S., Prabakaran, S., Pavithra, S.: Integrating artificial intelligence and educational technology for inclusive and ethical education practices. In: 2025 3rd International Conference on Intelligent Data Communication Technologies and Internet of Things (IDCIoT), pp. 1152–1157. IEEE, New York (2025)
35. Sergeeva, O.V., Zheltukhina, M.R., Shoustikova, T., Tukhvatullina, L.R., Dobrokhotov, D.A., Kondrashev, S.V.: Understanding higher education students' adoption of generative AI technologies: An empirical investigation using UTAUT2. Contemp. Educ. Technol. **17**(2), ep571 (2025). https://doi.org/10.30935/cedtech/16039

36. Bozkurt, A. (ed.): Unleashing the potential of generative AI, conversational agents and chatbots in educational praxis: a systematic review and bibliometric analysis of GenAI in education. Open Praxis **15**(4), 261–270 (2023). https://doi.org/10.55982/openpraxis.15.4.609
37. Chan, C.K.Y., Hu, W.: Students' voices on generative AI: perceptions, benefits, and challenges in higher education. Int. J. Educ. Technol. High. Educ. **20**(1), 43(2023). https://doi.org/10.1186/s41239-023-00411-8
38. Michel-Villarreal, R., Vilalta-Perdomo, E., Salinas-Navarro, D.E., Thierry-Aguilera, R., Gerardou, F.S.: Challenges and opportunities of generative AI for higher education as explained by ChatGPT. Educ. Sci. **13**(9), 856 (2023). https://doi.org/10.3390/educsci13090856
39. Pavlik, J.V.: Collaborating with ChatGPT: considering the implications of generative artificial intelligence for journalism and media education. Journalism Mass Commun. Educ. **78**(1), 84–93 (2023). https://doi.org/10.1177/10776958221149577
40. Redecker, C., Punie, Y.: Digital competence of educators—DigCompEdu (2017)
41. Cotton, D.R., Cotton, P.A., Shipway, J.R.: Chatting and cheating: ensuring academic integrity in the era of ChatGPT. Innov. Educ. Teach. Int. **61**(2), 228–239 (2024). https://doi.org/10.1080/14703297.2023.2190148
42. Duc, A.N., Lønnestad, T., Sundbø, I., Johannessen, M.R., Gabriela, V., Ahmed, S.U., El-Gazzar, R.: Generative AI in undergraduate information technology education—insights from nine courses. arXiv preprint. arXiv:2311.10199 (2023). https://doi.org/10.48550/arXiv.2311.10199
43. Qureshi, B.: Exploring the use of ChatGPT as a tool for learning and assessment in undergraduate computer science curriculum: opportunities and challenges. arXiv preprint. arXiv:2304.11214 (2023). https://doi.org/10.1145/3613944.3613946
44. Tsai, Y.S., Gasevic, D.: Learning analytics in higher education—challenges and policies: a review of eight learning analytics policies. In: Proceedings of the Seventh International Learning Analytics & Knowledge Conference, pp. 233–242 (2017). https://doi.org/10.1145/3027385.3027400
45. Tsai, Y.S., Moreno-Marcos, P.M., Tammets, K., Kollom, K., Gašević, D.: SHEILA policy framework: informing institutional strategies and policy processes of learning analytics. In: Proceedings of the 8th International Conference on Learning Analytics and Knowledge, pp. 320–329 (2018). https://doi.org/10.1145/3170358.3170367
46. Cheng, M.W., Yim, I.H.: Examining the use of ChatGPT in public universities in Hong Kong: a case study of restricted access areas. Discov. Educ. **3**(1), 1 (2024). https://doi.org/10.1007/s44217-023-00081-8
47. Xiao, P., Chen, Y., Bao, W.: Waiting, banning, and embracing: an empirical analysis of adapting policies for generative AI in higher education. arXiv preprint (2023). arXiv:2305.18617, https://doi.org/10.48550/arXiv.2305.18617
48. Plata, S., De Guzman, M.A., Quesada, A.: Emerging research and policy themes on academic integrity in the age of ChatGPT and generative AI. Asian J. Univ. Educ. **19**(4), 743–758 (2023). https://doi.org/10.24191/ajue.v19i4.24697
49. Creswell, J.W., Poth, C.N: Qualitative Inquiry and Research Design: Choosing Among Five Approaches. Sage Publications, Thousand Oaks (2016)
50. Yin, R.K.: Case Study Research and Applications: Design and Methods. Sage Publications, Thousand Oaks (2017)
51. Argyris, C., Schön, D.A.: Organizational Learning: A Theory of Action Perspective. Reis (77/78), pp. 345–348 (1997). https://doi.org/10.2307/40183951
52. March, J.G.: Exploration and exploitation in organizational learning. Organ. Sci. **2**(1), 71–87 (1991). https://doi.org/10.1287/orsc.2.1.71
53. Zhai, X.: ChatGPT for next generation science learning. XRDS: Crossroads ACM Mag. Students **29**(3), 42–46 (2023). https://doi.org/10.1145/3589649
54. Chan, C.K.Y.: A comprehensive educational framework for university learning and learning. Int. J. Educ. Technol. High. Educ. **20**(1), 38 (2023). https://doi.org/10.1186/s41239-023-00408-3

55. Luckin, R., Cukurova, M., Kent, C., du Boulay, B.: Permettere agli educatori di essere pronti per l'IA. Comput. Educ.: Artif. Intell. **3**, 100076 (2022). https://doi.org/10.1016/j.caeai.2022.100076
56. George, B., Wooden, O.: Managing the strategic transformation of higher education through artificial intelligence. Adm. Sci. **13**(9), 196 (2023). https://doi.org/10.3390/admsci13090196

Re-designing Higher Education in the Age of GenAI: A Case Study on the Transition to Design-Based Education

Laura-Maija Hero, Ulla-Maija Knuutti, Jari Jussila, Martti Aho, and Telle Hailikari

Abstract Universities of applied science are re-designing their pedagogies in line with generative AI. Individual innovation competence is a core learning target because it relates to creativity, social skills, future orientation, development project management skills, and more items that are not easily replaced with AI. Design-based pedagogies are applied in student-led research, development and innovation (RDI) learning in university–industry collaborations. Design thinking can significantly improve innovation outcomes. However, it is not yet known what hinders and enables a shift to design-based pedagogies in higher education institutions (HEI). This paper reports findings from a case study of one Finnish university of applied science aiming to transform the entire university for design-based education. The research question was: Based on current curricula and teachers' and students' experiences, does a total shift to design-based education (DBE) seem possible in HEI? What are the enabling and hindering factors? Current curricula were analyzed, semi-structured interviews with teachers of design thinking pilot projects were conducted, and students were surveyed. The findings indicate that most curricula need to be rewritten. According to the teachers, students' proactivity, the design thinking process model, and well-functioning communication with industry partners are the strongest DBE enablers. Problems in planning and the industry co-operation workload are among the biggest obstacles. Based on the student material, the strongest enabling factor is organizing in multidisciplinary teams. The most common obstacle is an unclear task. Improvements in teaching, teacher training and pedagogical management of DBE are suggested. This paper's originality is in highlighting an opportunity for applied science universities to promote student-led RDI for innovations by unfolding the opportunities for design thinking in this time of GenAI.

Keywords Artificial intelligence · Design-based education · Design thinking · University of applied science · Individual innovation competence · SDG 9 · Industry · Innovation · Infrastructure

---

L.-M. Hero (✉) · U.-M. Knuutti · J. Jussila · M. Aho · T. Hailikari
HAMK Häme University of Applied Science, Hämeenlinna, Finland
e-mail: laura-maija.hero@hamk.fi

A. Visvizi et al. (eds.), *Research and Innovation Forum 2025 Vol2*, Springer Proceedings in Complexity, https://doi.org/10.1007/978-3-032-09062-1_2

1 Introduction: The Effects of AI on Higher Education Institutions

Universities of applied sciences (UAS) are reinventing themselves in this time of generative AI (GenAI) [1, 2]. Generative AI (GenAI) tools, such as ChatGPT and Copilot, are increasingly being used in higher education [3, 4]. The use of GenAI tools in higher education has disrupted and transformed various educational practices in learning, teaching, and assessment [3, 5, 6]. With the rapid growth of GenAI, higher education institutions (HEIs) are obliged to use more developed pedagogies than reading and essay writing. Collecting information, memorizing written texts, participating in exams, and writing essays need to be replaced with more complex real-life problem-solving and collaborative pedagogies that use AI as a tool that does not replace the learner. Targeted competencies, such as individual innovation competence [7], have become a core learning target because they relate to creativity, social skills, future orientation, development project management skills, and making, productization, and innovation implementation skills [7–18] that are not so easy to replace with AI. These competencies are learned and acquired in authentic, design thinking supported innovation projects with work organizations and with teachers supporting the work [19, 20].

This paper presents findings from a case study of a Finnish university of applied science that aims to transform the entire university to an innovation pedagogy called design-based education [DBE; 21, 22, 23, 24, 25]. Häme University of Applied Science (HAMK) has started implementing DBE as a pedagogical strategy [26]. In HAMK, the DBE strategy has been collaboratively developed. Today, DBE is defined based on eight dimensions that describe the way of learning (Fig. 1).

Based on these eight dimensions, it is now essential to put DBE pedagogy to the test by piloting it. There may also be negative effects on students, teachers may find it useless, or it may be too difficult to implement in curricula [cf. 24, 25]. Thus, the main research question was: Based on current curricula, and teachers' and students' experiences, does a total shift to DBE seem possible? The sub-questions were as follows: Do the current curricula support the shift to DBE? Based on the pilot DBE projects, what hinders and enables the total shift to DBE? These questions must be answered to be able to plan pedagogical management actions in practice. To this end, the argument in this paper is structured as follows. First, we argue that, according to recent literature, there are positive and negative effects of generative AI, and aim to shift the pedagogies towards more authentic and design-based ways of learning. Third, we present the qualitative methods we have used and the findings from three sub-studies. Based on the findings, we discuss the enabling and hindering factors of shifting to DBE. We conclude by presenting the preconditions of a total shift to DBE, and finish with limitations and future research proposals.

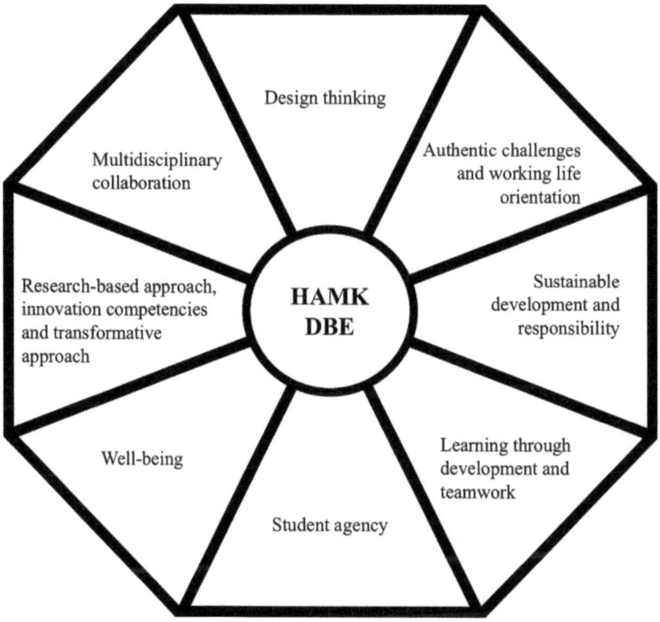

Fig. 1 HAMK design-based education (DBE) dimensions. *Source and permission*: HAMK Häme University of Applied Science, edit: Authors [26]

2 Generative AI and Design-Based Education

Generative AI comprises technologies that leverage natural language processing and deep learning models to generate human-like content e.g., words, images, audio, and video; and machine-readable content e.g., code and machine-readable data; in response to complex and varied prompts e.g., languages, instructions, and questions [3, 28, 29].

ChatGPT interventions are predominantly implemented in university teaching and learning and integrated into classroom environments as part of regular educational practices [27]. Recent research has highlighted the diverse capabilities of GenAI, such as analyzing student performance and providing suggestions for improvement [6], delivering detailed feedback [30], reflective writing of assignments [31], generating code [32], grading code [6], improving multimedia learning experiences [33], generating adaptive educational content, and offering tailored learning paths based on students' needs and preferences [3]. ChatGPT seems to improve academic performance, affective-motivational states, and higher-order thinking propensities. However, it reduces mental effort and has no significant effect on self-efficacy [27].

The use of GenAI has led to challenges of plagiarism and students completing various assignments either completely or partially using GenAI tools, raising concerns regarding critical thinking and learning outcomes [2]. Hence, GenAI is increasingly impacting the design of learning goals, teaching and learning methods, and assessment in higher education. Shifting from well-defined problems in post-intervention assessments to more complex, project-based assessments that require skill demonstration, adopting proctored assessments, or incorporating metrics such as originality alongside quality is suggested [25]. There is a need to redefine pedagogies, emphasizing the importance of interpersonal competences to tackle the new employability challenges [16]. However, it is not yet known what kind of pedagogies support these learning goals to tackle the challenges raised by GenAI.

Applying design thinking in university–industry innovation collaborations in the applied science university context extends learning outside the classroom—beyond learning from predefined learning materials and closed learning management systems to interaction with industry partners and their stakeholders. This activity has the potential to significantly improve innovation outcomes [34, 35]. Design-based education (DBE) is described as co-creative interaction between teachers, students, and industry [36]. UAS–industry innovation collaboration involves mutually beneficial knowledge and technology exchange aimed at spurring innovation [37, 38]. The creativity and innovation literature highlights that useful ideas often arise from the combination of very different viewpoints and the integration of knowledge among diverse actors [39].

Design-based pedagogies are applied in authentic, open challenge-based, and student-led RDI learning in UAS–industry collaborations [19, 25, 37, 40–42]. Students work in teams and facilitate their own co-creative RDI processes. DBE builds on the educational concepts of problem-based learning and competence-based education, but it places greater importance on the role of working life in the learning environment [37]. DBE as an educational concept focuses on solving authentic and complex problems via design processes [23], which empowers the learning processes of students, teachers, researchers, and the professional field [25].

Although the use of GenAI in preparing student homework is a threat to learning, it has essential advantages for DBE. These include accessing information and promoting creativity in the learning–teaching process by providing tools for various student co-creation activities [43]. DBE projects aim to develop concrete solutions to genuine challenges that a company itself would be unable to develop without innovative young people. Companies and other organizations can utilize the expertise of students from educational institutions in various projects that serve industry. For students, DBE enhances authenticity by providing access to real-life challenges, perspectives of professionals, materials, and tools that are not available in the university environment [44]. From the perspective of working life and organizations, the basic prerequisite for the development of a company and organization is that the actors strive for continuous development. The activities of organizations and enterprises include a built-in competitive mechanism that forces them to improve their activities. In DBE implementation, the actors form a useful network for the company, a development-oriented ecosystem where new solutions can be created. Regional student-centered co-creation aims to encourage students to network with working life and other fields of education and educational institutions during their studies [21–25, 37]. To conclude, while there is a wide body of literature on DBE, there is a research gap on factors that hinder and enable a shift to design-based pedagogies in HEI.

3 Methods and Materials

This research is a case study of one university of applied sciences in Finland. If a HEI plans to shift to DBE pedagogy, it is crucial to (1) analyze the state of the current curricula and (2) conduct a pilot study of several implementations, and (3) collect students' and teachers' experiences to avoid negative effects. Such studies help in understanding how an HEI can tackle the learning challenges raised by GenAI and transform to produce innovation competences in addition to unidisciplinary content knowledge.

3.1 Analysis of the Current Curricula

First, an analysis of the 2023–2024 curricula was conducted via theory-driven content analysis. The material consisted of 28 bachelor's degree curricula. One curriculum was typically a 10–50 page pdf document containing the description, content, aims, and assessment of a degree of study in HAMK. The curricula data were analyzed using a theory-driven thematic analysis [45]. As HAMK aims to implement DBE according to its eight dimensions (Fig. 1), the dimensions were used as an analytical framework to thematize the curricula. The analysis was conducted by two authors who first read the curricula twice and then by one author who deductively thematized the content piece by piece according to the themes named in the HAMK DBE model (Fig. 1). The Atlas.ti program was used to encode the reference units and track the stages of the analysis to ensure the rigor of the process. One reference unit was a course or module description (i.e., a group of sentences discussing the same content). In the second phase, the first author questioned the preliminary thematization of approximately 15% of the material, and the coding decisions were made in discussions with the third author [46]. The purpose of such negotiation in the coding phase was to guard against subjectivity bias and to avoid making premature conclusions.

3.2 Enabling and Hindering Factors: Teacher Interviews and Student Survey After DBE Pilots

After the curricular analysis and before the changes in the curricula and structures, eight pilot DBE implementations were conducted during spring 2024. The implementations were either multidisciplinary or involved students from one field of study. In total, 548 students and 39 teachers participated in the pilots (Table 1).

Semi-structured interviews were conducted with the teachers of the design thinking pilot projects (N = 11). The eighth dimension, "well-being," was left out from the teacher interviews because this concerned the students.

Table 1 Pilot implementations run with design-based pedagogy

Degree program	Students	Study year	Teachers	Partner companies and organizations	Open challenge	Multidiciplinary
International business	80	2	3	11 companies from Häme area	Several related to marketing	No
Sustainability, information and communication technology, bioeconomy	100	2	8	Finnish Environment Institute and Katumajärvi Conservation Association	Develop a product for lake water monitoring	Yes
Construction engineering	41	4	3	Flagmore Ltd.	How to renovate the area ? How to manage procurement?	No
Constructed environment, construction engineering, bioeconomy (masters)	62	3	8	SRV Group Plc	How to support biodiversity and sustainability through deck yards, green roofs and roof terraces?	Yes
Sustainable forest management	12	1	2	The Regional Council of Häme	How Kanta-Häme forestry expertise can be highlighted and promoted?	No

(continued)

Table 1 (continued)

Degree program	Students	Study year	Teachers	Partner companies and organizations	Open challenge	Multidiciplinary
Information and Communication Technology, Machine technology, Entrepreneurship and Business	53	2	8	Konesilta Ltd.	What can be achieved by developing business and increasing remote monitoring of construction machinery when a change in the operating environment leads to a shift from sales-oriented business to leasing and rental-oriented business?	Yes
Information and communication technology, ICT bio economy	140	1	3	Vauser Ltd., Envor Group Ltd., NOAC Ltd., Forssa City	Different challenges related to company datasets based on company strategy	Yes
Design	60	1	2	Forssa museum, HAMK Smart research unit	Develop new services for the museum. Develop new smart clothing	No (different design fields)
Total	548		37	23		4

Table Authors

3.3 Student Open Question Survey

Third, we conducted an open-question survey to find out what hinders and enables a total shift to DBE based on the pilot DBE projects. The survey was sent to students participating in the DBE implementation pilots. Altogether, 66 students from five different pilots answered the survey. The students were asked to think about their studying and learning during the design-based project course and what challenged and promoted their well-being during the project.

3.4 Analyses of the Teacher and Student Materials

We applied a deductive approach and conducted a theory-driven analysis of the teacher and student materials [47]. In this analysis, the identified themes are based on the HAMK DBE preliminary theoretical framework and the analytical interest in focusing on the DBE dimensions defined jointly in HAMK (Fig. 1). First, the fourth author thematized the teacher interview material, and the first author questioned the thematization. One coding unit was a sentence, a part of it, or a longer text, such as a chapter. The second author analyzed the enabling and hindering factors, and the first author questioned the thematization in several negotiation rounds. Second, the student material was collected by the fifth author and thematized by the second author, and the enabling and hindering factors were analyzed by the first author and discussed with the second author. This collaborative way of working allowed for many discussions and rounds, making the analysis more reliable [48].

4 Findings

In this section, we summarize the results from the qualitative analysis of the curricula and the teacher and student materials, focusing on the following research questions: Do the current curricula support the shift to DBE? Based on the pilot DBE projects, what hinders and enables the total shift to DBE?

4.1 Analysis of Current Curricula

We analyzed the frequencies of references to the HAMK DBE dimensions (Fig. 1) in HAMK curricula (N = 28 UAS curricula) in the 2023–2024 academic year. Sustainable development and responsibility, learning through development and teamwork, a research-based approach, and an innovation competence and transformative approach were well mentioned in the curricula, whereas student agency and well-being were very seldom mentioned (Table 2).

We found a large difference between the degree programs' current curricula. Some degrees barely mentioned the DBE dimensions (Table 3). The construction engineering, bio, and traffic engineering programs did not include much direct DBE-type learning in their curricula.

International business, design, sustainable development, and entrepreneurial nursing-related programs directed DBE-type learning the most (Table 4).

The frequencies of references in the curricula varied considerably. It was surprising that multidisciplinary collaboration was mentioned much more in entrepreneurial nursing than in any other curriculum. Notably, all five health care degrees had more emphasis on multidisciplinarity than the other degrees.

Table 2 Frequencies of references to the HAMK DBE dimensions in HAMK 2023–2024 curricula (N = 28 UAS curricula)

DBE dimension	Reference f = x
Sustainable development and responsibility	381
Learning through development and teamwork	299
Research-based approach, innovation competence and transformative approach	247
Authentic challenges and working life orientation	192
Design thinking	185
Multidisciplinary collaboration	110
Student agency	88
Well-being skills	36
Total	1538

Table Authors

Table 3 Lowest frequencies of references to HAMK DBE dimensions in HAMK 2023–2024 curricula

DBE dimension	Construction engineering FI	Bio and food technology FI	Construction and civil engineering FI	ICT FI	Traffic FI
Sustainable development and responsibility	4	4	6	1	6
Learning through development and teamwork	5	3	3	6	7
Research-based approach, innovation competence and transformative approach	2	4	3	4	5
Authentic challenges and working life orientation	1	1	4	4	5
Design thinking	2	2	2	11	6
Multidisciplinary collaboration	0	2	1	0	1
Student agency	1	0	0	0	1
Well-being skills	0	0	0	0	0
Total	15	16	19	26	31

Table Authors

Table 4 Highest frequencies of references to HAMK DBE dimensions in HAMK 2023–2024 curricula

DBE dimension	International business EN	Smart and sustainable design EN	Sustainable development FI	Smart and sustainable Design FI	Entrepreneurial nursing FI
Sustainable development and responsibility	14	23	31	19	24
Learning through development and teamwork	25	18	24	14	11
Research-based approach, innovation competence and transformative approach	28	10	11	10	13
Authentic challenges and working life orientation	20	11	16	12	7
Design thinking	8	40	4	37	4
Multidisciplinary collaboration	6	4	2	4	22
Student agency	5	0	12	0	3
Well-being skills	0	0	3	0	7
Total	106	106	103	96	91

Table Authors

4.2 *The Enablers and Hinderers According to Teachers*

The enablers and hinderers were monitored using the eight HAMK DBE dimensions (Fig. 1). Based on the pilot DBE projects' teacher interviews, there were many enabling and hindering factors for a shift to DBE (Table 5).

Among the most important factors were students' proactivity and smooth and well-functioning communication with the industry partner (i.e., the client).

> … they didn't come up with any half-baked solutions … in presentation events when there was a company's business activities spot … there, they showed responsibility … and pride in their own work in a completely different way … (Teacher B)

> And then I thought the communication was very smooth, and everyone was on the map all the time, what was going on, and what needed to be done. (Teacher C)

Building implementation around the design thinking process was also an important enabler. DBE implementation was seen to develop generic skills and content knowledge. The implementation raised awareness of sustainable development and responsibility.

Table 5 Hindering and enabling factors based on teachers' experiences in DBE pilot projects

DBE dimensions	Enabling factors	Frequency f = X	Hindering factors	Frequency f = X
Multidisciplinary collaboration	Multidisciplinarity led to thinking outside the box	5	Planning of the implementation	12
	Multidisciplinarity was achieved	4	Multidisciplinarity was not achieved	7
	Design factory was involved	4	Distance of campuses	6
	Intends to implement multidisciplinary teaching in the future	4	Curricula structure	5
	Complementarity of competences	2	Too big student amount	4
	Multidisciplinarity enabled the final product	2	The project did not bring the sectors closely together	2
Design thinking	Implementation built around design thinking process	8	Getting used to the freedom of method	3
	Offered tools for teams	7	Only some disciplines applied design thinking	1
	Customer profiles created by students	4		
	End-users have been involved	4		
	Iterative way of developing	4		
	Double diamond method	2		
	Teachers working together with students on solutions	2		
Authentic challenges and working life collaboration	Communication with the client was smooth	10	Too much work	11
	Cooperation with working life went well	7	The project was not the right choice	4
	Challenge from a company	3	Students had difficulties in understanding the aims	4
	Challenge from public sector	3		

(continued)

Table 5 (continued)

DBE dimensions	Enabling factors	Frequency f = X	Hindering factors	Frequency f = X
	Challenge from third sector	1		
Sustainable development and responsibility	The implementation raised awareness of sustainable development and/or responsibility	8	The working life partner did not operate in the region	4
	How were the teams formed?	8		
	The implementation promoted sustainable development and/or responsibility	6		
	The implementation had a regional impact in Häme	5		
Learning through development and teamwork	Multilingualism	6	One-discipline teams	9
	Teams were multidisciplinary	5	Difficulties in teamwork	8
	Teamwork went well	4		
	Teams were multilingual	3		
Student agency	Students were proactive	15	Student attendance	7
	Teachers gave boundaries	2	Students had difficulties in starting the project	6
			Students' proactivity and responsibility in the project	6
Research-based approach, innovation competencies and transformative approach	Developed generic skills	10		
	Developed content knowledge	7		
	End-users have been interviewed/tested	4		
	Students had to find the information themselves	3		
	Familiarisation with related literature	3		

(continued)

Table 5 (continued)

DBE dimensions	Enabling factors	Frequency f = X	Hindering factors	Frequency f = X
	Familiarisation with the working environment of the challenge	2		
Total		167		99

Table Authors

> The brief included the UN Sustainable Development Goals, which the students had to take into account or choose from there as a starting point for their own work. (Teacher C)

Hindering factors included problems in planning the implementation through teacher collaboration and the teachers' workload in working-life co-operation. Multidisciplinarity was not achieved in all projects. Teamwork difficulties were among the biggest hindering factors.

> ... Personally, I can say that it might have needed more of this kind of joint planning and preparation of teachers in advance... and going through things together … (Teacher A)

4.3 The Enablers and Hinderers According to Students

To answer the research question on what hinders and enables the total shift to DBE based on the pilot DBE projects, we analyzed the student survey's open question answers to detect students' experiences of the enabling and hindering factors during the DBE project course. The analyses revealed six different categories that enabled and hindered students' well-being during the DBE project course (Table 6).

Collaboration with team members and support and feedback from the teacher were the most frequently mentioned factors that promoted student well-being in DBE:

> .... The support from peers and mentors greatly contributed to my well-being during the design-based project course. Collaborating with team members provided opportunities for brainstorming, sharing ideas, and receiving feedback, which helped alleviate feelings of isolation and boosted confidence in my work. (Student 53)

> Group work and interdisciplinary collaboration. (Student 17)

> Additionally, having access to knowledgeable mentors who provided guidance and encouragement was invaluable in navigating challenges and staying motivated throughout the course. (Student 53)

> I liked this kind of work where the teachers were also a strong part of the team and brought their own views forward. (Student 4)

Table 6 Hindering and enabling factors based on students' experiences in the pilot projects

DBE dimensions	Enabling factors	Frequency f = X	Hindering factors	Frequency f = X
Multidisciplinary collaboration	Multidisciplinarity	2	Multidisciplinarity	3
			Foreign language	3
Design thinking	Support and feedback from teacher	12	New way of working	5
	Working at campus environment	4		
	Learning methods	2		
Authentic challenges and working life collaboration			Unclear task	14
Sustainable development and responsibility				
Learning through development and teamwork	Collaboration with team members	29	Challenges in teamwork	13
	Support from classmates from own degree	4	Team communication	9
	Communication	3		
Student agency	Autonomy and flexibility	6	Time management	12
	Experiences of success	5	Challenges in personal life	6
	Own motivation	2		
Research-based approach, innovation competencies and transformative approach				
Nothing	Nothing	1	No challenges	5
Total		70		70

Table Authors

Unclear tasks challenged student well-being in DBE the most:

During the design-based project course, one aspect that challenged my well-being was the high level of ambiguity inherent in the projects. Often, the objectives were not clearly defined, requiring extensive research and exploration to establish a direction. This ambiguity sometimes led to feelings of uncertainty and stress, as it was challenging to gauge progress and ensure alignment with the project. (Student 53)

Some students felt that time management, problems in teamwork, and team communication challenged their well-being:

> Some team members don't make contributions. (Student 32)

> ... managing interpersonal dynamics within the team... (Student 55)

Working in a multidisciplinary team challenged the well-being of three students and promoted the well-being of two students. A research-based approach, an innovation competence and transformative approach, and sustainable development and responsibility were not mentioned by the students as hindering or enabling factors.

5 Discussion and Conclusions

This study aimed to address the difficult pedagogical challenge in HEI in the time of GenAI [2, 27]. Learning tasks can be easily completed by AI if they test only unidisciplinary content knowledge via reading and writing. One possible solution is a total shift to DBE. To be able to educate for student innovation competence [7], we have proposed design-based pedagogies to be applied in UAS. However, does it seem possible in practice? Do the current curricula support the shift? What hinders and enables this shift? These questions must be answered to be able to plan teacher training and pedagogical management actions in practice. Based on these findings, there are many implications for practice that we can discuss and conclude on. Do the current curricula support the shift to DBE? The findings imply that the current curricula do not support a shift in all degree programs. Those degrees that emphasize a multidisciplinary phenomenon in their headline, not only one industry, directed DBE-type learning more (e.g., sustainable development, international business, and entrepreneurial nursing). More traditional disciplines, such as construction and civil engineering, seemed to promote DBE-type learning the least.

Based on the teachers' experiences in the DBE pilot projects, there are many more enabling than hindering factors that support the total shift to DBE (Table 5). According to the teachers, students' proactivity, the design thinking process model, and smooth and well-functioning communication with the industry partner are the strongest enablers of DBE. Problems in planning, the workload in industry co-operation, and teamwork difficulties are among the largest hindering factors. Based on these findings, we recommend that teachers be well trained in DBE-type work-oriented, team-based collaborative teaching and innovation pedagogy. They already seem to know how to promote student agency, and most can apply design thinking. In teacher training, more emphasis should be devoted to (1) planning the DBE curricula and implementation for more clarity in learning targets and assignments that need to be returned and (2) how to efficiently co-operate with industry partners and collaboratively design the open challenge and offer student support.

Based on the student material, the strongest enabling factor is the (multidisciplinary) team itself, in which a single student works to co-create new solutions for the client (Table 6). Teachers should invest in supporting teamwork, especially when the multidisciplinary team formation offers a more challenging learning environment than a unidisciplinary team. The most common hindering factor is an unclear task. In DBE, students are challenged with an authentic, open task from the industry. The task is inherently "unclear," and design thinking as such promotes an unclear, open starting point [23, 25, 37]. Students are supposed to trust their collaborative agency to co-create a concrete solution to it. However, the feeling of unclarity might affect motivation and engagement in the project work and thus affect the learning experience.

To conclude, a total shift to DBE seems possible if the hindering factors like problems in collaborative planning, the workload and training of teachers, challenges in teamwork, and students' feeling of unclarity are solved. Even if the teachers found student agency to be a key enabling factor (Table 5), the teacher's role is to monitor, support, and balance the experience and ease the pain if needed [19]. The tolerance of uncertainty should grow during the four years of studying at UAS so that students are more prepared for an unclear and uncertain working life requiring employee agency and proactiveness without affecting employee well-being. If the design thinking way of working is too vague, the GenAI tools offered along with the design thinking tools may offer help, especially in creative work [43]. The assignments that should be returned during the project should allow for creativity, but the form of returning them should be clear to avoid unnecessary unclarity.

There are limitations in this study. The study concerns only one institution, and the analysis frame was context specific. However, based on literature review, the analysis frame (Fig. 1) reflects the characteristics of design thinking and DBE as a form of innovation pedagogy. The teacher and student samples were small. Thus, the applicability of the results is limited. While some DBE implementations had already been running for some years, others were new. Some teachers were well prepared to implement DBE, while others were new to it. However, for students, this type of learning was quite new to all, even though the students came from different grades and degree programs (Table 1). Future research should focus on monitoring learning outcomes, the benefits for the participating industry partners, the potential added value of AI tools in design thinking processes, and the effects of the proposed curricula renewal and teacher training.

References

1. Buselic, V., Rajkovic, I.: Teaching generic skills with/or against Chat GPT. In: 47th MIPRO ICT and Electronics Convention (MIPRO), Opatija, Croatia, 2024, pp. 707–712. IEEE (2024). https://doi.org/10.1109/MIPRO60963.2024.10569431
2. Michel-Villarreal, R., Vilalta-Perdomo, E., Salinas-Navarro, D.E., Thierry-Aguilera, R., Gerardou, F.S.: Challenges and opportunities of generative AI for higher education as explained by ChatGPT. Educ. Sci. **13**(9), 856 (2023). https://doi.org/10.3390/educsci13090856
3. Chukwuere, J.E., Handoko, B.L.: The future of generative AI chatbots in higher education. J. Emerg. Technol. **4**(1), 36–44 (2024). https://doi.org/10.57040/y62k8d79

4. Hofmann, P., Brand, A., Späthe, E., Lins, S., Sunyaev, A.: AI-based tools in higher education: a comparative analysis of university guidelines. In: Proceedings of Mensch und Computer 2024 (MuC '24), pp. 665–673. Association for Computing Machinery (2024). https://doi.org/10.1145/3670653.3677513
5. Jin, Y., Yan, L., Echeverria, V., Gašević, D., Martinez-Maldonado, R.: Generative AI in higher education: a global perspective of institutional adoption policies and guidelines. Comput. Educ. Artif. Intell. **8**, Article 100348 (2025). https://doi.org/10.1016/j.caeai.2024.100348
6. Saari, M., Rantanen, P., Nurminen, M., Kilamo, T., Systä, K., Abrahamsson, P.: Toward guiding students: exploring effective approaches for utilizing AI tools in programming courses. In: Generative AI for effective software development, pp. 331–346. Springer Nature Switzerland (2024). https://doi.org/10.1007/978-3-031-55642-5_16
7. Hero, L.-M., Lindfors, E., Taatila, V.: Individual innovation competence: a systematic review and future research agenda. Int. J. High. Educ. **6**(5), 103–121 (2017). https://doi.org/10.5430/ijhe.v6n5p103
8. Arvanitis, S., Stucki, T.: What determines the innovation capability of firm founders? Ind. Corp. Chang. **21**(4), 1049–1084 (2012). https://doi.org/10.1093/icc/dts003
9. Avvisati, F., Jacotin, G., Vincent-Lancrin, S.: Educating higher education students for innovative economies: what international data tell us. Tuning J. High. Educ. **1**(1), 223–240 (2013). https://doi.org/10.18543/tjhe-1(1)-2013pp223-240
10. Bruton, D.: Learning creativity and design for innovation. Int. J. Technol. Des. Educ. **21**(3), 321–333 (2011). https://doi.org/10.1007/s10798-010-9122-8
11. Chatenier, E.D., Verstegen, J.M., Biemans, H.A., Mulder, M., Omta, O.F.: Identification of competencies for professionals in open innovation teams. R&D Manag. **40**(3), 271–280 (2010). https://doi.org/10.1111/j.1467-9310.2010.00590.x
12. Edwards-Schachter, M., García-Granero, A., Sánchez-Barrioluengo, M., Quesada-Pineda, H., Amara, N.: Disentangling competences: interrelationships on creativity, innovation and entrepreneurship. Thinking Skills Creativity **16**, 27–39 (2015). https://doi.org/10.1016/j.tsc.2014.11.006
13. Hero, L.-M.: Learning to develop innovations. Individual competence, multidisciplinary activity systems and student experience, Annales universitatis Turkuensis, 475, Doctoral dissertations, University of Turku (2019). https://www.utupub.fi/handle/10024/147038
14. Montani, F., Odoardi, C., Battistelli, A.: (Individual and contextual determinants of innovative work behaviour: Proactive goal generation matters. J. Occup. Organisational Psychol. **87**(4), 645–670 (2014). https://doi.org/10.1111/joop.12066
15. Nielsen, J.A.: Assessment of innovation competency: a thematic analysis of upper secondary school teachers' talk. J. Educ. Res. **108**(4), 318–330 (2015). https://doi.org/10.1080/00220671.2014.886178
16. Ojan, M.P., Lara-Navarra, P., Sánchez-Navarro, J.: Profiling soft skills in the labour market: the case of communication. In: Visvizi, A., Troisi, O., Corvello, V. (eds.) Research and Innovation Forum 2023. RIIFORUM 2023. Springer Proceedings in Complexity. Springer, Cham (2024). https://doi.org/10.1007/978-3-031-44721-1_325
17. Vila, L.E., Pérez, P.J., Coll-Serrano, V.: Innovation at the workplace: do professional competencies matter? J. Bus. Res. **67**(5), 752–757 (2014). https://doi.org/10.1016/j.jbusres.2013.11.039
18. Waychal, P., Mohanty, R., Verma, A.: Determinants of innovation as a competence: an empirical study. Int. J. Bus. Innov. Res. **5**(2), 192–211 (2011). https://doi.org/10.1504/IJBIR.2011.038781
19. Hero, L.-M., Lindfors, E.: Students' learning experience in a multidisciplinary innovation project. Educ. + Training **61**(4), 500–522 (2019). https://doi.org/10.1108/ET-06-2018-0138
20. Manetti, A., Lara-Navarra, P., Serradell-López, E.: Design thinking innovation and trends foresighting. In: Visvizi, A., Troisi, O., Grimaldi, M. (eds) Research and Innovation Forum 2022. RIIFORUM 2022. Springer Proceedings in Complexity. Springer, Cham (2023). https://doi.org/10.1007/978-3-031-19560-0_2
21. Assen, H., Benhadda, L., Losekoot, E., van Diggelen, M.: Design thinking in hospitality education: lessons learned and future opportunities. J. Hospitality Leisure Sport Tourism Educ. **32**, Article 100439 (2023). https://doi.org/10.1016/j.jhlste.2023.100439

22. Geitz, G., Donker, A., Parpala, A.: Studying in an innovative teaching–learning environment: design-based education at a university of applied sciences. Learning Environ. Res. (2023). https://doi.org/10.1007/s10984-023-09467-9
23. Geitz, G., de Geus, J.: Design-based education, sustainable teaching, and learning. Cogent Educ. **6**(1), Article 1647919 (2019). https://doi.org/10.1080/2331186X.2019.1647919
24. Guaman-Quintanilla, S., Everaert, P., Chiluiza, K., Valcke, M.: Impact of design thinking in higher education: a multi-actor perspective on problem solving and creativity. Int. J. Technol. Des. Educ. **33**(1), 217–224 (2023). https://doi.org/10.1007/s10798-021-09724-z
25. van Gameren, M.J., van Diggelen, M.R., Evers, A.T.: Art and design teachers' personal rules of thumb while designing instruction for studios. Int. J. Technol. Des. Educ. 1–41 (2024). https://doi.org/10.1007/s10798-024-09943-0
26. HAMK Häme University of Applied Science: Design-based education (DBE). Retrieved 20 Feb 2025, from https://www.hamk.fi/en/about-hamk/strategy/design-based-education-dbe/
27. Deng, R., Jiang, M., Yu, X., Lu, Y., Liu, S.: Does ChatGPT enhance student learning? A systematic review and meta-analysis of experimental studies. Comput. Educ. **227**, Article 105224 (2025). https://doi.org/10.1016/j.compedu.2024.105224
28. Lim, W.M., Gunasekara, A., Pallant, J.L., Pallant, J.I., Pechenkina, E.: Generative AI and the future of education: Ragnarök or reformation? A paradoxical perspective from management educators. Int. J. Manag. Educ. **21**(2), Article 100790 (2023). https://doi.org/10.1016/j.ijme.2023.100790
29. Iorliam, A., Ingio, J.A.: A comparative analysis of generative artificial intelligence tools for natural language processing. J. Comput. Theor. Appl. **1**(3), 311–325 (2024). https://doi.org/10.62411/jcta.9447
30. Dai, W., Lin, J., Jin, H., Li, T., Tsai, Y.S., Gašević, D., Chen, G.: Can large language models provide feedback to students? A case study on ChatGPT. In: 2023 IEEE International Conference on Advanced Learning Technologies (ICALT), pp. 323–325. IEEE (2023). https://doi.org/10.1109/ICALT58122.2023.00100
31. Li, Y., Sha, L., Yan, L., Lin, J., Raković, M., Galbraith, K., Lyons, K., Gašević, D., Chen, G.: Can large language models write reflectively. Comput. Educ. Artif. Intell. **4**, Article 100140 (2023). https://doi.org/10.1016/j.caeai.2023.100140
32. Liu, X.: Navigating uncharted waters: teachers' perceptions of and reactions to AI-induced challenges to assessment. Asia-Pac. Educ. Res. 1–12 (2024). https://doi.org/10.1007/s40299-024-00890-x
33. Vartiainen, H., Tedre, M.: Using artificial intelligence in craft education: crafting with text-to-image generative models. Digit. Creativity **34**(1), 1–21 (2023). https://doi.org/10.1080/14626268.2023.2174557
34. Liedtka, J.: Evaluating the impact of design thinking in action. Acad. Manag. Proc. **2017**(1), Article 10264 (2017). https://doi.org/10.5465/AMBPP.2017.177
35. Prud'homme van Reine, P.: The culture of design thinking for innovation. J. Innov. Manag. **5**(2), 56–80 (2017). https://doi.org/10.24840/2183-0606_005.002_0006
36. van der Meer, R., van der Moolen, M.: How to support faculty in educational innovation. ETH Learn. Teach. J. **2**(2), 378–381 (2020). https://doi.org/10.16906/lt-eth.v2i2.168
37. Lahdenperä, J., Jussila, J., Järvenpää, A., Jumisko-Pyykkö, S.: Developing students' innovation capability through interdisciplinary product development projects for industry. In: Teoksessa 1st Design Factory Global Network Research Conference 'Designing the Future' 5–6 Oct 2022, pp. 14–17 (2023). http://urn.fi/URN:ISBN:978-952-64-9627-6
38. Rajalo, S., Vadi, M.: University-industry innovation collaboration: reconceptualization. Technovation **62**, 42–54 (2017). https://doi.org/10.1016/j.technovation.2017.04.003
39. Caccamo, M., Pittino, D., Tell, F.: Boundary objects, knowledge integration, and innovation management: a systematic review of the literature. Technovation **122**, Article 102645 (2023). https://doi.org/10.1016/j.technovation.2022.102645
40. Konst, T., Kairisto-Mertanen, L.: Developing innovation pedagogy approach. On the Horizon **28**(1), 45–54 (2020). https://doi.org/10.1108/OTH-08-2019-0060

41. Raij, K.: Summarising the basis of LbD for further development—review. In: Juvonen, S., Marjanen, P., Meristö, T. (Eds.) Learning by Developing 2.0: Case Studies in Theory and Practice, pp. 16–27. Laurea Publications (2018)
42. Leijon, M., Gudmundsson, P., Staaf, P., Christersson, C.: Challenge-based learning in higher education–a systematic literature review. Innov. Educ. Teach. Int. **59**(5), 609–618 (2022). https://doi.org/10.1080/14703297.2021.1892503
43. Saritepeci, M., Yildiz Durak, H.: Effectiveness of artificial intelligence integration in design-based learning on design thinking mindset, creative and reflective thinking skills: an experimental study. Educ. Inf. Technol. **29**, 1–35 (2024). https://doi.org/10.1007/s10639-024-12829-2
44. Nachtigall, V., Shaffer, D.W., Rummel, N.: Stirring a secret sauce: a literature review on the conditions and effects of authentic learning. Educ. Psychol. Rev. **34**(3), 1479–1516 (2022). https://doi.org/10.1007/s10648-022-09676-3
45. Krippendorff, K.: Content Analysis: An Introduction to its Methodology, 3rd edn. Sage, Thousand Oaks, California (2013)
46. Bowden, J.A., Green, P.J.: Relationality and the myth of objectivity in research involving human participants. In: Higgs, J., Cherry, N., Macklin, R., Ajjawi, R. (eds.) Researching Practice: A Discourse on Qualitative Methodologies, pp. 105–121. Sense, Rotterdam (2010)
47. Braun, V., Clarke, V.: Using thematic analysis in psychology. Qual. Res. Psychol. **3**(2), 77–101 (2006). https://doi.org/10.1191/1478088706qp063oa
48. Elo, S., Kyngäs, H.: The qualitative content analysis process. J. Adv. Nurs. **62**(1), 107–115 (2008). https://doi.org/10.1111/j.1365-2648.2007.04569.x

Global Research Adoption of AI in Education: Trends and Applications

Ali Mohamad Mouazen, Ana Beatriz Hernández Lara, and Jawad Chahine

Abstract Artificial Intelligence (AI) is rapidly reshaping educational systems, influencing pedagogical practices, learning environments, and institutional decision-making. While scholarly interest in AI in education has grown exponentially, the field remains fragmented across disciplines, regions, and thematic orientations. To address this gap, the present study conducts a comprehensive bibliometric analysis of 5676 peer-reviewed articles indexed in Scopus. The study looks at global research trends, collaboration patterns, and new topic groups by analyzing how often articles are cited together, mapping keywords that appear together, and visualizing networks of authors who work together. Findings reveal a marked surge in research activity over the past five years, with heightened interdisciplinary collaboration and the rise of themes such as personalized learning, intelligent tutoring systems, ethical considerations, and generative AI. Despite this growth, significant gaps persist in the areas of equity, policy integration, and the inclusion of underrepresented regions and perspectives. This study adds to academic discussions by providing a comprehensive and worldwide perspective on how AI is being used in education, while also pointing out important areas for future research, ethical guidelines, and inclusive educational improvements.

Keywords Digital transformation · Education systems · Educational technology · Innovation 5.0 · Artificial intelligence in education · Educational governance · SDG 4—quality education

A. M. Mouazen (✉)
Business Management Department, Lebanese International University, Beirut, Lebanon
e-mail: ali.mouazen@liu.edu.lb

A. B. H. Lara
Department of Business Management, Universitat Rovira I Virgili, Reus, Spain
e-mail: anabeatriz.hernandez@urv.cat

J. Chahine
College of Business, Australian University, Kuwait City, Kuwait
e-mail: j.chahine@au.edu.kw

A. Visvizi et al. (eds.), *Research and Innovation Forum 2025 Vol2*, Springer Proceedings in Complexity, https://doi.org/10.1007/978-3-032-09062-1_3

1 Introduction

The integration of Artificial Intelligence (AI) into education marks a transformative shift, fundamentally altering contemporary educational systems. Beyond its early uses in automation, AI has evolved to encompass predictive analytics, intelligent tutoring systems, adaptive learning platforms, and real-time feedback mechanisms [1]. These innovations are reshaping instructional practices, enhancing student engagement, and influencing institutional decision-making at multiple levels [2].

Academic interest in AI in education has grown substantially, expanding across disciplines such as computer science, cognitive psychology, education, learning analytics, and educational ethics [3]. While early research primarily emphasized technological development and algorithmic efficiency, more recent contributions have begun to address broader concerns, ranging from AI-assisted assessment and personalized learning to educational recommender systems, ethical implications, and educational innovation [4].

Despite these advances, the field remains fragmented across thematic, geographic, and disciplinary boundaries. Existing reviews often adopt narrow lenses, focusing on higher education or STEM domains, or confining analysis to specific regional contexts [1, 5–7]. This fragmentation limits our ability to synthesize the field's intellectual evolution and inhibits the formulation of cohesive strategies for inclusive AI integration in education. Moreover, although AI is increasingly recognized as a key driver of educational innovation, the global distribution of scholarly leadership and collaboration remains unequal, raising important questions about equity, access, and epistemic representation [3, 8].

This research contributes to ongoing debates on the digital transformation of education and knowledge systems, particularly in relation to institutional responsiveness, pedagogical adaptability, and resilience in post-COVID contexts [9]. Recent studies have also emphasized the implications of uncertainty and crisis for educational systems, highlighting the potential influence of Innovation 5.0 frameworks—especially the integration of artificial intelligence—on governance structures and learning models [10].

By advancing this line of inquiry through a global bibliometric perspective, the present study contributes empirical insight into the structural and thematic development of AI in education. In particular, it interrogates research output distribution, identifies collaborative patterns and regional asymmetries, and highlights conceptual shifts—offering a basis for more informed policy dialogue and scholarly engagement around the future of AI-supported education. The research is guided by the following question:

- *How has the global adoption of AI in education evolved in terms of research output, thematic trends, and patterns of international collaboration?*

To answer this question, the study analyzes at 5676 peer-reviewed articles listed in Scopus, using methods like analyzing co-authorship and country connections, mapping keyword occurrences, and examining publication trends over time.

The remainder of the paper is structured as follows: Sect. 2 reviews the existing literature, focusing on both bibliometric studies and the conceptual evolution of AI in education. Section 3 outlines the methodological approach and data collection process. Section 4 presents the results of the bibliometric analyses, while Sect. 5 offers a critical discussion of the findings. Section 6 concludes with reflections on future research priorities, ethical considerations, and policy implications.

2 Literature Review

The integration of Artificial Intelligence (AI) in education has developed into a substantial area of multidisciplinary research, bridging fields such as computer science, psychology, pedagogy, learning analytics, and educational ethics. This convergence is reshaping instructional methodologies and educational systems globally, driven by technological acceleration and a growing demand for personalized, scalable learning solutions. AI-driven tools, including adaptive learning platforms, intelligent tutoring systems, and Natural Language Processing (NLP) applications, are increasingly recognized as promising methods for reducing educational disparities and promoting access and achievement among marginalized learners [11].

Since 2018, there has been a significant increase in global research on AI in education, primarily driven by contributions from countries such as America and China [12, 13]. However, collaboration among researchers remains uneven, and regional implementations of AI vary significantly. In higher-income regions such as North America, East Asia, and Europe, AI applications frequently target medical education, institutional analytics, and digital assessment [14]. In contrast, initiatives in the Global South tend to focus on addressing infrastructural inequalities through gamified platforms and mobile-adaptive technologies—though these remain constrained by challenges related to digital literacy, access, and policy support [15]. In this context, the role of local knowledge and culturally grounded educational practices is often is often ignored in general AI research, despite its importance for designing sustainable, context-sensitive learning solutions in areas with limited resources [16]. The overall change in teaching methods towards using data and personalized instruction is shown by the growing number of AI tools in education, such as learning analytics dashboards and real-time feedback systems [17–19]. These systems align with models of personalized learning and promise increased engagement and performance. However, they also prompt pedagogical and ethical concerns, including algorithmic bias, diminished learner autonomy, opaque decision-making, and questions about the evolving roles of educators and students [20, 21].

Recent bibliometric work has sought to map this evolving field. Zawacki-Richter et al. [1] highlighted an overreliance on computer science methodologies at the expense of educational theory, exposing a key interdisciplinary gap. Chen et al. [3] and Gao et al. [22] showed how ideas have changed over time, starting from basic topics like "student modeling" and "intelligent tutoring systems" to more advanced and socially relevant areas like "educational data mining," "deep learning," and "AI ethics." These findings emphasize an emerging scholarly interest in integrating AI more responsibly and equitably into educational systems.

Despite these developments, critical gaps remain. Many existing studies are temporally or geographically narrow and fail to account for longitudinal shifts or cross-regional asymmetries. Furthermore, questions of **epistemic inequality**—who produces, shapes, and benefits from AI in education—are seldom addressed. Scholarship from the Global South remains underrepresented in leading publication venues, reinforcing structural imbalances in how educational futures are conceptualized and governed [2, 5, 6]. Furthermore, few studies engage extensively with educational theory. Fundamental frameworks like constructivism, critical pedagogy, and the Technological Pedagogical Content Knowledge (TPACK) model are seldom used to explore how AI technologies influence teaching methods or learning experiences [23].

The absence of these viewpoints hinders the field's ability to examine how AI reshapes conceptions of teacher agency, learner identity, and the classroom as a social space. In the same way, ethical guidelines like IEEE's Ethically Aligned Design or the AI4K12 program offer useful ideas, but they are not used enough in real studies about how AI is adopted and impacts education [21]. Moreover, while ethics is increasingly referenced, much of the discussion remains abstract or procedural. Few studies engage with **critical AI ethics**—which emphasizes power, accountability, inclusion, and social justice—as a foundation for educational AI governance [20, 24]. As AI systems begin to shape decision-making processes in areas such as student evaluation, resource allocation, and curriculum design, it becomes essential to move beyond technological optimism and toward nuanced, context-sensitive evaluations of their educational implications.

To address these conceptual, methodological, and geographic limitations, this study conducts a large-scale bibliometric analysis of 5676 Scopus-indexed journal articles. Through co-authorship and country network analysis, keyword co-occurrence mapping, and longitudinal trend analysis, the study provides a data-driven assessment of how AI in education has evolved. In doing so, it contributes to ongoing discussions on the future of educational innovation, with particular attention to research equity, collaborative structures, and the ethical governance of AI-enhanced learning systems.

3 Methodology

To examine global trends in AI and education research, a comprehensive bibliometric analysis was conducted using the Scopus database. Scopus was selected due to its broad multidisciplinary coverage, inclusion of peer-reviewed literature, and robust metadata structure—making it especially well-suited for large-scale trend and network analysis. However, we acknowledge that exclusive reliance on Scopus may introduce certain limitations, including regional bias, exclusion of grey literature, and lag in indexing newly published articles. These limitations are considered in the interpretation of results.

The literature search was systematically structured using keyword combinations designed to capture thematic breadth while maintaining conceptual focus. The query used was:

"(TITLE-ABS-KEY (generative AND ai AND in AND education) OR TITLE-ABS-KEY (chatgpt AND in AND education) OR TITLE-ABS-KEY (ai-powered AND learning AND tools) OR TITLE-ABS-KEY (global AND adoption AND of AND ai AND in AND education) OR TITLE-ABS-KEY (educational AND technology AND trends)) AND (LIMIT-TO (DOCTYPE, "ar"))"

This search yielded an initial dataset of 6608 articles, covering all publication years up to—but not including—2025. To ensure dataset relevance and accuracy, a rigorous filtering process was undertaken. Articles that lacked key terms in titles, abstracts, or keywords (826); those missing essential metadata such as author affiliations (103); and duplicate entries (3) were removed. The final dataset included 5676 peer-reviewed journal articles.

To analyze geographic patterns, author affiliation data was cleaned and standardized using a hybrid approach combining text-pattern recognition, city-to-country matching, and manual verification. Countries were classified according to the UNICEF regional framework [25], then further categorized into geographic regions such as Western and Eastern Europe, the Middle East, North and South America, Gulf countries, Southeast Asia, and Africa. This granular classification allowed for a detailed examination of regional collaboration dynamics and publication trends.

We initially retrieved 6608 articles that covered all research categories and publication years, with the exception of 2025, to ensure a comprehensive overview of the topic. A careful review was done to check for consistency and relevance; as a result, 826 articles without important keywords, 103 articles missing author information or affiliations, and 3 duplicate entries were removed, leaving a final total of 5676 articles for analysis.

Author-level data was also extracted and processed to identify the most prolific and influential contributors. Additionally, keyword data was normalized by consolidating semantic variants (e.g., merging "AI" with "artificial intelligence") and correcting inconsistencies to improve thematic clustering accuracy.

The refined dataset was analyzed using VOSviewer, a well-known tool for bibliometric visualization and clustering, to perform keyword co-occurrence and co-authorship network analysis [26, 27]. SPSS was used for descriptive statistical analysis to aid in the quantitative study of publishing trends, authorship patterns, and geographical production.

By combining these techniques, the study aims to deliver a robust, data-driven overview of the global research landscape on AI in education, while also recognizing the limitations of bibliometric methods—particularly in capturing conceptual nuance and contextual meaning.

4 Findings

This section provides a multi-layered bibliometric and thematic analysis of the AI in education research landscape, focusing on publishing patterns, authorship structures, cooperation networks, and topic themes. Beyond mapping volume and structure, the findings also reflect the underlying asymmetries in research productivity, visibility, and leadership.

4.1 Publication Trends: An Accelerating Field

The trajectory of AI in education research demonstrates exponential growth, particularly in the past five years. From fewer than 150 annual publications between 2006 and 2018, output accelerated beginning in 2020—likely due to the COVID-19 pandemic's catalyzing effect on digital education. Publications rose from 253 in 2020 to 301 in 2022 and then surged to 850 in 2023. In 2024, output tripled to 2771 articles—a 226% increase from the previous year (see Fig. 1).

This dramatic expansion reflects increased interest in educational AI and signifies a broader structural transformation. The growth of tools like ChatGPT has changed the conversation from just improving algorithms to considering their effects on society and teaching, moving AI in education from a minor focus to a major topic that involves many fields.

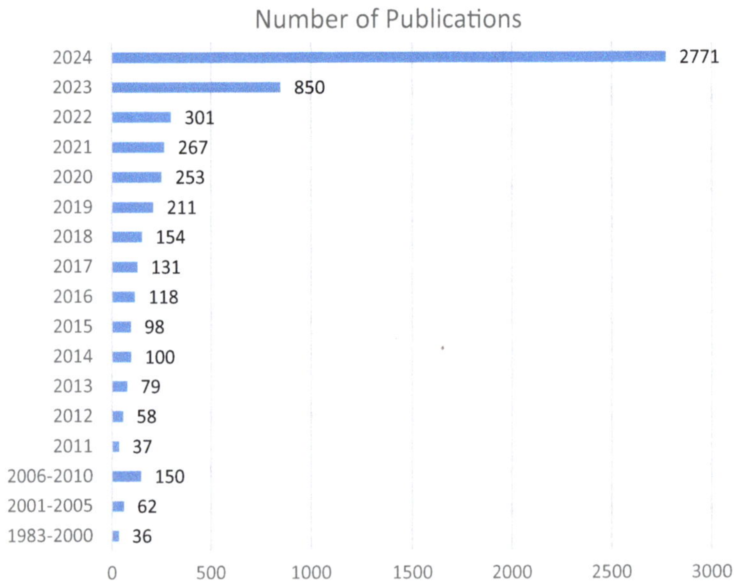

Fig. 1 Publications trend. *Source* Authors

Yet, this growth has not been globally balanced. While volume has increased, visibility and influence remain concentrated in high-output nations and institutions. The quick growth runs the risk of strengthening current hierarchies in educational innovation in the absence of more equitable infrastructure.

4.2 Co-authorship Structure: High-Impact Authors and Collaborative Clusters

We identified 81 influential authors from 17,955 studies, forming eight co-authorship clusters that reflect thematic alignment, collaboration frequency, and regional ties (See Fig. 2).

- **Red and Green clusters** are centered around Chinese-speaking scholars (e.g., Zhang X., Wang J., Liu J.), revealing strong intra-regional collaboration networks, especially in East Asia.

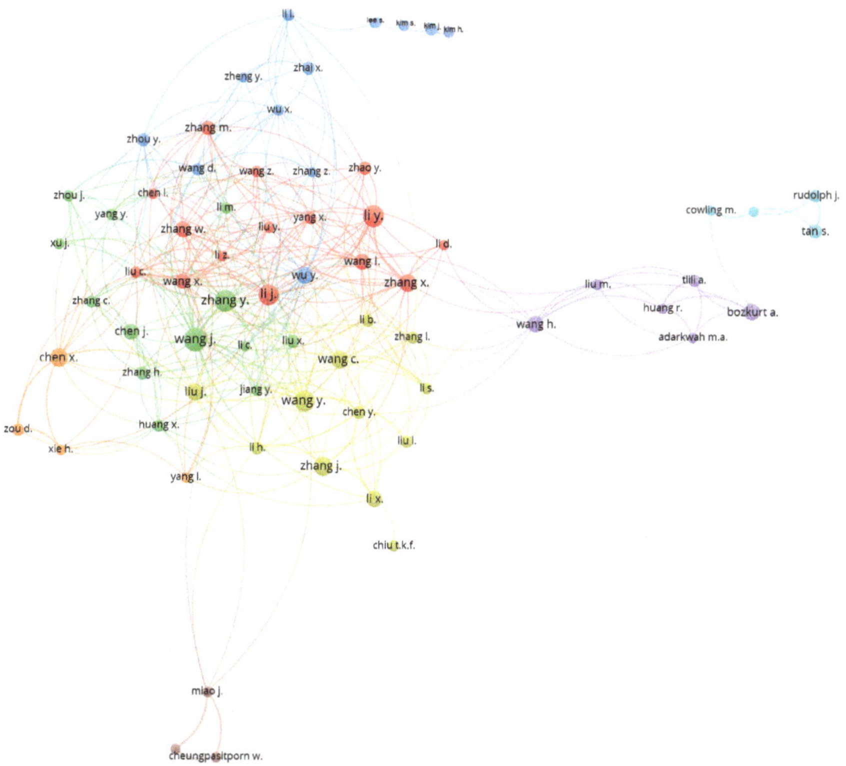

Fig. 2 Co-authorship structure-generated by VOSviewer

- **Blue cluster** (e.g., Kim H., Lee S.) appears more self-contained, likely reflecting Korean research circles.
- **Purple cluster** (e.g., Bozkurt A., Huang R., Liu D.) displays genuine interdisciplinary and international linkages, suggesting increasing East–West engagement.
- **Yellow, Cyan, Brown, and Orange clusters** represent niche technical areas or peripheral collaborations with varied degrees of thematic integration.

Authors such as Wang Y., Chen X., and Zhang X. emerge as structural bridges connecting otherwise insular networks. However, their connectivity often replicates regional patterns rather than dismantling them. While signs of globalization are present, true integrative collaboration across continents—especially involving the Global South—remains limited.

The findings point to a persistent imbalance in scholarly power: East Asian institutions lead in output volume, but leadership is often concentrated within closed networks. Scholars from underrepresented regions are more likely to serve in co-authoring or supporting roles rather than leading research agendas. This echoes broader critiques about **epistemic inequality** in global education research.

Authorship roles further illustrate the dynamics of scholarly leadership and participation. Wang Y. and Chen X. frequently occupy first-author positions, suggesting topic-setting influence. In contrast, figures such as Tan S. consistently appear in supporting roles, highlighting the structural division between intellectual leadership and collaborative labor.

The co-existence of centralized leadership and peripheral participation reinforces the need for more inclusive research ecosystems. Bridging isolated authorship clusters and enabling reciprocal collaborations—especially across regions with disparate resources—would foster a more globally responsive field.

These authorship patterns also reflect institutional investment in AI research, where certain universities or national systems strategically support publication productivity and citation visibility. However, the limited presence of African, Latin American, or Southeast Asian scholars in leadership roles suggests that institutional efforts to globalize AI research must be more inclusive and collaborative in nature (See Table 1).

4.3 International Collaboration in AI and Education Research

Country-level co-authorship networks provide a macro-level view of research connectivity in AI and education. Using co-authorship data, we found eight international groups, each showing different levels of teamwork within the group and connections with other regions (see Fig. 3).

As expected, high-income nations dominated global collaboration structures. China, the US, the UK, and Australia have emerged as important hubs, participating in regular bilateral and multilateral research interactions. The **Red cluster**,

Table 1 Authorship pattern—number of publications per author

Author	1st author	2nd author	3rd author	Total publications
Wang Y	9	6	0	15
Wang C	6	6	3	15
Chen X	11	3	0	14
Zhang X	6	7	1	14
Zhang X	6	7	1	14
Wang J	4	6	4	14
Li J	9	4	0	13
Zhang Y	6	2	4	12
Zhang J	2	6	4	12
Hwang G.-J	6	3	2	11
Tan S	0	6	4	10
Bozkurt A	5	3	1	9
Lee S	2	6	1	9
Perkins M	5	2	0	7

Source Authors

led by China, included Taiwan, Malaysia, Singapore, and Hong Kong—reflecting strong East Asian and ASEAN intra-regional ties. The **Green cluster**, anchored by the United States, included Canada, Mexico, and Brazil, with collaborations often reflecting shared institutional funding streams and continental proximity.

The **Blue cluster** represented Western European partnerships, especially between the UK, Germany, and the Netherlands. The **Yellow cluster**, composed of Australia, New Zealand, and South Korea, suggested more trans-regional academic ties. Meanwhile, the **Purple cluster** comprised South Asian countries like India, Pakistan, and Iran—regions often characterized by internal collaboration rather than global integration. The **Brown and Orange clusters**, led respectively by Spain–Portugal and Saudi Arabia–Gulf states, formed geographically adjacent constellations. Lastly, the **Cyan cluster**, containing South Africa, Nigeria, and other Sub-Saharan African nations, appeared loosely connected and more peripheral.

While these clusters reveal some regional diversity, the collaborative landscape remains structurally uneven. Institutions in the Global North lead in output and control the terms of collaboration—often acting as intellectual gatekeepers. Often, researchers from the Global South appear in secondary authorship roles, suggesting partnerships that are functionally asymmetric rather than co-designed or co-led.

These patterns mirror broader **epistemic disparities** in global research: access to funding, infrastructure, and publication channels tends to be concentrated in well-resourced academic systems. Although South–South collaboration is visible (e.g., within Africa, or between India and Pakistan), such interactions remain limited in scale and visibility. Furthermore, regions such as Sub-Saharan Africa and parts of Latin America remain underrepresented in both research leadership and co-authoring networks.

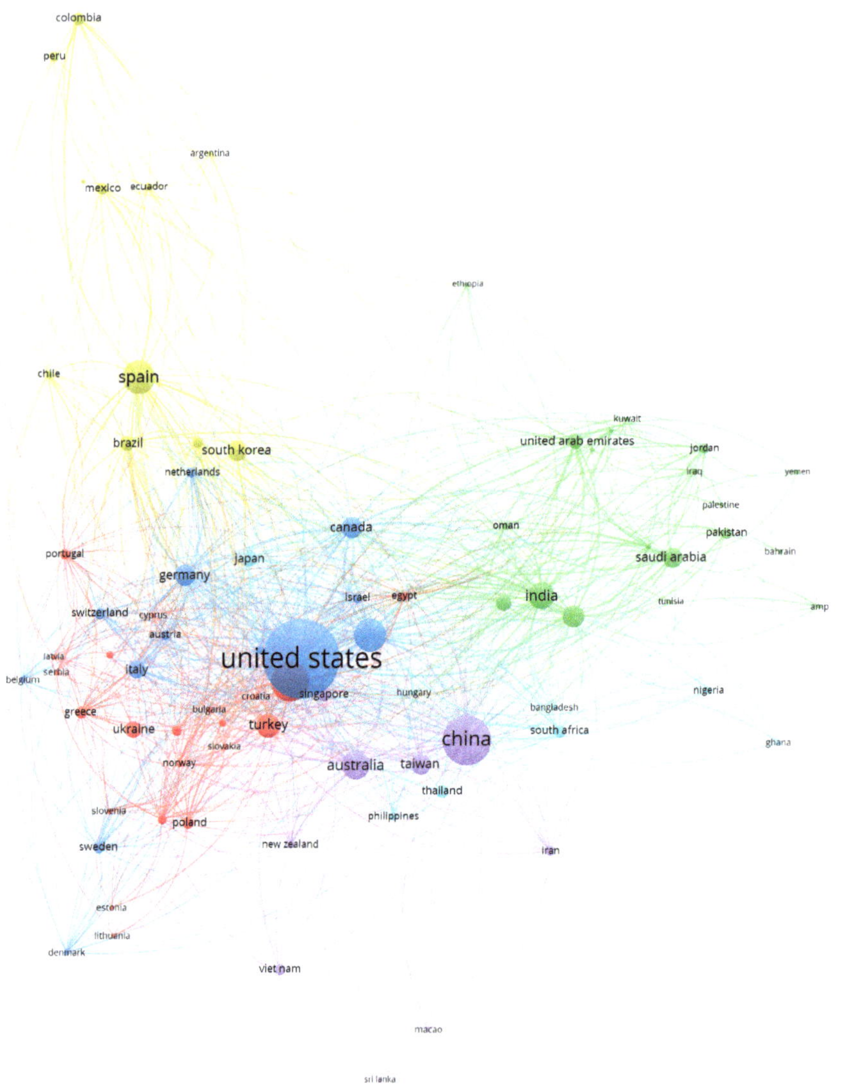

Fig. 3 Countries collaboration-generated by VOSviewer

This asymmetry raises concerns about the inclusiveness and representativeness of the emerging AI in education knowledge base. Without intentional efforts to diversify participation and facilitate equitable collaboration, the expansion of AI scholarship risks reproducing the same global inequities it claims to address through technological innovation. Future initiatives must therefore not only enhance access but also restructure governance and authorship norms to empower marginalized voices.

While these clusters highlight productive regional specialization, they also expose pronounced structural asymmetries in the global research landscape. A small group of dominant countries—most notably the United States, China, and the United Kingdom—continues to shape research priorities, set thematic agendas, and define collaboration norms. Their centrality enables them to influence not only publication trends but also the intellectual framing of the field. In contrast, many countries in Latin America, Sub-Saharan Africa, and South Asia remain situated at the periphery of these networks. While international collaborations may support capacity building, they often reflect asymmetrical partnerships in which Global South scholars contribute empirical data or contextual expertise yet have limited influence over conceptual development or methodological design.

At the continental level, Asia leads global research output in AI and education, accounting for 38.8% of publications, followed by Europe (29.6%) and North America (20.2%). Most collaborations within Asia and Europe remain intraregional, reflecting strong internal research ecosystems but limited cross-continental integration. Africa, by contrast, contributes just 1.7% of publications as lead authors and 2.6% as collaborators. Notably, over half of African-led studies involve no international partners, pointing to regional isolation, limited access to collaborative networks, and persistent capacity constraints (see Fig. 4).

Subcontinental patterns reveal deeper disparities. Eastern and Southeastern Asia—driven by China, South Korea, Taiwan, and Malaysia—produces nearly one-third of global publications. These regions exhibit strong domestic ecosystems, paired

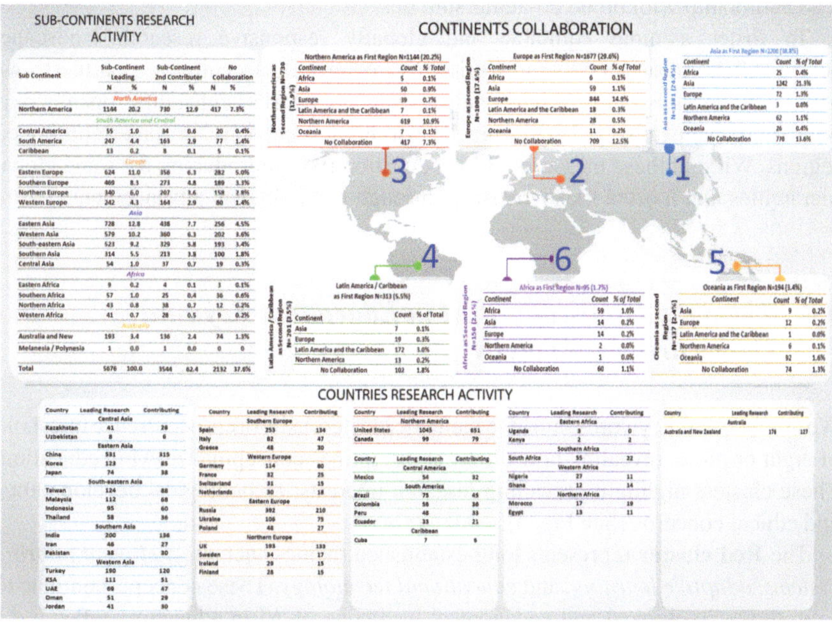

Fig. 4 Continent, sub-continents and countries collaboration—author's own work, map from freepik.com

with selective external engagement, particularly with Europe and North America. Southern Asia, led by India, is gaining visibility but remains underrepresented in global leadership roles. Western Asia, encompassing Turkey and the Gulf states, shows increasing engagement—likely the result of sustained investment in EdTech and research infrastructure.

In Europe, internal asymmetries are also apparent. Northern and Southern European countries (e.g., the UK and Spain) display strong leadership and collaborative density, while Eastern Europe (e.g., Russia and Ukraine) shows a more insular research posture. Western Europe—including Germany, France, and Switzerland—plays a dual role, functioning both as a hub for continental integration and as a bridge to global networks.

Sub-Saharan Africa presents the starkest contrast. Outside of South Africa and Nigeria, most countries in the region appear primarily as co-authors or data sources, with minimal representation in lead authorship or agenda-setting roles. These patterns underscore the persistent underrepresentation of African voices in shaping global discourse on AI in education.

At the country level, the United States leads with 1045 first-author publications, followed by China (531), Russia (392), Spain (253), and India (200). The U.S. exhibits a globally distributed collaboration profile, whereas China's model is more internally focused, with relatively fewer co-authored papers given its overall volume. Countries such as Pakistan, Iran, and Egypt contribute meaningfully but often in supporting roles. Meanwhile, smaller research economies—such as Taiwan, Malaysia, and Turkey—demonstrate a high-impact engagement strategy, balancing lead authorship with broad co-authorship ties.

To foster a more equitable and globally responsive research landscape, future collaboration models must prioritize reciprocal, decentralized partnerships. This includes bolstering South-South research networks, enhancing open-access publishing pathways, and investing in institutional capacity in underrepresented regions. Without these measures, AI in education risks reproducing the same global hierarchies that it often seeks to disrupt through technological advancement.

4.4 Thematic Landscape of AI in Education: Keyword Co-occurrence Analysis

We grouped 407 keywords into nine thematic clusters using a co-occurrence threshold of eight or more, revealing how different disciplines conceptualize AI in education. These clusters highlight prevailing research interests, technological developments, and ethical concerns (See Fig. 5).

The **Red cluster** represents long-established themes such as *intelligent tutoring systems*, *adaptive learning*, and *educational technology*. These concepts continue to form the technological and pedagogical backbone of AI in education, reflecting a sustained focus on personalization, engagement, and learning analytics. The **Green cluster** captures a newer wave of interest in *generative AI*, *ChatGPT*, and *large language models*. These keywords highlight the field's recent shift toward more

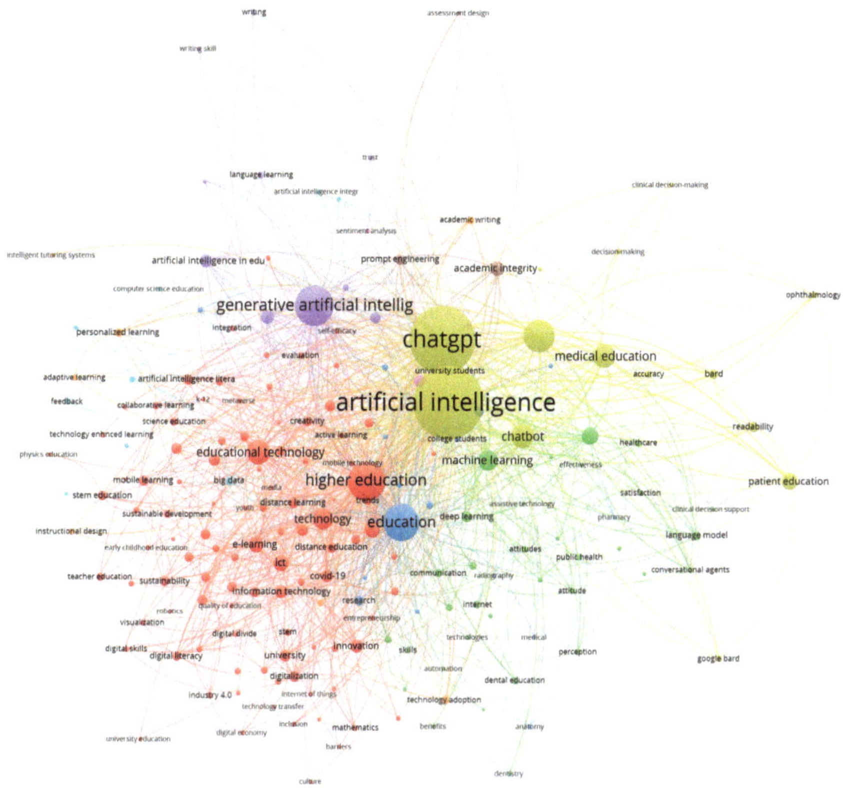

Fig. 5 Keyword co-occurrence—generated by VOSviewer

socially visible and disruptive technologies. Their emergence reflects not only technological advancement but also a growing societal awareness of AI's role in reshaping communication, assessment, and authorship in educational contexts. The **Blue cluster** focuses on *ethical concerns*, including *bias*, *transparency*, and *trust*. This signals a necessary maturation of the field—one in which technological progress is increasingly accompanied by reflective discourse on accountability, fairness, and governance. The **Yellow cluster** includes keywords such as *educational policy*, *leadership*, and *digital transformation*. These terms position AI not merely as a classroom tool but as a force influencing institutional change, policy-making, and systemic reform. The presence of this cluster underscores the field's expansion beyond individual learning outcomes toward broader educational ecosystems. The **Purple cluster** connects AI to *higher education*, *STEM learning*, and *medical education*, emphasizing domain-specific applications and reinforcing the view that AI adoption often follows sector-specific needs and institutional capacities.

Collectively, these clusters illustrate a field that is both consolidating and diversifying. Foundational technologies remain central, but newer themes—particularly those related to generative AI and ethics—are rapidly gaining prominence. This suggests an evolving research landscape in which practical applications and critical reflections are increasingly interwoven. However, keyword co-occurrence analysis is not without limitations. It depends on author-selected keywords, which may vary in specificity or terminology. Additionally, such mappings offer structural insights but may underrepresent deeper theoretical nuances or context-specific meanings embedded within articles. Even with these limitations, the grouping of themes shows that the research area is growing up—it's starting to look beyond just technology to think about wider social, ethical, and institutional issues. Understanding these thematic trajectories can guide future research and assist researchers, educators, and policymakers in navigating the complex intersection of AI and education.

5 Discussion

This study provides a comprehensive and data-driven view of the emerging AI landscape in education. The dramatic increase in publishing output, notably in the aftermath of the COVID-19 epidemic and the advent of generative tools like ChatGPT, demonstrates not just greater scholarly interest but also a more profound reconfiguration of the discipline. AI is no longer understood as a solitary technical tool but as a transformative system that reshapes institutional logic, assessment paradigms, and pedagogical design.

While collaboration patterns have expanded, they remain structurally uneven. A small group of highly productive scholars—largely based in East Asia—continues to anchor the core of the field, whereas researchers from emerging regions remain mostly peripheral. Nevertheless, boundary-spanning contributors are beginning to bridge geographic and disciplinary divides, a positive sign of an evolving and potentially more inclusive research ecology.

Country-level collaboration patterns reinforce this asymmetry. Research leadership is concentrated in the United States, China, and the United Kingdom, while emerging economies in Latin America, Africa, and the Middle East still often engage in asymmetrical partnerships. These collaborations, though valuable, frequently reflect dependency structures in which Global South scholars provide data or implementation contexts but are underrepresented in conceptual framing or lead authorship. This aligns with wider concerns about epistemic inequality and the risk of reinforcing hierarchical knowledge flows.

Subcontinental analysis reveals further regional stratification. East and Southeast Asia dominate both in volume and leadership, driven by strategic national investments. Southern and Western Asia demonstrate growing engagement but remain modest in global visibility. Europe reveals a split between highly networked Northern/Southern nations and a more independent Eastern bloc. In Sub-Saharan Africa and parts of Latin America, a few countries are emerging as leaders, yet limited regional infrastructure continues to inhibit broader participation.

Thematic clustering confirms a field in intellectual transition. Core technologies—such as intelligent tutoring systems and adaptive learning platforms—remain central, but new clusters have emerged around generative AI, academic integrity, and prompt engineering. Significantly, themes related to trust, transparency, ethics, and user perception suggest a growing shift toward human-centered AI frameworks in education.

This shift underlines the importance of aligning AI implementation with local values, social contexts, and pedagogical goals. Human-centered principles such as fairness, explainability, and user trust are no longer peripheral—they are essential to the success and acceptance of AI systems. However, participatory research methods, particularly those engaging educators and learners as co-designers, remain under-utilized. Without these, the field risks designing systems that optimize performance metrics while neglecting educational meaning.

Another critical theme emerging from the analysis is the enduring gap between innovation and implementation. Clusters addressing barriers to adoption—such as infrastructure, teacher preparedness, and digital literacy—underscore the challenges facing under-resourced contexts. Unless these structural issues are addressed, AI may unintentionally exacerbate educational inequalities rather than reduce them.

These findings echo themes raised in recent RII Forum proceedings, particularly regarding the unequal distribution of digital opportunity and the urgent need to rethink innovation through the lens of equity, inclusion, and sustainability (Visvizi et al. 2023, 2024). Where previous Rii Forum contributions have explored the disruptive and transformative potential of educational technologies, this study adds a bibliometric perspective on how those transformations are being distributed, led, and interpreted across regions and institutions.

In synthesizing these insights, it becomes clear that the field of AI in education stands at a pivotal juncture. It is simultaneously expanding, converging, and reconfiguring in response to global technological shifts and educational imperatives. While progress in system development and instructional design is evident, the next phase of research must prioritize ethical accountability, inclusive participation, and cross-sector collaboration. The future of AI in education will depend not only on technical sophistication but also on the collective will to ensure that these technologies advance human learning in transparent, equitable, and socially meaningful ways.

6 Conclusion

This study offers a comprehensive bibliometric analysis of 5676 peer-reviewed articles, mapping the structural and thematic evolution of AI in education. The findings confirm that the field is expanding rapidly, with increasing attention to both technical innovation and broader pedagogical, ethical, and institutional implications. The increase in publishing output reflects not just the impetus generated by the COVID-19 pandemic and the advent of generative AI, but also a shift to the way AI is perceived—as a systemic force affecting educational models rather than a separate collection of tools.

Despite encouraging signs of interdisciplinary engagement and global interest, the distribution of research leadership and collaboration remains uneven. Dominant regions, particularly East and Southeast Asia, continue to drive much of the field's output. However, many nations in Latin America, Sub-Saharan Africa, and South Asia continue to operate on the margins, often contributing to studies without participating in agenda-setting or conceptual development. These asymmetries risk reinforcing global dependencies and limiting the contextual relevance of AI-enabled educational innovations.

The thematic analysis suggests that the field is undergoing an important conceptual transition. While intelligent tutoring systems and adaptive learning technologies remain central, newer concerns—such as algorithmic fairness, trust, explainability, and academic integrity—are gaining prominence. These themes reflect a broader shift toward human-centered perspectives that recognize the importance of aligning AI with educational values, learner agency, and socio-cultural diversity. However, there remains a gap between innovation and implementation, especially in under-resourced settings where digital infrastructure, institutional capacity, and teacher preparedness remain significant challenges.

Responding to these realities requires a recalibration of research priorities. AI in education must move beyond a purely innovation-driven paradigm toward more inclusive, interdisciplinary, and socially responsive approaches. Collaborative models that involve scholars from underrepresented regions not only as contributors but as co-leaders, and partnerships that center equity and mutual learning, are essential. Likewise, the field must continue to integrate educational theory, critical ethics, and policy relevance into its design and evaluation frameworks.

As recent RII Forum proceedings have emphasized, the challenge ahead is not merely to advance AI technologies but to govern and guide their development in ways that reflect shared human values [9, 10]. These global conversations frame AI not as an isolated innovation but as a transformative domain that intersects with policy, pedagogy, and equity.

References

1. Zawacki-Richter, O., Marín, V.I., Bond, M., Gouverneur, F.: Systematic review of research on artificial intelligence applications in higher education—where are the educators? Int. J. Educ. Technol. High. Educ. **16**, 39 (2019). https://doi.org/10.1186/s41239-019-0171-0
2. Holmes, W., Bialik, M., Fadel, C.: Artificial intelligence in education promises and implications for teaching and learning. Cent. Curriculum Redesign (2019). https://doi.org/10.58863/20.500.12424/4276068
3. Chen, L., Chen, P., Lin, Z.: Artificial intelligence in education: a review. IEEE Access **8**, 75264–75278 (2020). https://doi.org/10.1109/ACCESS.2020.2988510
4. Woolf, B.P., Lane, H.C., Chaudhri, V.K., Kolodner, J.L.: AI grand challenges for education. AI Mag. **34**, 66–84 (2013). https://doi.org/10.1609/aimag.v34i4.2490
5. Hwang, G.-J., Tu, Y.-F.: Roles and research trends of artificial intelligence in mathematics education: a bibliometric mapping analysis and systematic review. Mathematics **9** (2021). https://doi.org/10.3390/math9060584

6. Barbu, C., Sbughea, C.: Artificial intelligence in education: global trends, clusters, and perspectives. Ann. Univ. Dunarea de Jos of Galati Fascicle I. Econ. Appl. Inform. **30** (2024). https://doi.org/10.35219/eai1584040945
7. Zafari, M., Bazargani, J.S., Sadeghi-Niaraki, A., Choi, S.-M.: Artificial intelligence applications in K-12 education: a systematic literature review. IEEE Access **10**, 61905–61921 (2022). https://doi.org/10.1109/ACCESS.2022.3179356
8. Guo, S., Zheng, Y., Zhai, X.: Artificial intelligence in education research during 2013–2023: a review based on bibliometric analysis. Educ. Inf. Technol. **29**, 16387–16409 (2024). https://doi.org/10.1007/s10639-024-12491-8
9. Visvizi, A., Troisi, O., Grimaldi, M.: Research and Innovation Forum 2022: Rupture, Resilience and Recovery in the Post-Covid World. Springer Cham, Cham (2023). https://doi.org/10.1007/978-3-031-19560-0
10. Visvizi, A., Troisi, O., Corvello, V.: Research and Innovation Forum 2023: Navigating Shocks and Crises in Uncertain Times—Technology, Business, Society. Springer Cham, Cham (2024). https://doi.org/10.1007/978-3-031-44721-1
11. Bushuyev, S., Bushuyeva, N., Murzabekova, S., Khusainova, M., Saidullayev, R.: Transformation of the Education Landscape in an AI Environment, pp. 57–64 (2024). https://doi.org/10.56889/strd5315
12. Durak, G., Çankaya, S., Özdemir, D., uleyman Can, S.: Artificial intelligence in education: a bibliometric study on its role in transforming teaching and learning. Int. Rev. Res. Open Distrib. Learn. **25**, 219–244 (2024). https://doi.org/10.19173/irrodl.v25i3.7757
13. Afzaal, M., Xiao, S., Yan, D., Younas, M.: Mapping artificial intelligence integration in education: a decade of innovation and impact (2013–2023)-a bibliometric analysis. IEEE Access **1** (2013). https://doi.org/10.1109/access.2024.3443313
14. Tsekhmister, Y., Konovalova, T., Tsekhmister, B., Пушкарьова, T., NAHORNIAK, S.: Contemporary education: globalization and transformation process under the influence of artificial intelligence. Int. J. Eval. Res. Educ. **13**, 3443 (2024). https://doi.org/10.11591/ijere.v13i5.29016
15. D'Souza, R.: Scope of artificial intelligence in education. Int. J. Emerg. Knowl. Stud. **3**, 670–678 (2024). https://doi.org/10.70333/ijeks-03-09-035
16. Rucavado Rojas, D., Postigo, J.C.: The cycle of underrepresentation: structural and institutional factors limiting the representation of Global South authors and knowledge in the IPCC. Clim. Change **178**, 19 (2025). https://doi.org/10.1007/s10584-025-03857-z
17. Al-Tarawneh, A.M.A., AlOmoush, R.A., ul Islam, T., Janjua, J.I., Abbas, T., Ihsan, A.: Current trends in artificial intelligence for educational advancements. In: 2024 International Conference on Decision Aid Sciences and Applications (DASA), pp. 1–6. IEEE (2024). https://doi.org/10.1109/DASA63652.2024.10836340
18. Das, S. Anowar, S. Ghosh, B.: The rise of artificial intelligence in education: current trends and future prospects. In: Life as Basic Science: An Overview and Prospects for the Future, vol. 1, pp. 57–67. International Academic Publishing House (IAPH) (2024). https://doi.org/10.52756/lbsopf.2024.e01.006
19. Rashid, S.F., Duong-Trung, N., Pinkwart, N.: Generative AI in education: technical foundations, applications, and challenges. In: S. Kadry (ed.) Artificial Intelligence for Quality Education—Shaping the Future of Learning. IntechOpen (2024). https://doi.org/10.5772/intechopen.1005402
20. Williamson, B., Eynon, R., Potter, J.: Pandemic politics, pedagogies and practices: digital technologies and distance education during the coronavirus emergency. Learn. Media Technol. **45**, 107–114 (2020). https://doi.org/10.1080/17439884.2020.1761641
21. Smuha, N.A., Ahmed-Rengers, E., Harkens, A., Li, W., MacLaren, J., Piselli, R., Yeung, K.: How the EU can achieve legally trustworthy AI: a response to the European commission's proposal for an artificial intelligence act. SSRN Electron. J. (2021). https://doi.org/10.2139/ssrn.3899991
22. Gao, F., Jia, X., Zhao, Z., Chen, C.-C., Xu, F., Geng, Z., Song, X.: Bibliometric analysis on tendency and topics of artificial intelligence over last decade. Microsyst. Technol. **27**, 1545–1557 (2021). https://doi.org/10.1007/s00542-019-04426-y

23. Fisser, P., Rosenberg, J., Teske, J., Koehler, M., Mishra, P., Voogt, J., Tondeur, J., van Braak, J., Harris, J., Angeli, C.: Technological pedagogical content knowledge (TPACK): revision and rethinking. In: Society for Information Technology & Teacher Education International Conference. Association for the Advancement of Computing in Education (AACE), pp. 907–911 (2015)
24. Filgueiras, F.: Artificial intelligence and education governance. Educ. Citizsh. Soc. Justice **19**, 349–361 (2024). https://doi.org/10.1177/17461979231160674
25. UNICEF: JME Regional Classification. https://data.unicef.org/wp-content/uploads/2018/05/JME_Regional-Classifications.xlsx. Last accessed: 2 Apr 2025
26. Mouazen, A.M., Hernández-Lara, A.B.: The role of sustainability in the relationship between migration and smart cities: a bibliometric review. Digit. Policy Regul. Gov. **23**, 77–94 (2021). https://doi.org/10.1108/DPRG-04-2020-0051
27. Mouazen, A.M., Hernández-Lara, A.B.: Visualising the quality and the evolution of transactional and transformation leadership research: a 16-year bibliometric review. Total Qual. Manag. Bus. Excell. **34**, 148–182 (2023). https://doi.org/10.1080/14783363.2022.2034497

Optimizing the Feedback Process Through Instructor and AI-Generated Feedback

Anna Moni

Abstract This study explores how artificial intelligence (AI) can support feedback practices in undergraduate English for Academic Purposes (EAP) blended courses. It focuses on the challenge of providing timely and varied feedback that helps students understand goals, assess progress, and improve in the online asynchronous component of learning environments. Using a framework addressing different types and levels of learning support, the study compares instructor- and AI-generated feedback through content analysis of archived comments on student writing. Results show that instructor feedback was often personalized and context-specific but mainly addressed surface-level issues, offering limited support for learning strategies or independent skill development. In contrast, AI-generated feedback provided more balanced support and clearer forward-looking guidance but lacked personalization. Combining both forms of feedback may lead to a more comprehensive learner-centered experience. These findings offer insights for developing scalable and sustainable feedback systems that foster student engagement, autonomy, and long-term progress in higher education.

Keywords Instructor-generated feedback · AI-generated feedback · Content analysis · Feedforward · Self-regulation · SDG 4—quality education

1 Introduction

Feedback is essential to student learning, yet in online asynchronous settings, instructors struggle to provide timely, detailed responses that promote engagement [6, 38]. English for Academic Purposes (EAP) courses, which require complex written production and revision, make this challenge especially visible, as studies have shown that instructor feedback often emphasizes surface-level issues such as grammar and structure, with limited attention to broader learning strategies or self-regulation [12,

A. Moni (✉)
UOC (Universitat Oberta de Catalunya) & Deree College–The American College of Greece, Athens, Greece
e-mail: amoni@acg.edu

A. Visvizi et al. (eds.), *Research and Innovation Forum 2025 Vol2*, Springer Proceedings in Complexity, https://doi.org/10.1007/978-3-032-09062-1_4

17, 21, 25, 26]. The Matrix of Feedback for Learning [5] (see Fig. 1) offers a structure for supporting task performance, strategic thinking, and learner reflection, but its application with AI tools in asynchronous contexts remains underexplored.

This study advances the central argument that a hybrid feedback model—integrating instructor-generated and AI-generated feedback—offers a more pedagogically effective and scalable approach to supporting learner engagement, autonomy, and long-term skill development in asynchronous academic writing contexts. The rationale is developed in four stages. First, the paper situates the inquiry within existing scholarship on feedback types, levels, and the emerging role of AI in higher education, emphasizing persistent challenges in achieving timely, balanced, and actionable feedback in online learning environments [6, 10, 15, 21, 36].

Second, it details the methodological design, which involves a two-phase content analysis of over 14,000 feedback segments, guided by the Matrix of Feedback for Learning [5] and informed by established content analysis protocols [18, 27]. Third, the paper presents comparative findings on the distribution, alignment, and contextualization of instructor- and AI-generated feedback across feedback types (feed-up, feed-back, feed-forward) and levels (task, process, self-regulatory).

Finally, it critically discusses the pedagogical implications of these findings, proposing a human–AI collaborative feedback model that addresses the limitations of each modality and supports learner agency and pedagogical integrity [28, 35, 37, 38, 41]. In doing so, the study contributes to ongoing scholarly and policy debates on the responsible use of AI in education and explicitly supports quality education, which emphasizes inclusive, equitable, and effective learning for all [32, 37–39].

As a preliminary investigation, this study also lays the foundation for a broader research design. The hybrid feedback approach developed here will be implemented in future course cycles, accompanied by the collection of additional data—including student and instructor perceptions and control group comparisons—to enable triangulated evaluation and refinement. This qualitative exploratory study compares instructor-generated and AI-generated feedback within the asynchronous component of a blended undergraduate EAP course. It investigates the following research questions (RQs):

- RQ1. What are the affordances and limitations of instructor-generated versus AI-generated feedback in maintaining alignment with the Matrix of Feedback for Learning?
- RQ2. What potential implications does the combined use of instructor and AI-generated feedback have for enhancing learner and instructor experiences?

2 Literature Review

Effective feedback in higher education is a cyclical process involving students in understanding, interpreting, and responding to comments [6, 10]. It addresses three questions—where students are going (feed-up), how they're performing (feed-back), and how to improve (feed-forward)—across task, process, and self-regulation

Learner Stage	Feedback Level	Feeding Up: Where am I going?	Feeding Back: How am I going?	Feeding Forward: What do I have to do next?
Novice Proficient Advanced	Task	**Prompts:** - Today we are learning… - Success in this task will look like… (exemplar/model) - The key criteria for success are… - We are looking for…	**Prompts:** - You have/haven't met the learning intention by… - You have/haven't met the success criteria by… - Your answer/work is/isn't what we are looking for because…	**Prompts:** - To fully meet the learning intention you could… - Addressing the following success criteria would improve your work… - Adding/removing ___ would improve your work…
		Strategies: - Reduce complexity - Use exemplars/models - Identify misconceptions - Use diagnostic assessment for goal setting	**Strategies:** - Avoid overemphasis of error analysis - Feedback must be immediate - Match feedback to success criteria	**Strategies:** - Use language from the success criteria - Use scaffolding - Feed Forward must be timely - Use challenge - Refer to goals
	Process	**Prompts:** - The key ideas/concepts in this task are… - These ideas/concepts are related by… - Key questions you could ask about this task	**Prompts:** - Your understanding of the ideas/concepts within this task is… - Your thinking about this task is…	**Prompts:** - You could improve your understanding of ___ concepts by… - Thinking further about ___ could improve your work by…

Fig. 1 The matrix of feedback for learning. *Source* Brooks et al. [5] [Permission attached]

Advanced	**Process**	are… - Skills you will need in this task are… - Strategies you will need in this task are… **Strategies:** - Use graphical organisers - Reduce scaffolding - Increase complexity - Use mastery goals	- You demonstrated ___ skills to a ___ level - You used ___ strategies to a ___ level **Strategies:** - Feedback amount can start to increase - Feedback complexity can increase - Use prompts or cues	- You could improve your ___ skills by… **Strategies:** - Feed Forward amount can start to increase - Feed Forward complexity can increase - Use prompts or cues - Use challenge
	Self-Regulatory	**Prompts:** - How will you use the learning intention? - How could you use the success criteria? - Which other ways could you monitor your work? **Strategies:** - Reduce emphasis of exemplars - Mastery and performance goals	**Prompts:** - Are you on track with your work? - How do you know? - To which level are you satisfying the success criteria? - Are you on track to achieving your goal? - How do you know? **Strategies:** - Delay feedback - May only require verification feedback	**Prompts:** - How could you deepen your understandings? - How could you improve your work? - What is the next step for your learning? - How do you know? **Strategies:** - Delay feedback - Reduce teacher reliance - Develop self-regulated learners

Fig. 1 (continued)

levels [5, 15]. These levels align with learner proficiency. Task-level feedback focuses on accuracy and meeting expectations [15]; process-level feedback helps students improve how they plan and complete tasks [5, 15]; self-regulation feedback fosters reflection and strategy adjustment, supporting autonomy and growth [15, 28]. Research shows this feedback improves outcomes, supports independence, and enables learners to take control of their learning [6, 10, 15, 28, 40, 42]. Yet teachers often fail to provide balanced feedback due to large classes, workload, poor training, personal preferences [8, 17, 19, 23], and institutional reliance on summative assessment [6, 28]. In academic writing and EFL/L2 classes, instructors often miss opportunities to provide balanced feedback across all levels [5, 15, 21], reducing its effectiveness [1, 20]. Only by integrating process- and self-regulation-level feedback with feedforward guidance can future performance and stronger writing skills develop [5, 15, 20, 30, 42].

AI-generated feedback offers a promising strategy to support instructor feedback in developing learner autonomy and self-regulation [13]. It is particularly effective in addressing surface-level issues like mechanics, vocabulary, and grammar, in line with rubrics [9, 24, 35], although it may overcorrect [22]. Tools like ChatGPT generate feedback with greater clarity and readability than that of instructors [7, 9]. Studies show that AI-based feedback prompts more learner revisions of their work because it is timely, specific, and actionable [14, 35]. Unlike time-limited instructor feedback, AI responses are instant, allowing for frequent, iterative improvement. However, AI lacks the ability to personalize feedback, essential for developing argumentation, creativity, and critical thinking [31, 36, 44]. Instructors address this through context-rich guidance and discussion [2, 24, 42], though their feedback can be inconsistent and time-consuming [34, 43]. A blended approach helps balance these issues: AI supports grammar and structure; instructors guide content and reasoning [22].

The sequence of feedback also matters. Research shows that AI feedback followed by instructor feedback and revision is more effective for learners [35]. This approach aligns with cognitive load theory, which recommends starting with task-level feedback before advancing to process or self-regulation [11]. Research confirms that AI cannot replace human feedback, but combining both maximizes writing improvement and self-regulation [3, 12]. Still, optimal strategies for integrating AI and instructor feedback across disciplines remain underexplored [9].

Recent interdisciplinary research highlights AI's potential to support sustainability, transparency, and learner agency in education [32, 37]. In alignment with a call for quality-focused integration of digital tools in education [38], this study proposes a human-AI feedback model grounded in pedagogical coherence and learner agency. While this study does not directly address AI ethics, it contributes to responsible AI use by emphasizing the limits of automation and the continued need for instructor judgment and pedagogical integrity [32, 39, 39].

3 Method

This qualitative exploratory study was conducted within the online asynchronous component of an undergraduate blended EAP course. The research applied content analysis to compare instructor-generated and AI-generated feedback using the Matrix of Feedback for Learning [5] as the primary coding framework.

3.1 Phase 1: Data Collection and Coding

Institutional approval was obtained to analyze anonymized archived data, including formative and summative assignments and instructor comments. The full dataset consisted of 1026 student documents, comprising 14,223 feedback segments and 281,579 words. Purposeful sampling was used to reflect the study's focus on embedded feedback design. Coding units were segmented into meaningful sentences, and analysis was conducted using MAXQDA 2022. Two independent coders followed an established content analysis protocol [18, 33]. Intercoder reliability was calculated using Kappa coefficients [4], improving from 0.43 to 0.83 across five iterative rounds, indicating a high level of agreement and coding consistency.

3.2 Phase 2: Subsampling and AI-Generated Feedback

To ensure representation across the full instructional and learner landscape, a stratified random subsample comprising 20% of the coded dataset (n = 205 documents) was selected [29]. Stratification was based on three criteria: (1) inclusion of all course sections during the sampling period, (2) representation of all instructors involved in the course, and (3) a wide distribution of student performance levels, based on final assignment grades (from high to low).

This approach was designed to capture variation in instructional style, feedback practices, and student engagement, enabling a more nuanced comparison between instructor and AI-generated feedback across a diverse set of real-world learning contexts. AI-generated feedback was produced using ChatGPT-4, which was prompted using task rubrics, assignment guidelines, exemplar submissions, and the Matrix of Feedback for Learning. This multi-layer prompting strategy followed established principles designed to enhance alignment between AI feedback and pedagogical criteria [16]. Both instructor- and AI-generated feedback were coded independently and then compared along four dimensions: feedback type (feed-up, feedback, feed-forward), feedback level (task, process, self-regulatory), rubric alignment, and contextualization.

4 Findings

The findings are presented in two phases, aligned with the study's two research questions. In Phase 1, content analysis of 14,223 instructor-generated feedback segments revealed a dominant focus on feed-back (62.6%), comments about current performance, followed by feed-forward (34.8%) and minimal use of feed-up (2.6%). As shown in Table 1, feedback was primarily situated at the task level (64.4%), with much less emphasis on process-level strategies (23.8%) or self-regulation (11.9%).

In phase 2, the comparison between instructor and AI-generated feedback, from the 20% sample, revealed interesting similarities and differences:

a. Instructor feedback mainly consisted of feed-back commentaries (72%) with moderate use of feed-forward suggestions (27.2%) and very low instances of feed-up comments (0.8%). In contrast, AI-generated feedback presented a more balanced distribution of feedback types between feed-back (55.3%) and feed-forward (43%), while feed-up was minimal (1.7%);
b. In relation to the level of feedback, instructors' feedback was heavily task-oriented (75.51%), with significantly less emphasis on process (19.33%), and very low focus on self-regulation (5.2%). In contrast, AI feedback showed a more even distribution across task (37.10%), process (32.44%), and self-regulatory levels (30.47%) (see Table 2 and 3);

Table 1 Instructor feedback segments by feedback types and levels

Segments by feedback type			Segments by feedback level		
	Frequency (type)	Percentage (type) (%)		Frequency (level)	Percentage (level) (%)
Feed-up	372	2.60	Task level	9154	64.36
Feed-back	8901	62.60	Process level	3379	23.76
Feed-forward	4950	34.80	Self-regulatory level	1690	11.88
Total	14,223	100	Total	14,223	100

Source Author

Table 2 Summary of instructor feedback segments (20% sample)

Source	Feedback type	Frequency	Percentage (%)	Feedback level	Frequency	Percentage (%)
Instructor	Feed-up	20	0.80	Task level	1860	75.51
	Feed-back	1773	72.00	Process level	475.8	19.33
	Feed-forward	670	27.20	Self-regulatory level	127	5.16
	Total	2463	100		2463	100

Source Author

Table 3 Summary of AI feedback comments (20% sample)

Source	Feedback type	Frequency	Percentage (%)	Feedback level	Frequency	Percentage (%)
AI	Feed-up	49	1.7	Task level	1074	37.10
	Feed-back	1622	55.30	Process level	939	32.44
	Feed-forward	1224	43.00	Self-regulatory level	882	30.47
	Total	2895	100		2895	100

Source Author

Table 4 Rubric alignment and contextualization

Source	Rubric alignment	Percentage (%)	Contextualization	Percentage (%)
Instructor	2300	95.00	1569	98.00
AI	2600	99.00	1987	45.60

Source Author

c. In relation to rubric alignment, both AI (99%) and instructors' feedback (95%) reflected consistency in evaluation criteria adherence (see Table 4);
d. Contextualization (see Table 4) was higher in instructor feedback (98%), while on AI feedback it was much lower (45.6%).

Tables 5 and 6 show examples of coding for both instructor-generated and AI-generated feedback. As shown on Table 3, instructor-generated feedback primarily emphasizes detailed corrections (task-level feedback), coupled with context-specific, personalized remarks (high contextualization). However, instructors rarely engage learners in reflective or metacognitive (self-regulatory) strategies. AI-generated feedback, in contrast (see Table 4), frequently provides rubric-aligned, balanced feedback across task, process, and self-regulatory levels, though it notably lacks the deep contextualization and personalized approach present in instructor-generated feedback.

5 Discussion

This study examined the strengths and limitations of instructor- and AI-generated feedback (RQ1) and how combining the two might improve learner and instructor experiences in an asynchronous EAP course (RQ2). Instructor feedback was personalized and grounded in course context but focused mostly on providing corrections, confirming earlier research on EFL and L2 writing [5, 21]. While it aligned with grading criteria, it offered less support for how students complete tasks (process) or manage their own learning (self-regulation). These results reflect broader concerns that time and workload often prevent instructors from offering higher-level feedback

Table 5 Examples of coding for instructor-generated feedback

Code	Feedback segment example
Feed-back (task level)	You need to check verb tense consistency in paragraph 2 where you shift from past to present
Feed-forward (process level)	In future essays, consider outlining your thesis statement in your introduction to help structure your argument more effectively. Remember: we need to guide the reader from general to specific
Feed-up (task level)	As discussed in the asynchronous online week 9, as requested in the essay instructions, your essay must adhere to APA referencing standards to demonstrate your engagement with credible academic sources
Feed-back (self-regulation)	I noticed some online tasks are incomplete! Have you reviewed your progress in the Week 3 to ensure you're on track?
Rubric alignment	Good job consistently formatting your essay according to the rubric requirements, especially the headings and in-text citations
Contextualization	Good comments about academic integrity, you integrated concepts from our Week 9 VoiceThread discussion, highlighting your deeper understanding of the topic

Source Author

Table 6 Examples of coding for AI-generated feedback

Source	Feedback segment example
Feed-back (task-level)	Your introduction lacks a clear thesis statement as required by the rubric criterion 'Thesis Clarity'
Feed-forward (process-level)	For improved coherence, try using transitional phrases to explicitly show the connections between your paragraphs
Feed-forward (self-regulatory level)	Consider reviewing the provided exemplar essays to enhance your understanding of effective argument structuring, and reflect on your own strengths and areas for improvement
Feed-up	To successfully complete this task, refer closely to the rubric criteria for argumentation and coherence
Contextualization	Your thesis statement effectively outlines your central argument and aligns with the essay prompt. To further strengthen your introduction, revisit the funnel structure
Rubric alignment	You have successfully formulated a thesis that accurately reflects the purpose of your essay and aligns directly with the assignment guidelines

Source Author

[6, 17], and underscore the value of combining human and AI feedback to meet different student needs.

In contrast, AI-generated feedback demonstrated a more balanced distribution across all feedback levels, particularly in providing feed-forward suggestions and promoting self-regulatory practices. This aligns with research suggesting that generative AI (GenAI) tools can support learner autonomy and revision through clear, structured, and rubric-aligned commentary [9, 12, 35]. However, the lack of contextualization in AI feedback, particularly its inability to reference course-specific material, prior student work, or instructor expectations, reinforces critiques about its limitations in supporting deeper rhetorical and analytical development [2, 31, 44].

A hybrid model may offer the most effective solution. It supports feedback diversity and distributes cognitive load by scaffolding revision from surface-level issues toward deeper learning [11]. Moreover, starting with AI-generated feedback followed by instructor input may increase agency, uptake and engagement, as learners move from basic corrections to more reflective and independent revision [35]. This approach may also help reduce instructor workload, allowing faculty to focus on more meaningful pedagogical interactions without compromising feedback quality [9]. More importantly, combining instructor and AI-generated feedback supports the practical use of the Matrix of Feedback for Learning [5] in asynchronous course design. It promotes clearer alignment between task instructions, learning strategies, and student self-regulation, essential components of feedback that drives long-term academic growth [6, 28].

This hybrid model aligns with current calls for responsible and pedagogically grounded uses of AI in education, emphasizing sustainability, transparency, and student autonomy [37, 38]. Rather than viewing AI as a value-neutral technological enhancement [37, 38], this study frames human–AI collaboration as a deliberate design decision that serves educational goals. The integration of AI is positioned not as innovation for its own sake, but as a means to foster equitable and scalable feedback systems that support long-term academic growth and learner empowerment. In this sense, the model responds to concerns about superficial applications of educational technology [38] and contributes to the broader shift toward ethical, conceptually sound, and learner-centered AI practices in line with sustainable development objectives [37].

6 Conclusions, Future Research Directions, and Limitations

This study supports an emerging educational philosophy that positions human–AI collaboration not as substitution but as a complementary process. This hybrid feedback approach aligns with current calls for ethical and pedagogically grounded uses of AI in learning design [38], emphasizing AI's strengths—clarity, immediacy, and scalability—while preserving the contextual sensitivity and dialogic value of instructor feedback [2, 44]. In this approach, AI acts as a scaffold for promoting learner autonomy and increasing engagement with formative feedback.

The hybrid approach relevance extends beyond language education to writing-intensive discipline contexts (e.g., STEM, teacher training, and business education), where formative assessment is central. This application aligns with the Matrix of Feedback for Learning [25, 26], which—though not discipline-specific—is grounded in cognitive and pedagogical principles such as task complexity, self-regulation, and formative cycles. These foundations make this approach adaptable across subject areas [7].

As a preliminary investigation, this study focused on multiple sections of a single undergraduate EAP course and a large corpus of archived feedback. While this allowed for methodological depth, it constrained the exploration of feedback's relational and dynamic dimensions. These limitations will inform the next phase of research design. The hybrid feedback model will be implemented in future EAP course cycles and an undergraduate Psychology course, enabling initial cross-disciplinary validation. Data collection will include student and faculty perceptions, control group comparisons, and focus groups with both learners and instructors, supporting a triangulated evaluation of the model pedagogical value, scalability, and adaptability.

References

1. Abdullah, N., Aziz, A: Between preaching and practicing. The incongruities of teachers' written feedback beliefs and practices: a case study. J. Educ. Res. Indigenous Stud. **1**(1), 1–19, (2020)
2. Barrot, J.S.: Using ChatGPT for second language writing: pitfalls and potentials. Assessing Writ. **57** (2023). https://doi.org/10.1016/j.asw.2023.100745
3. Bitchener, J., Storch, N.: Written corrective feedback for L2 development. Multilingual Matters, Bristol (2016). https://doi.org/10.21832/9781783095056
4. Brennan, R.L., Prediger, D.J.: Coefficient kappa: some uses, misuses, and alternatives. Educ. Psychol. Meas. **41**(3), 687–699 (1981). https://doi.org/10.1177/001316448104100307
5. Brooks, C., Carroll, A., Gillies, R.M., Hattie, J.: A matrix of feedback for learning. Aust. J. Teach. Educ. **44**(4), 14–32 (2019). https://doi.org/10.14221/ajte.2018v44n4.2
6. Carless, D., Boud, D.: The development of student feedback literacy: enabling uptake of feedback. Assess. Eval. High. Educ. **43**(8), 1315–1325 (2018). https://doi.org/10.1080/02602938.2018.1463354
7. Dai, W., Lin, J., Jin, F., Li, T., Tsai, Y.-S., Gašević, D., Chen, G.: Can large language models provide feedback to students? A case study on ChatGPT. Comput. Educ. **201** (2023)
8. Er, E., Dimitriadis, Y., Gašević, D.: Collaborative peer feedback and learning analytics: theory-oriented design for supporting class-wide interventions. Assess. Eval. High. Educ. **46**(2), 169–190 (2021). https://doi.org/10.1080/02602938.2020.1764490
9. Escalante, J., Pack, A., Barrett, A.: AI-generated feedback on writing: Insights into efficacy and ENL student preference. Int. J. Educ. Technol. High. Educ. **20**(57) (2023). https://doi.org/10.1186/s41239-023-00425-2
10. Esterhazy, R.: Re-conceptualizing feedback through a sociocultural lens. In: Henderson, M., Ajjawi, R., Boud, D., Molloy, E. (eds.) The impact of feedback in higher education. Palgrave Macmillan (2019). https://doi.org/10.1007/978-3-030-25112-3_5
11. Evans, P., Vansteenkiste, M., Parker, P., Kingsford-Smith, A., Zhou, S.: Cognitive load theory and its relationships with motivation: a self-determination theory perspective. Educ. Psychol. Rev. **36**(1) (2024). https://doi.org/10.1007/s10648-023-09841-2
12. Ferris, D.R.: Responding to student writing: teachers' philosophies and practices. Assess. Writ. **19**, 6–23 (2014). https://doi.org/10.1016/j.asw.2013.09.004

13. Ghafouri, M., Hassaskhah, J., Mahdavi-Zafarghandi, A.: From virtual assistant to writing mentor: exploring the impact of a ChatGPT-based writing instruction protocol on EFL teachers' self-efficacy and learners' writing skill. Lang. Teach. Res. (2024). https://doi.org/10.1177/13621688241239764
14. Guo, K., Wang, D.X.: To resist it or to embrace it? Examining ChatGPT's potential to support teacher feedback in EFL writing. Educ. Inform. Technol. **29**(7), 8435–8463 (2024)
15. Hattie, J., Timperley, H.: The power of feedback. Rev. Educ. Res. **77**(1), 81–112 (2007). https://doi.org/10.3102/003465430298487
16. Ingley, C., Pack, C.: Prompting frameworks for AI feedback in education. Educ. Technol. Rev. **31**(3), 142–155 (2023)
17. Junqueira, L., Payant, C.: I just want to do it right, but it's so hard: a novice teacher's written feedback beliefs and practices. J. Second. Lang. Writ. **27**, 19–36 (2015). https://doi.org/10.1016/j.jslw.2014.11.001
18. Krippendorff, K.: Content analysis: an introduction to its methodology, 4th edn. SAGE (2019)
19. Lee, I.: L2 writing teachers' perspectives, practices, and problems regarding error feedback. Assess. Writ. **8**(3), 216–237 (2003). https://doi.org/10.1016/j.asw.2003.08.002
20. Lee, I.: Ten mismatches between teachers' beliefs and written feedback practice. ELT J. **63**(1), 13–22 (2009). https://doi.org/10.1093/elt/ccn010
21. Lee, I.: Understanding teachers' written feedback practices in Hong Kong secondary classrooms. J. Second. Lang. Writ. **17**(3), 144–164 (2008). https://doi.org/10.1016/j.jslw.2007.10.001
22. Lin, S., Crosthwaite, P.: The grass is not always greener: teacher vs. GPT-assisted written corrective feedback. System **127** (2024). https://doi.org/10.1016/j.system.2024.103529
23. Mao, Z., Crosthwaite, P.: Investigating written corrective feedback: (mis)alignment of teachers' beliefs and practices. J. Second Lang. Writ. **45**, 46–60 (2019)
24. Mizumoto, A., Eguchi, M.: Exploring the potential of using an AI language model for automated essay scoring. Res. Methods Appl. Linguis. **2**(2) (2023). https://doi.org/10.1016/j.rmal.2023.100050
25. Moni, A.: Learner perceptions of the feedback process in the online component of a blended course. Online Learn **28**(2), 1–25 (2024). https://doi.org/10.24059/olj.v28i2.3967
26. Moni, A., Martínez-Argüelles, M.-J., Serradell-López, E.: Exploring feedback design perceptions and relationships with scores in the online component of an EAP-blended course. Appl. Sci. **14**(24), 11554 (2024). https://doi.org/10.3390/app142411554
27. Neuendorf, K.A.: The content analysis guidebook, 2nd edn. SAGE (2017)
28. Nicol, D., Macfarlane-Dick, D.: Formative assessment and self-regulated learning: a model and seven principles of good feedback practice. Stud. High. Educ. **31**(2), 199–218 (2006). https://doi.org/10.1080/03075070600572090
29. Patton, M.Q.: Qualitative research and evaluation methods: Integrating theory and practice, 4th edn. SAGE (2015)
30. Sadler, D.R.: Beyond feedback: developing student capability in complex appraisal. Assess. Eval. High. Educ. **35**(5), 535–550 (2010). https://doi.org/10.1080/02602930903541015
31. Shabara, R., ElEbyary, K., Boraie, D.: Teachers or ChatGPT: The issue of accuracy and consistency in L2 assessment. J. Teach. Engl. Technol. **24**(2), 71–92 (2024). https://doi.org/10.56297/vaca6841/LRDX3699/XSEZ5215
32. Siddiquee, T.A.R., Islam, M.M., Farid, A., Hossain, M.A.: Architecture education: rubrics in Google classroom as a tool of improving the assessment and learning. In: Visvizi, A., Troisi, O., Corvello, V. (eds.) Research and Innovation Forum 2023. RIIFORUM 2023. Springer Proceedings in Complexity, pp. 373–386. Springer, Cham (2024). https://doi.org/10.1007/978-3-031-44721-1_28
33. Strijbos, J.W., Martens, R.L., Prins, F.J., Jochems, W.M.G.: Content analysis: what are they talking about? Comput. Educ. **46**(1), 29–48 (2006). https://doi.org/10.1016/j.compedu.2005.04.002
34. Teng, M.F.: A systematic review of ChatGPT for English as a foreign language writing: opportunities, challenges, and recommendations. Int. J. TESOL Stud. **6**(3), 36–57 (2024)

35. Tran, T.T.T.: Enhancing EFL writing revision practices: the impact of AI-and teacher-generated feedback and their sequences. Educ. Sci. **15**(2), 232 (2025). https://doi.org/10.3390/educsci15020232
36. Varghese, N.N., Jose, B., Cleetus, A.: The power duo: unleashing cognitive potential through human-AI synergy in STEM and non-STEM education. Front. Educ. (2025). https://doi.org/10.3389/feduc.2025.1534582
37. Visvizi, A.: Artificial intelligence (AI) and sustainable development goals (SDGs): exploring the impact of AI on politics and society. Sustainability **14**, 1730 (2022). https://doi.org/10.3390/su14031730
38. Visvizi, A., Daniela, L., Chen, Ch.W.: Beyond the ICT- and sustainability hypes: a case for quality education. Comput. Hum. Behav. **107** (2020). https://doi.org/10.1016/j.chb.2020.106304
39. Visvizi, A., Kozłowski, K., Nawaz, R.: Higher education institutions and the global digital transformation: beyond the question of which skills for the 21st century. In: Visvizi, A., Kozłowski, K., Nawaz, R. (eds.) Higher Education Institutions and Covid-19. Toward Resilience and Sustainability Through Emergencies. Routledge, London & New York (2023). https://www.routledge.com/Higher-Education-Institutions-and-Covid-19-Toward-Resilience-and-Sustainability/Visvizi-Kozlowski-Nawaz/p/book/9781032380476
40. Winstone, N., Carless, D.: Designing effective feedback processes in higher education: a learning-focused approach, 1st edn. Routledge (2020)
41. Winstone, N.E., Nash, R.A., Parker, M., Rowntree, J.: Supporting learners' agentic engagement with feedback: a systematic review and a taxonomy of recipience processes. Educ. Psychol. **52**(1), 17–37 (2017). https://doi.org/10.1080/00461520.2016.1207538
42. Wisniewski, B., Zierer, K., Hattie, J.: The power of feedback revisited: a meta-analysis of educational feedback research. Front. Psychol. **10**, 3087 (2020). https://doi.org/10.3389/fpsyg.2019.03087
43. Yu, S., Zheng, Y., Jiang, L., Liu, C., Xu, Y.: "I even feel annoyed and angry": teacher emotional experiences in giving feedback on student writing. Assessing Writ. **48** (2021). https://doi.org/10.1016/j.asw.2021.100528
44. Zhu, Y., Zhang, L.: Exploring AI-generated feedback on writing: a case study of ChatGPT in EFL contexts. Lang. Teach. Res. **28**(3), 345–362 (2024). https://doi.org/10.17265/1539-8080/2024.03.002

The Trajectory of Educational Reform in Chinese Classrooms During the GenAI Era

Chenxiang Long and Simian Zhu

Abstract A new wave of technological revolution and knowledge has been initiated with the emergence and pervasive application of generative AI, which has had a profound impact on all sectors. This paper focuses on the bidirectional impacts of generative AI on four key educational elements—teaching objectives, content, methodologies, and assessment—proposing a novel educational partnership framework centered around the "teacher-AI-student triadic interaction model". Through a mixed-methods approach combining literature analysis, case studies, and model construction, the research reveals that while generative AI enhances teaching efficiency, it may simultaneously induce challenges including fragmentation of educational objectives, content accuracy issues, and weakening of teacher-student emotional bonds. Building on these findings, the study proposes systematic implementation pathways: achieving balance between technological empowerment and educational values through restructuring classroom collaboration patterns, strengthening digital literacy cultivation, designing critical thinking training strategies, and establishing multi-stakeholder collaborative governance frameworks. This paper suggests implementation pathways for classroom reform in China, providing educators with valuable references and insights as they explore practical strategies for classroom reform in the era of generative AI. This is achieved through an analysis of the challenges involved in classroom teaching reform in China.

Keywords Generative AI · Chinese education · Teaching reform · Classroom teaching · Digital literacy · Critical thinking · Quality education

1 Introduction

The report of the 20th National Congress of the Communist Party of China, which was held in October 2022, proposed the "promotion of education digitalization and the establishment of a learning society and learning-oriented country for lifelong

C. Long · S. Zhu (✉)
College of Biology and Engineering, Guizhou Medical University, Guiyang, China
e-mail: simianzhu@gmc.edu.cn

A. Visvizi et al. (eds.), *Research and Innovation Forum 2025 Vol2*, Springer Proceedings in Complexity, https://doi.org/10.1007/978-3-032-09062-1_5

education for all" in response to the rapid advancement of artificial intelligence (AI) technology. The "14th Five-Year Plan for the Development of the Digital Economy" also underscores the significance of "efficiently deploying AI infrastructure and enhancing industry empowerment capabilities to support the development of 'intelligent +' industries."

A technology known as Generative Artificial Intelligence is capable of generating original content, including text, images, and audio, in response to its input. The capacity of the system to produce customized content suggests that it has the potential to infuse new energy into educational reform by generating personalized learning materials that are tailored to the needs and abilities of learners [1]. Generative AI, a novel element of education digitalization, broadens its scope by introducing innovation into the dissemination of knowledge and the teaching process [2]. However, its unregulated applications have sparked controversies concerning academic integrity, data security, and diminished creativity [3]. The core research question addressed in this paper is: How can systematic optimization of instructional design harness generative AI to enhance teaching efficiency while circumventing its technical limitations and ethical risks?

The paper is structured as follows: Part 2 elaborates the research methodology; Part 3 analyzes generative AI's benefits and challenges for classroom teaching; Part 4 The Path to Implementation for Establishing New Educational Collaborations; Part 5 presents specific strategies for cultivating digital literacy and critical thinking; Part 6 discusses research limitations and practical implications; Part 7 concludes with key findings and future directions.

2 The Research Methodologies

This study employs a mixed-methods research approach:

Literature analysis: By searching major academic databases, we systematically screened core journal articles and policy documents related to "generative AI + education" published between 2018 and 2023.

Case studies: In the implementation pathways section, relevant cases are cited to analyze observed behavioral patterns of teachers and students using AI tools and their impact on teaching effectiveness.

Model construction: Grounded in generative AI's impacts on the "four fundamental elements" of classroom teaching, we propose a framework for establishing new educational partnerships in the digital intelligence era. This model comprehensively explores implementation pathways through three dimensions: restructuring classroom collaboration patterns, constructing next-generation classroom environments, and developing collaborative governance and guidance mechanisms for stakeholders.

3 Generative AI's Benefits and Challenges for Classroom Teaching

The education industry is undergoing a rapid transformation in the digital-intelligent era. The traditional teaching and learning processes have been enhanced and information has become more accessible as a result of the integration of generative AI. It has profoundly penetrated all facets of education and has introduced new tools to address real-world teaching challenges. The subsequent section provides a concise summary of its contributions in relation to the four components of classroom instruction.

The traditional method of teaching content was constrained by the knowledge of individual teachers and textbooks, resulting in a relatively static exposure of students to knowledge. Interdisciplinary and cross-cultural exchange and collaboration are facilitated by the efficacy of Generative AI in personalized content generation and delivery. Dialogues and Q&A sessions between students and AI-powered "assistants" have become common examples of AI + education applications. These tools considerably improve the efficiency and adaptability of content delivery, thereby expanding the scope of learning beyond textbooks, broadening students' perspectives, and facilitating a truly personalized education.

Traditional classrooms were primarily focused on the one-way transmission of knowledge by teachers, with students passively receiving information. This was the case in terms of teaching methodologies. The transition to discovery-based and participatory teaching models is expedited by generative AI [4]. Generative AI serves as a teaching assistant for educators and a learning companion for students, facilitating the rapid retrieval, organization, extraction, and review of knowledge. It generates plausible virtual scenarios, delivers content that is appropriate for students' levels, and customizes learning paths. This significantly reduces the burden of teachers, enabling them to concentrate on the improvement of their teaching abilities. Additionally, it provides students with increased autonomy, which in turn enhances their motivation and initiative [5]. Traditional methods of evaluating teaching rely largely on exam results and the subjective judgment of teachers, which frequently fail to sufficiently reflect the learning status of students. Generative AI can provide personalized learning resources that are tailored to the requirements and interests of students by efficiently comparing their learning outcomes both horizontally and vertically through natural language processing. Generative AI also facilitates the effective evaluation of teaching outcomes for educational administrators by integrating and analyzing data from online learning, practical teaching, student feedback, and teaching results. This enhances the objectivity, comprehensiveness, and efficacy of teaching evaluations.

Nevertheless, the utilization of generative AI tools does not solely provide advantages. Their irrational abuse can result in a variety of issues, as is the case with all emergent technologies. This is particularly apparent when generative AI is employed casually by teachers and students as a teaching instrument without proper organization, which exposes a variety of drawbacks and controversies.

Classroom teaching encompasses four fundamental components: teaching objectives, teaching content, teaching methods, and teaching evaluation. They are interconnected and constitute a comprehensive curriculum system. Classroom teaching efficacy can only be significantly enhanced when these components are in harmony [6].

The following is a comprehensive examination of the manner in which generative AI tools affect the four components of teaching.

3.1 *Influence on Educational Objectives*

It is imperative that educators possess a comprehensive comprehension of the teaching objectives when employing AI tools for instructional design. AI tools may be unable to accurately understand the teacher's teaching intentions if explicit instructions are not carefully designed, resulting in discrepancies between the generated learning objectives and the actual teaching goals [7]. For pupils, the casual use of AI tools may result in fragmented teaching objectives that lack systematization and coherence. The current "conversational" use logic of generative AI tools may cause students to become distracted during interactions, causing their attention to transfer to a variety of different knowledge points. This can make it difficult to form a comprehensive knowledge system and skill set, thereby impacting the achievement of teaching goals [8].

The cultivation of ability goals and holistic development may be neglected by generative AI tools, which may unduly emphasize knowledge transmission and skill training as teaching objectives. Human educators are capable of guiding students in the development of good moral values and cooperation through daily interactions, whereas certain AI tools are unable to do so. In addition, the inability to accurately establish personalized teaching objectives may result from the fact that generative AI's comprehension of students is primarily based on data and algorithms, which may not completely capture the unique needs and interests of those students.

3.2 *Influence on the Content of Instruction*

The accuracy, systematization, and diversity of AI-generated instructional content may be compromised as a result of certain inherent flaws in generative AI tools.

Errors, falsehoods, or information that deviates from reality in AI-generated content are referred to as the "hallucination" problem in AI. These outputs may appear plausible, but they are actually inaccurate or nonexistent, which potentially misleads users. This issue has the potential to lead to substantial factual errors and misinformation in the educational content produced by AI.

The "context" problem arises when AI encounters difficulty in comprehending and maintaining logical coherence in its generated content, resulting in inconsistencies or a lack of alignment with the encompassing context. The "short-term memory" issue is a result of AI's limited capacity to store and recall information, which causes it to rapidly forget prior content. Additionally, the absence of long-term memory can impede its capacity to utilize past experiences to produce more insightful and precise content. The AI may produce superficial or repetitive content as a result of these memory issues. Additionally, the AI may not provide accurate or comprehensive answers when confronted with tasks that necessitate in-depth background knowledge or complex logical and mathematical reasoning, which can impact the depth of the generated content.

The coherence and logical flow of learning can be compromised by the combination of these two issues, necessitating that instructors or students interrupt the process to make prompt adjustments or restart the dialogue, resulting in fragmented teaching [9]. Moreover, generative AI is unable to evoke emotional resonance in students as effectively as teachers do with evocative stories and personal experiences when it comes to social hot topics, history, culture, or any content that requires emotional and value-based explanations. This absence of emotional connection can impede students' comprehension and acceptance of the instructional material.

3.3 *Effect on Teaching Methods*

The "only" mode of interaction between teachers and pupils is no longer face-to-face communication due to the widespread penetration of AI tools in classroom teaching. This change not only enables learning to transcend the confines of the classroom but also somewhat undermines the emotional bond and interaction between instructors and students.

There are two primary disadvantages to instructors' excessive dependence on AI tools: On the one hand, teachers may incline to prioritize the knowledge delivery methods that AI tools excel at during the preparation process, neglecting diversified teaching methods such as group discussions or experimental activities that are crucial for the development of students' comprehensive abilities. Alternatively, students may not be able to satisfy their emotional requirements during classroom interactions, which could lead to a lack of the social emotional skills necessary for success in the real world [10].

The over-reliance of students on AI tools for learning also has two significant drawbacks: On the one hand, it may impede their capacity to think critically and learn independently, as they depend on AI tools for solutions. Conversely, the ease with which students can acquire knowledge and complete assignments undermines their intellectual growth and profound insights. This is due to the instantaneous satisfaction that AI tools offer, which diminishes their capacity for profound comprehension and wisdom.

3.4 Influence on the Evaluation of Teaching

Generative AI provides personalized support for classroom assessments and real-time dynamic analysis; however, its technical limitations and over-reliance can result in a variety of risks, such as imbalances in educational fairness, academic integrity crises, and compromised teacher evaluation capabilities.

From the teacher's standpoint: One potential consequence of AI "hallucinations" and biases in training data is that the accuracy of evaluations may be compromised, resulting in erroneous or biased assessments. Teachers will experience difficulty accurately understanding students' learning situations if they excessively depend on these evaluations and neglect their own observation, analysis, and judgment. Consequently, their own evaluative abilities will be compromised. Conversely, an overreliance on generative AI for evaluation, while disregarding teacher observations and student self-assessments, can result in rigid, restricted evaluation methods that hinder the comprehensive and effective assessment of students, thereby distorting the evaluation results [11].

From the student's standpoint: On the one hand, the overreliance on AI-generated content can result in academic integrity crises by fostering plagiarism, scamming, and other forms of academic dishonesty [12]. Furthermore, the development of critical thinking and analytical abilities is restricted by the absence of profound cognitive engagement with the material. Conversely, privacy violations and security risks are a cause for concern, as AI tools collect and analyze students' learning data for evaluation. The personal privacy and information security of students may be jeopardized if this data is abused.

Education administrators (schools) are also affected. Initially, the high costs associated with the implementation of generative AI teaching tools may prevent certain schools from affording them, resulting in an unequal distribution of educational resources and exacerbating educational inequality. Secondly, the inability of certain students to fully utilize AI tools is a result of the disparities in technological infrastructure between schools, which impedes the attainment of educational equity and impacts learning outcomes.

4 The Path to Implementation for Establishing New Educational Collaborations

4.1 Comply with the Objective of Instruction

The long-term, fundamental core and purpose of the course should be the teaching objectives, and new instruments should be designed and utilized in conjunction with the achievement of these objectives. The teaching objectives of a course in a Chinese classroom are typically delineated into "four objectives." These objectives are as follows: the knowledge objective, which is dedicated to the transmission of knowledge and the development of skills; the ability objective, which is designed to foster the capacity to solve complex problems; the quality objective, which is designed to improve thinking literacy; and the ideological objective, which is designed to cultivate future leaders of socialism.

The teaching team develops a teaching plan that explicitly delineates the teaching objectives that each teaching task serves, based on these objectives. This is the basis for the development of a knowledge map and teaching activities that are designed to achieve these objectives. The consistency and scientific character of the teaching objectives should be preserved when employing generative AI tools in teaching design. Additionally, the materials and content generated by the generative AI tools should be organized using the knowledge map.

4.2 *Revise the Method of Collaboration*

A binary structure (Fig. 1) is the traditional Chinese classroom teaching paradigm, which emphasizes bidirectional interaction as a critical component of the educational process. This interaction model underscores that teaching is not a one-way transmission, but rather a process of active communication and joint progress between the instructor and students.

The architecture of deep learning models is the foundation of generative AI. It is capable of recognizing and internalizing patterns and laws within it through a thorough examination of a vast amount of training data. Upon receiving precise instructions, it is capable of "comprehending" and independently producing content that aligns with the training objectives. This encompasses topics such as video creation, programming applications, and course design [13]. It is capable of self-improvement through linguistic feedback, which enables it to transcend the fragmented characteristics of conventional AI and more seamlessly integrate into the various facets of classroom instruction as a peer or assistant [14]. Teachers supervise and train the generative AI in this model, which establishes a connection through the instructional content by transferring knowledge to them. Generative AI transmits knowledge to students through teaching activities, and students also provide feedback to the generative AI.

The "Teacher-AI-Student" triadic interactive model (Fig. 2) incorporates generative AI in addition to the knowledge transfer and feedback between students and

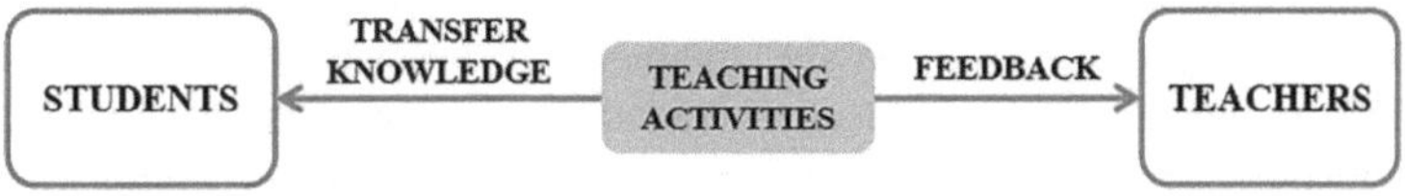

Fig. 1 Binary structure model for classroom subjects. *Source* Authors

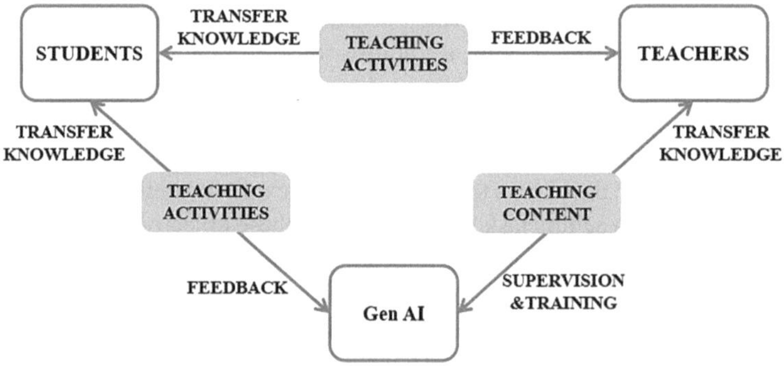

Fig. 2 Interactive model of classroom disciplines with ternary framework. *Source* Authors

instructors through instructional activities. In this model, educators can more effectively supervise and direct the application of generative AI to guarantee its safety and efficacy in the classroom. On the other hand, students can improve their learning outcomes by utilizing more interactive methods in a more comprehensive learning environment, which also fosters digital literacy and critical thinking abilities. After the educational reform in the generative AI era, this new triadic interactive model of "Teacher-AI-Student" will serve as the cornerstone of the primary teaching relationship in the classroom.

4.3 Changing the Classroom Environment

The classroom serves as the primary arena for cultivating and instructing talent. Education is characterized by its extensive implementation, while student learning is characterized by its unique individualization [15]. The transformation of roles and the reshaping of relationships among instructors, students, and generative AI are closely linked to the construction of the classroom teaching environment. Our objective is to establish a learning environment that is both enriching and comprehensive by incorporating generative AI into classroom instruction.

Teachers are no longer limited to the role of knowledge transmitters; they are now assuming the responsibilities of guiding wisdom and cultivating abilities. This encourages educators to proactively lead the future of education, perpetually improve themselves, and transition from the transmission of knowledge to the development of students' independent thinking, innovation, and critical thinking abilities. In turn, students must transition from passive recipients of knowledge to active creators, thereby forming well-rounded personalities and closely integrating personal development with the realization of social value. Students must become autonomous learners. In terms of generative AI, it functions similarly to the blackboard or slide projector in history—a modern teaching instrument that enhances human education but cannot supplant the central role of human beings [16].

Classroom teaching necessitates the integration of generative AI into the realm of intelligent education in the new era, as it is a distinctive form of interpersonal communication. This is not intended to diminish the fundamental status of instructors, but rather to improve the precision of education and foster personalized student development [17]. Teachers can capitalize on the potent capabilities of generative AI's "chain of thought" to incorporate a variety of teaching tools and models into the classroom, thereby promoting personalized learning while simultaneously preserving both scale and individualization. Teachers can intelligently generate personalized learning paths and resources by conducting a thorough analysis of academic data, including grades, routines, and interests, using generative AI technology [18]. Simultaneously, the utilization of AI to acquire profound insights into the personalities and shared characteristics of students enables the monitoring of learning states in real time and the provision of scientific support for personalized instruction. This promotes the growth of digital intelligence in education, enabling students to accomplish comprehensive development by engaging in both active knowledge construction and gradual progression [19].

Education is significantly influenced by the emotions and interactions of students, particularly in the classroom. Teachers can enhance the effectiveness of their instruction by comprehending the emotional states and patterns of interaction of their students, thereby guaranteeing that they remain engaged and derive value from the course.

As an illustration, if a teacher observes students crossing their arms, lowering their heads, or maintaining distracted gazes during class, these body language signals may indicate confusion, disinterest, or resistance to the current lesson content. Students may develop a negative attitude toward learning if the teacher disregards these signals and proceeds with the original plan, resulting in a loss of interest due to difficulty keeping up. The teacher can, however, make immediate adjustments by slowing their speech, simplifying explanations, or re-engaging students through inquiries and interactions, provided that they can sensitively detect these emotional cues. For example, the instructor may hesitate and inquire, "Is this concept somewhat challenging to comprehend?" Is anyone interested in exchanging their opinions? This type of emotional exchange can not only assist students in overcoming perplexity, but also enable them to feel acknowledged and supported, thereby increasing motivation and engagement. Additionally, emotional interaction facilitates the evaluation of students' cognitive states by educators. Teachers can determine that the content is well comprehended by observing students' focused eye contact, nodding, or smiling, and subsequently provide more detailed explanations. Conversely, teachers may modify their approach by employing examples or analogies to facilitate learning, as evidenced by frequent frowning or avoidance of eye contact, which are indicators of anxiety [20].

4.4 Guidelines and Supervision

Academic Integrity Monitoring. The education sector is experiencing the negative consequences of generative AI's continued penetration, including pupils utilizing AI tools to complete assignments or write papers [21]. To address these significant obstacles, educational administrators and teachers must implement a variety of evaluation methods to reform assessment systems. For instance, they should strictly prohibit the use of AI during examinations, mandate the honest disclosure of personal and AI contributions, and utilize tools like Copyleaks or ZeroGPT to confirm whether content was generated by AI, particularly for tasks that assess creativity and writing ability, such as essays or assignments.

Improving Digital Literacy. Digital literacy is the comprehensive capacity to effectively utilize information technology to acquire essential information and generate new knowledge in order to identify, analyze, and resolve issues. This encompasses dimensions such as technical skills, knowledge accumulation, moral conduct, and digital awareness. This capability is especially important for students in the digital era. In the education sector, the pressing issue is how to improve students' digital literacy and instruct them on the proper use of generative AI tools. The cultivation of digital literacy requires a systematic, phased curriculum design and practical training framework integrated with dynamic assessment mechanisms, constructing a comprehensive cultivation system covering "cognition-skills-ethics." At the basic education level, generative AI application education could be embedded through information technology curriculum reforms. For instance, adding modules like "Fundamental Applications of AI Tools" to teach students effective prompt engineering, identification of hallucination issues in AI-generated content, and enhancement of information discernment capabilities through cross-verification of sources. In higher education, interdisciplinary courses such as "AI and Digital Ethics" could guide students to deeply explore data biases, privacy risks, and copyright controversies in generative AI, strengthening ethical decision-making awareness through case analyses. Practically, project-based learning activities should be designed to facilitate technology-to-reality skill transfer. For example, organizing students to analyze local environmental data using generative AI tools and generate research reports, followed by verifying AI conclusions' reliability through field interviews and questionnaire surveys. Such projects not only require technical operational proficiency but also demand understanding of the connections between technological outputs and real societal needs, thereby cultivating problem-solving abilities and social responsibility. Regarding teacher professional development, an "AI Teaching Mentor" system could be established, where technologically proficient teacher teams develop school-based training resources including discipline-specific AI Tool Pedagogical Guidelines and "Ethical Review Workshops." Role-playing simulations of academic misconduct scenarios could help instructors design anti-cheating strategies.

Ensuring the Sustainability of Teacher-Student Relationships. Amidst the profound integration of generative AI into classrooms, maintaining teacher-student relationships requires reconstructing pedagogical interactions centered on cultivating critical thinking. The traditional knowledge-transmission paradigm faces challenges: teachers' knowledge renewal pace struggles to match AI's resource integration capabilities, while overreliance on knowledge stock risks diminishing educators' agency. Consequently, AI-era educational objectives should prioritize teachers' emotional value and higher-order thinking cultivation, achieving "co-creation of cognition" through strategic design [22]. First, designing critical interactions. Teachers can transform AI-generated content into critical examination objects using "questioning-verification" pedagogy. For instance, in history classes, students compare AI-generated event narratives with authoritative historical sources, annotating logical flaws and tracing verification; in social science courses, debates on "how algorithmic biases influence gender equality discourses" stimulate deep reflections on technological ethics. Such activities simultaneously train information verification skills and expose AI's inherent limitations. Second, developing systematic thinking training tools. Deconstruct AI outputs into three-tiered "fact-inference-conclusion" frameworks, requiring students to annotate unsourced claims and visualize argumentation logic through mind maps. Students first obtain AI-generated grammar corrections for assignments, then teachers pose probing questions targeting argument depth and value expression, driving cognitive restructuring. Third, co-constructing emotional bonds and roles. Embed "role-reversal" tasks in group projects: students act as teachers, designing lesson plans using AI and defending their choices against peer challenges like "Why select this AI case?" or "Why not use traditional materials?" This process fosters mutual understanding through empathetic engagement. Teachers must also attune to classroom emotional dynamics, promptly converting AI outputs into discussion catalysts. Collaborative inquiry rebuilds knowledge construction, positioning AI as a medium for deep learning rather than an authoritative substitute. Ultimately, the essence of AI-era teacher-student relationships lies in educators anchoring emotional connections and cognitive guidance, ensuring technological tools serve human growth rather than displacing human value.

5 Discussion

5.1 Research Limitations

While this study systematically explores generative AI's impacts on the four core elements of classroom teaching, several limitations persist. First, case samples primarily focus on K-12 education, leaving the application patterns of AI tools in higher education and vocational training underexplored. Second, rapid technological advancements—such as GPT-4's evolution into multimodal models—may constrain the temporal validity of certain findings. Third, the "teacher-AI-student" interaction model, constructed within China's educational context, requires cross-cultural adaptability verification through international comparative studies. Fourth, the long-term effects of AI tools on students' critical thinking and academic integrity remain unmonitored, necessitating longitudinal experiments to address this gap.

5.2 Practical Implications

This research offers multidimensional insights for educational digital transformation. At the policy level, a regulatory-empowerment dual governance framework should be established, exemplified by drafting Guidelines for Generative AI Applications in Education to clarify technical准入 standards and ethical boundaries. For teacher development, integrating "AI-augmented instructional design" and "academic integrity identification" into in-service training programs can enhance educators' technological integration and risk management competencies. Curriculum reform could adopt a dual-track "AI + critical thinking" model, such as adding modules like "Authenticity Analysis of AI-Generated Texts" in language courses to strengthen digital literacy. Resource allocation strategies should leverage national cloud platforms to share high-quality AI tools, bridging the urban–rural digital divide and advancing educational equity.

5.3 Future Research Directions

Future studies should deepen exploration in three domains: technological adaptability, ethical mechanisms, and cross-cultural comparisons. Technologically, multimodal AI's potential in experimental and arts education—such as visual-language models guiding chemistry lab operations—warrants investigation. Ethically, developing blockchain-based learning behavior traceability systems could address academic integrity oversight challenges. Cross-culturally, comparative analyses of AI acceptance patterns in Eastern and Western classrooms may elucidate how cultural values mediate technology integration. Additionally, longitudinal studies spanning 3–5 years could map generative AI's developmental trajectories in shaping students' higher-order cognitive abilities, providing empirical foundations for sustainable educational reform.

6 Conclusion and Prospects

In summary, this paper provides a thorough examination of the substantial acceleration of classroom teaching reform that generative AI tools have brought about, as well as the new issues they have introduced. It concentrates on the four critical components of classroom teaching: teaching objectives, teaching content, teaching methods, and teaching evaluation. It emphasizes the new expectations for both students and educators. In this context, the paper underscores the importance of educators utilizing the powerful capabilities of generative AI tools to reform the collaborative approach to the curriculum, transform the classroom environment, and maintain teaching objectives by proposing the establishment of a new "Teacher-AI-Student" educational collaboration model. Simultaneously, it is imperative to consistently oversee and direct this innovative educational collaboration model. The paper also offers specific recommendations for utilizing AI as a catalyst to promote the creativity of students, foster lifelong learning abilities, and drive educational system reforms by ensuring academic integrity, enhancing digital literacy, and preserving healthy teacher-student relationships. This is of substantial practical significance and reference value in the context of helping to reform education in the digital era and in the development of individuals with a greater capacity for innovation and practical skills that will benefit society.

In the future, generative AI has the potential to further integrate deeply into the educational ecosystem, thereby moving away from tool-assisted classroom instruction and toward ecological reconstruction. The current technological constraints are anticipated to be overcome by generative AI, which is anticipated to gradually achieve a precise understanding and dynamic response to complex teaching scenarios through continuous algorithm optimization and enhanced multimodal interaction capabilities.

Future research should further investigate the synergistic mechanisms between generative AI and teacher professional development. Particularly in rural areas with scarce educational resources, empirical evidence remains crucial to determine whether AI can effectively bridge teacher resource gaps. The "Teacher-AI-Student" model proposed in this study serves as a preliminary framework, whose global applicability necessitates refinement through cross-cultural comparative research.

Lastly, this paper establishes the new "Teacher-AI-Student" educational collaboration model by examining the four components of Chinese-style classroom instruction. However, its effectiveness and significance should be globally validated and promoted. Inherently possessing internationality and universality, generative AI, which was born from the free flow of internet information, ensures its seamless integration and self-adaptation in educational collaboration relationships across various cultural contexts. The author aspires to establish a practical and forward-thinking exploration paradigm for global educational reform based on the new "Teacher-AI-Student" educational collaboration model and its guidelines, which will foster mutual trust, complementarity, and a healthy symbiosis between teachers, students, and AI. This will contribute to the global consensus on educational fairness, personalized learning, and individualized teaching in the era of generative AI.

References

1. Benjamin Luke Moorhouse: Generative artificial intelligence and ELT. ELT J. **78**(4), 378–392 (2024)

2. Espín-León, A., Pertegal-Felices, M.L., Jimeno-Morenilla, A., Molina-Carmona, R.: The Evolution of Education in Amazonian Indigenous Communities: A Post-Covid Perspective. RIIFORUM 2023, pp. 327–341. Springer, Cham (2024)
3. Weng, W.: Classroom teaching reform in first-class undergraduate education: what needs to be done, what is difficult and what can be done. J. National Inst. Educ. Adm. **03**, 58–64 (2020)
4. Wang, Z.: New constructivist pedagogy. Mod. Educ. Technol. **24**(05), 5–11 (2014)
5. Skrzek-Lubasińska, M., Malik, R., Çekani, V.: Critical Thinking Skills on the Labour Markets—Comparative Analysis of Online Job Postings in Italy and Poland. RIIFORUM 2023, pp. 415–425. Springer, Cham (2024)
6. Establish an integrated approach to curriculum thinking that unifies teaching, learning, and assessment. http://www.moe.gov.cn/jyb_xwfb/moe_2082/2021/2021_zl53/tsj/202112/t20211216_587719.html. 16 Dec 2021
7. Chiu, T.K.F.: The impact of Generative AI (GenAI) on practices, policies and research direction in education: a case of ChatGPT and Midjourney. Interact. Learn. Environ. **32**(10), 6187–6203 (2024)
8. Wong, G.K.W., Ma, X., Dillenbourg, P., Huan, J.: Broadening artificial intelligence education in K-12: where to start? ACM Inroads **11**(1), 20–29 (2020)
9. Biggs, J., Tang, C.: Constructive Alignment: An Outcomes-Based Approach to Teaching Anatomy, pp. 31–38 (2015)
10. West, M.R., Buckley, K., Krachman, S.B., Bookman, N.: Development and implementation of student social-emotional surveys in the CORE Districts. J. Appl. Dev. Psychol. **55**, 119–129 (2018)
11. Hounsell, D.: Commenting Constructively: Feedback to Make a Difference (2015)
12. Nicol, D.J., Macfarlane-Dick, D.: Formative assessment and self-regulated learning: a model and seven principles of good feedback practice. Stud. High. Educ. **31**(2), 199–218 (2006)
13. Baidoo-Anu, D., Ansah, L.: Education in the era of generative artificial intelligence (AI): understanding the potential benefits of ChatGPT in promoting teaching and learning. J. AI **7** (2023)
14. Wu, Q., Guo, L.: Teaching reform in higher vocational colleges in the era of generative artificial intelligence: challenges, frameworks and paths. Univ. Educ. Sci. **06**, 112–120 (2023)
15. Kolb, A., Kolb, D.: Learning Styles and Learning Spaces: Enhancing Experiential Learning in Higher Education, vol. 4 (2003)
16. Li, S., Zheng, L.: Challenges and responses of generative artificial intelligence to classroom teaching. Curriculum Teach. Mater. Teach. Methods **44**(01), 39–46 (2024)
17. Wang, Z., Wu, Y., Wang, Y.: ChatGPT and educational transformation—how education should transform in the intelligent age. J. Dist. Educ. **41**(04), 27–36 (2023)
18. Loia, F., Capolupo, N., Gravili, G., Adinolfi, P.: Bridging Sustainable Higher Education and Metaverse During the Post-pandemic Era: Highlights from a Multiple Case Study. RIIFORUM 2023, pp. 363–371. Springer, Cham (2024)
19. Qiu, Y., Li, Z.: Challenges, integration, and change: a review of the conference "ChatGPT and Future Education". Mod. Distance Educ. Res. **35**(03), 3–12 (2023)
20. Lujan, H.L., Raizada, A., DiCarlo, S.E.: Critical skill of teaching: learning the cognitive and emotional states of our students during class. Adv. Physiol. Educ. **45**(01), 59–60 (2021)
21. Zhang, R.: The impact of generative artificial intelligence technology on education—an interview on ChatGPT. E-Educ. Res. **44**(02), 5–14 (2023)
22. Lei, X., Shao, B.: Research on embedding artificial intelligence-generated content into digital literacy education under a large model. Mod. Intell. **43**(06), 99–107 (2023)

Smart Engagement: The Role of AI in University Communication

Giuliano Gaveglia

Abstract Artificial intelligence (AI), while being an innovative tool, operates in certain respects within boundaries defined by human intelligence, replicating established models and patterns available online. This characteristic offers an opportunity to explore new modalities for public engagement (PE) within the framework of academia's third mission. The paper presents a reflection on how AI can enhance public participation through digital platforms and interactive tools, such as chatbots and virtual assistants, which amplify citizens' voices, stimulate co-creation of knowledge, and foster stronger connections between academia and society. However, it is essential to adopt a critical approach that integrates AI with inclusive decision-making processes, rooted in an abductive reasoning process that is distinctly human. Such an approach must be capable of understanding the unique characteristics of the territories and communities involved, avoiding the mere reproduction of existing dynamics and distortions. The analysis adopts a descriptive methodology to map the use of AI in academic institutional communication, with a focus on Italian university websites and an assessment of the significance and impact of introducing such technological innovation.

Keywords Public engagement · Artificial intelligence · Chatbots · Website · Administration · Innovation (SDG 9)

1 Introduction and Theoretical Framework

In addition to its traditional roles of teaching and research, the contemporary university is increasingly called to play an active role in society, directly contributing to the cultural, social and economic development of the surrounding community. This set of practices, commonly referred to as third mission, reflects the growing awareness that these institutions must take responsibility for interacting with the public beyond their internal borders [1]. In this framework, the public engagement (PE) emerges as

G. Gaveglia (✉)
University of Salerno, Fisciano, Salerno, Italy
e-mail: ggaveglia@unisa.it

A. Visvizi et al. (eds.), *Research and Innovation Forum 2025 Vol2*, Springer Proceedings in Complexity, https://doi.org/10.1007/978-3-032-09062-1_6

one of the most significant expressions of the third mission. More precisely, it refers to a wide range of initiatives through which universities establish open, inclusive and dialogic relationships with organizations and individuals that have no formal affiliation with the academic world. Through accessible tools, languages and platforms, the PE therefore aims to promote participation and knowledge sharing, helping to strengthen the civic mission of the university in support of inclusive and sustainable development [2]. In recent years, this dimension is intersecting with the rapid expansion of AI, which is transforming the way in which institutions communicate and interact with their audiences. AI is not simply a technical innovation; it alters the very logic of public communication, as it introduces new forms of interaction. However, these phenomena of great transformation raise crucial questions, first of all: to what extent can AI support—or weaken—the participatory dimension of academic engagement?

Currently, AI use in PE practices is still limited, fragmented, and often lacks a coherent strategic direction. However, this research intends to explore the potential of AI in improving engagement, understood in the broadest sense that the third mission interprets, for example by also incorporating the administrative dimension. And it is precisely in these two peculiar segments of the third mission—where the effectiveness of bidirectionality and the non-rigidly hierarchical characterization of the communicative phenomenon are of crucial importance [3]—that the most innovative technological tools can offer theoretical and empirical insights of extreme interest. Therefore, this emerging frontier is explored by analyzing how AI technologies are implemented, if implemented, in Italian university websites, with a particular focus on the sections dedicated to the PE and administrative services. The aim is twofold: to assess the actual diffusion and functionality of these tools and to critically reflect on their potential contribution to a more open, inclusive and contextualized institutional communication. The article is structured in five sections: the first outlines the theoretical framework; the second describes the methodology; the third presents the empirical results; the fourth offers a critical discussion; and the fifth concludes with recommendations and future directions.

It is worth highlighting some preliminary considerations: AI is defined here as a set of tools capable of replicating, through algorithmic models, some of the generative functions of human intelligence. Second, since we are still at the beginning of a profound technological and cultural transformation, this article constitutes a first step in a broader research program aimed at understanding the relationship between digital innovation, public communication and the civic role of universities.

Although there are critical and contrasting evaluations of AI, its use is increasingly frequent and there are virtuous cases in the academic field, both international and Italian, in which its use is at the center of several projects.

The University of Pisa, for example, has started a collaboration with OpenAI to integrate AI into teaching, research and administrative activities [4]. It is also interesting to note how recent studies show a rather clear direction: AI helps to lighten the workload of teachers by automating repetitive activities, especially in the areas of administrative management [5]. In the first and second university missions, studies and research aimed at exploring the potential of AI are proliferating [6] and, almost as a natural consequence, an impact is expected also on the third mission and, in particular, in the area of PE.

In this perspective, the research question that animates this work is the following: how can AI contribute to reducing information inequalities for stakeholders and communities, ensuring effective equal access to the services and opportunities offered by universities?

With this research question we want to highlight the importance of understanding how AI can act as a lever for social innovation and how, at the same time, it can support users with less social, cultural or economic capital.

Based on Merton's concept of "functional alternative" [7], the underlying hypothesis is that chatbots and virtual assistants can replace—or at least integrate—traditional forms of interaction and academic "consulting", facilitating access to information and allowing greater participation to those who, otherwise, would be disadvantaged by complex and cumbersome bureaucratic systems.

2 Methods and Materials

This research describes how (and if) official university websites in Italy use AI-based tools and how these tools are integrated, in particular, in the context of the PE.

The hypothesis is that such technologies can strengthen the relationship between universities and communities, going beyond traditional communication models in favor of more interactive and "smart" approaches. At the same time, the study examines whether AI is also used in those sections of university websites that deal with administrative matters, in particular student services. The research design is based on a qualitative approach, supported by a reasoned sampling to identify the most representative universities in terms of size and geographical location. The unit of analysis is constituted by the web pages dedicated to the PE and administrative functions. Ten Italian universities located in different parts of the country (North, Center and South) were selected. This sample was considered sufficient to offer the most heterogeneous perspective possible. The analysis was conducted in February 2025.

As for the data sources, the information was collected by consulting the sections of the university websites dedicated to PE and administrative activities, with the aim of identifying pages with features related to the adoption of AI. Internal search engines of each institutional portal were also used, with keywords such as "virtual assistant" or "chatbot". Although from a qualitative perspective, the analytical criteria partially refer to the model proposed by Marino and Lo Presti [8], adapted to the dimension of PE identified by Hart [9], called "Institutional relationship and partnership building". This perspective also takes into account the user experience in evaluating the effectiveness of navigation [10] [cf. 37]. Initially, the data collection phase simply involved checking the presence of chatbots, virtual assistants or dynamic FAQs within the relevant sections of the pages dedicated to PE and administrative practices. This initial binary check (presence = 1, absence = 0) made it possible to determine whether the universities were already using forms of AI or were still tied to static communication systems. Once the existence of AI-based tools was confirmed, the study proceeded with a qualitative evaluation using a five-point Likert scale (from 1 = very functional to 5 = excellent functionality). This scale was used to assign scores related to communication effectiveness and navigability: the first refers to the clarity and quality of written content, while the second concerns the ease of use of the specific web section.

3 Results

In the case of PE sections, most universities have not yet implemented chatbots or virtual assistants. The situation appears better in the administrative field, where there is a greater tendency to use such tools to manage bureaucratic activities.

The Table 1 summarizes the values assigned to communication effectiveness and navigability for the websites examined. The ratings cover the entire range of the Likert scale from 1 to 5, indicating a heterogeneous scenario. The University of Naples Federico II stands out for the integration of a virtual assistant specifically aimed at PE, a feature that significantly improves the clarity of content and ease of consultation [40]. Some universities, such as those of Milan, Florence, Turin and Ca' Foscari, still show a high level of clarity and organization of content, although they do not include any real-time AI or chatbots dedicated to this specific area. Other institutions, on the contrary, have less developed PE sections, where information is limited to simple reports of past initiatives.

The scenario appears significantly richer and more complex in the Table 2 (administrative section), where several universities have implemented chatbots and virtual assistants. Ca' Foscari Venezia has introduced a service developed in collaboration with IBM, clearly visible in the sections dedicated to registrations, fees and student careers, providing continuous support and immediate answers [38]. Sapienza University of Rome has taken a similar path, although its virtual assistant, as stated on the website, is only accessible via the *Infostud app* [42]. The University of Milan has adopted similar solutions, integrating a virtual assistant in the student services and *Informastudenti* section to provide timely answers to bureaucratic questions [39]. The situation is different at the University of Salerno, where the implementation of a chatbot (called *Matteo*) has not materialized. Finally, the University of Palermo has a section specifically designed for future students and has activated a virtual assistant that makes navigating questions on orientation and registration procedures quicker and easier [41]. A combined analysis of these two areas, PE and administrative services, reveals different communication strategies within the same universities. On the one hand, the use of AI tools remains relatively limited in promoting PE or in fostering innovative dialogue with the community. On the other hand, several universities are already quite advanced in their administrative applications. The growing number of universities adopting more sophisticated AI systems represents a significant development in terms of simplifying procedures and improving user experience,

Table 1 Public engagement section

University	AI tools or chatbots (1/0)	Communicative effectiveness (1–5)	Navigability (1–5)
Università degli Studi di Milano	0	4	4
Università degli Studi di Firenze	0	3	4
Università degli Studi di Bari Aldo Moro	0	2	2
Università degli Studi di Salerno	0	2	3
Università di Bologna—ALMA MATER STUDIORUM	0	2	3
Università di Roma Sapienza	0	1	1
Università degli Studi di Torino	0	4	4
Università degli Studi di Napoli Federico II	1	5	5
Università Ca' Foscari Venezia	0	4	4
Università degli Studi di Palermo	0	3	3

Source Author

especially to respond to recurring requests or to promptly manage administrative tasks.

4 Discussion

Communication processes are a fundamental element of the PE. For this reason, it was decided to examine university websites which, although virtual, act as privileged places for academic communication. They should be seen not only as repositories of information, but as strategic platforms capable of facilitating dialogue with the public and amplifying engagement between different categories of users [8]. The adoption of increasingly sophisticated tools that allow the functioning of chatbots or virtual assistants could further improve institutional engagement [11], promote inclusiveness and encourage co-participation between the academic community and citizens. However, as several theoretical reflections indicate, although AI can be considered an "impersonal" tool, it is certainly not neutral. In fact, the deepest essence of technology is not merely technical [12]. AI is, in fact, the result of models and algorithms of human

Table 2 Administrative section

University	AI tools or chatbots (1/0)	Communicative effectiveness (1–5)	Navigability (1–5)
Università degli Studi di Milano	1	5	5
Università degli Studi di Firenze	0	3	4
Università degli Studi di Bari Aldo Moro	0	2	2
Università degli Studi di Salerno	0	2	3
Università di Bologna—ALMA MATER STUDIORUM	0	2	3
Università di Roma Sapienza	0	2	3
Università degli Studi di Torino	0	3	3
Università degli Studi di Napoli Federico II	0	3	3
Università Ca' Foscari Venezia	1	5	5
Università degli Studi di Palermo	1	5	5

Source Author

knowledge, which can incorporate biases, limitations and even stochastic elements [13], as well as potentially reproducing, if not carefully guided, exclusive or even discriminatory dynamics [14]. It is therefore essential to adopt a critical approach that highlights both the opportunities and risks associated with the use of these technologies, particularly in university PE. In order to be effective, it requires a preliminary reflection that recalls Marshall McLuhan's well-known formulations [15] regarding the medium and the message: communication is profoundly influenced by the ways in which information is conveyed. Consequently, if AI were treated as a medium comparable to traditional ones, it becomes plausible to hypothesize that it could facilitate forms of co-creation and knowledge sharing that favor dialogic approaches, going beyond the mere vertical transmission of information [16]. In other words, feedback, both negative and positive [17], occurs at an unprecedented speed, making AI a key tool to promote dialogue and broaden the scope of PE initiatives. Therefore, AI can radically transform online interactions, adapting the information offer to users' needs more effectively and rapidly, integrating knowledge and facilitating proactive participation processes [cf. 18]. Some works [19], in describing the historical evolution of this area, underline how the need for transparency and greater citizen involvement in governance processes has become increasingly pressing. Law 150/2000, which regulates the communication responsibilities of public administrations, has made public relations offices (URP) mandatory. In short, the civic need to reduce barriers between citizens and institutions is an initiative that has been underway for decades. In this sense, within university "administrations"—with particular reference to the segment of communication provided on their websites—the adoption of AI could, by virtue of the values represented by academic institutions, act as a catalyst for the start of a process of social transformation that effectively responds to civic demands. In addition to this: it is hoped that the use of these new technologies can generate the latent function [7] of bridging that cultural gap [20] that distinguishes Italian academies with respect to third mission activities. The latter, and in particular the PE, has not yet reached a satisfactory level of implementation. According to data collected in 2022, only 46% of Italian universities and research centers have monitoring systems for PE activities; 23% are in the process of implementing such tools and 25% have not yet adopted specific measures in this regard.

Ultimately, it can be said that the priority given to these activities within Italian universities is still insufficient [21]. It is necessary, however, to underline that the evaluation of PE initiatives is based not only on visibility or communication, but also on their ability to respond to real social needs, which are not always easily verifiable.

In line with what has just been said, the results presented in the tables show a complex and multifaceted panorama. While the integration of AI into administrative services seems to be progressing in several institutions, the adoption of such tools within PE sections is still in its infancy. The presence of chatbots or virtual assistants specifically dedicated to interacting with the wider public represents, at this stage, more of an exception than a common standard.

This disparity calls a broader reflection on the role that AI can effectively play in reshaping the way universities interact with society. As Beaulieu [22] point out, engagement is not simply a technique, but an epistemic positioning: it implies the responsibility of institutions to co-produce knowledge [cf. 36] and contribute to social transformation [23]. In this context, AI can act as a "weak enabler", a tool that improves engagement only if integrated into a broader strategy based on active listening, civic empowerment and impact assessment. In any case, the data reveal significant heterogeneity between institutions, confirming a two-speed digital transformation. This carries the risk of institutional polarization, whereby some universities progress through partnerships and technological investments, while others remain tied to traditional communication paradigms.

The University of Naples Federico II represents a virtuous example: the integration of a virtual assistant specifically aimed at the PE has improved the accessibility and clarity of information, thus promoting interaction.

In the administrative field, the diffusion of AI appears more homogeneous. Universities such as Ca' Foscari Venezia and the University of Palermo have implemented advanced chatbots that, as highlighted by other studies, should significantly lighten the workload of university offices by providing continuous and immediate responses [24]. This confirms that the adoption of AI goes beyond technical functions and is part of a broader redefinition of the relationship between institutions and the public.

Therefore, it is not just about transmitting information, but about creating communication environments in which participation can be effective and inclusive. In this context, AI can facilitate the co-creation of knowledge and amplify feedback loops, promoting the real-time adaptation of information to user needs [18].

This reflection also relates to Merton's concept of "functional alternative" [7]. Chatbots and virtual assistants can complement or even replace traditional forms of academic counseling, thus providing access to users who would otherwise be excluded by complex bureaucratic systems [25]. This suggests that AI should be interpreted not simply as a technical tool, but as an opportunity to reconfigure social structures and reduce information inequalities [cf. 20].

However, critical issues remain: digital barriers and skills shortages can limit access to AI-based tools; furthermore, the alleged neutrality of algorithms is often misleading, as they reproduce human cognitive models permeated by prejudices, epistemic limits and exclusion dynamics [13, 14].

The risk of reinforcing technocratic logics or information asymmetries makes participatory governance essential, capable of involving both academic communities and civil society [26].

International experiences—including those of the University of Pisa and the University of Murcia—show that the integration of AI into university processes can generate significant impacts only when accompanied by clear data governance policies, defined responsibilities and interdisciplinary training [24]. The adoption of 24/7 operational services, capable of supporting multiple languages, constitutes a concrete measure of inclusion, especially for users with lower cultural or linguistic capital [27]. On the employment side, although there are fears of staff replacement, evidence indicates that AI tends to transform—rather than eliminate—jobs [28], freeing human resources from repetitive activities and starting an evolution similar to that observed in previous technological revolutions [24]. Universities manage huge volumes of data—from enrolments to academic results, from the use of spaces to interactions on online services—and AI tools allow them to be valorized to optimize internal processes and improve services. Dashboards and advanced metrics are useful for both accountability and continuous improvement [24]. This push towards increasingly data-driven management must be supported by robust data governance practices that clarify responsibility, intended use, quality, security and privacy of data. A significant example is offered by the University of Murcia, which introduced a chatbot capable of answering questions about the campus and the educational offer, managing over 38,000 queries with an accuracy of 91% and thus lightening the workload of the offices [24].

5 Conclusions

The analysis presented here is a step in a relatively new line of research. However, it is appropriate to recognize some methodological limitations. First, the identification of the true nature of AI tools implemented on university websites. Not all chatbots or virtual assistants use advanced natural language processing (NLP) technologies; they often rely on predefined scripts, thus offering rather limited interactions, which do not significantly improve either the accessibility to information or the user experience. Second, the lack of a clear taxonomy for AI tools risks leading to inaccurate assessments of their real potential.

It should be noted that, in addition to future research based on sample surveys of user experience, other significant elements for a more complete investigation of chatbot performance on the pages considered, such as: the average length of stay in the chat, the questions most frequently asked by the user, any time slots with the highest attendance and any other data useful for better understanding the interaction between users and the tool; will have to be the subject of further investigations also conducted through qualitative and quantitative analyses. These investigations, in agreement with the companies that own the chatbots (once the data are made available), will be necessary to fully understand the effectiveness and potential of these tools in university contexts.

Another challenge concerns the standardization of the analysis sample for data comparability: significant economic and infrastructural disparities among universities directly impact their capacity to develop and adopt advanced digital tools. Based on these considerations, it can be stated that the integration of AI in academic PE processes remains an open challenge. While on the one hand it presents significant opportunities, on the other it raises several issues. One of these concerns the generational gap in interaction with digital technologies. Younger generations tend to be attracted by interactive experiences, while older ones prefer more familiar ways of engaging with the institution [29]. This highlights the need for differentiated engagement strategies, which guarantee equal access to university resources to all age groups.

A further critical issue is the possibility that the role of technicians coincides in practice with that of policy makers. The introduction of AI systems in public institutions raises complex issues of governance and transparency in decision-making processes: there is a concrete risk that emerging technologies will be adopted without adequate involvement of the academic community and civil society, thus calling into question the democratic principles that should guide the digital transformation of universities [cf. 26]. This requires a political and cultural stance: universities must operate as "open" institutions, limiting access only to bad practices that threaten the rights of the individual [30]. Ultimately, the academic world, through a conscious use of the media, must be able to promote freedom of expression, develop inclusive citizen science and adopt a transparent communication style [cf. 31]. It must always be remembered that the implementation of AI also involves challenges related to the amplification of pre-existing biases and the risk of compromising users' privacy [32]. Some studies, inspired by utopian-dystopian reflections [33] or critical digital literacy approaches, paint a scenario in which AI acts as a catalyst for both potentials and risks, especially when technical innovation is introduced in social contexts characterized by strong inequalities, restrictions on freedom and lack of pluralism [34].

Another widespread fear, in the context of the adoption of AI, concerns the possible replacement of staff: reduction of staff in the secretariat, among librarians or even among teaching assistants. However, studies and surveys conducted among university managers outline a less drastic future: according to an international survey, 42% of academic leaders do not foresee any layoffs due to AI in the next decade; indeed, they believe that AI could foster an increase in qualified personnel, thanks to the new opportunities generated by innovation [24]. Rather than eliminating jobs, AI tends to redefine them. As already observed in previous technological revolutions, efficiency gains do not necessarily translate into a drop in employment, but rather into a reallocation of personnel towards new or different tasks. Carroll and colleagues [28] argue that AI does not necessarily have to destroy jobs, but can instead "enhance employment by augmenting human capabilities", provided that the approach to the adoption of such technologies is not alarmist and remains balanced.

Ultimately, within the framework of the third mission, universities are increasingly configuring themselves as engines of social, economic and cultural development of the territory in which they operate, promoting the transfer of knowledge and technologies not only towards the industrial sector but also towards civil society as a whole [16].

This perspective further increases the importance of careful management of new technologies, including the use of AI systems, so that they can effectively contribute to collective well-being, reducing imbalances and ensuring full protection of rights.

Although this study represents a small step towards the analysis of the use of AI in academic PE, it can be considered as a preliminary exploration within a broader line of investigation, which will necessarily have to make use of an interdisciplinary approach. As highlighted, the complexity of the phenomenon requires the involvement of communication experts, computer scientists and sociologists, among others, to fully understand the implications and the role that these tools play in the relationship between universities and civil society [cf. 35] [cf. 43]. Only by moving in this direction will it be possible to acquire a complete vision of the processes of academic engagement in the era of AI.

References

1. ANVUR—Agenzia Nazionale di Valutazione del Sistema Universitario e della Ricerca: I Criteri e gli indicatori di struttura di "terza missione". In: Valutazione della Qualità della Ricerca 2004–2010 (VQR 2004-2010), anvur.it, p. 15 (2011)
2. Martino, V.: Terza Missione e cultura delle università. Note per una socio-logia del patrimonio accademico, Rivista Trimestrale di Scienza dell'Amministrazione, 1 (2018). https://doi.org/10.32049/RTSA.2019.1.4
3. Sandmann, Lorilee R.: Conceptualization of the scholarship of engagement in higher education: a strategic review, 1996–2006. J. Higher Educ. Outreach Engag. **12**(1), 91–104 (2008). ISSN: 15346102, 21648212
4. Università degli Studi di Pisa: Artificial Intelligence, Partnership between Università di Pisa and OpenAI. In: News. Available online at https://www.unipi.it/index.php/news/item/28885-intelligenza-artificiale-alleanza-tra-universita-di-pisa-e-openai. Last accessed 19 Feb 2025
5. Sessa, C., Aiello, P.: Can My Co-teacher be a Chatbot?. In: ICERI2024 Proceedings. IATED (2024). https://doi.org/10.21125/iceri.2024.2049

6. Okonkwo, C.W., Ade-Ibijola, A.: Chatbots applications in education: a systematic review. Comput. Educ.: Artif. Intell. **2** (2021). https://doi.org/10.1016/j.caeai.2021.100033
7. Merton, R.K.: Teoria e struttura sociale. Il Mulino, Bologna (1970). ISBN: 978-88-150-7654-0
8. Marino, V., Lo Presti, L.: Strumenti di comunicazione digitale e public engagement: il caso delle università europee. Esperienze d'impresa **1**, 63–77 (2015). https://doi.org/10.3280/EI2015-001003
9. Hart, E., Northmore, S.: Auditing and evaluating university-community engagement: lessons from a UK case study. High. Educ. Q. **65**(1), 34–58 (2011). https://doi.org/10.1111/j.1468-2273.2010.00466.x
10. Polillo, R.: Il check-up dei siti web delle organizzazioni non profit. I Quaderni di THINK!, 3, Think! Foundation, pp. 1–55, (2013). ISBN: 978-88-907047-2-7
11. Lin, X., Chan, R.Y., Sharma, S., Bista, K.: ChatGPT and Global Higher Education: Using Artificial Intelligence in Teaching and Learning. STAR Scholars Press, Baltimore (2024). https://doi.org/10.32674/rh27qv16
12. Heidegger, M.: La questione della tecnica, trad. it. In: Saggi e discorsi, Mursia (1976). ISBN: 9788842555070
13. Esposito, E.: Comunicazione Artificiale. Come gli algoritmi producono intelligenza sociale. Egea, Milano (2022). ISBN: 9788899902971.
14. Joyce, K. et al.: Toward a sociology of artificial intelligence: a call for research on inequalities and structural change. Socius 7 (2021). https://doi.org/10.1177/2378023121999581
15. McLuhan, M.: Gli strumenti del comunicare. Il Saggiatore, Milano (1967). ISBN: 9788842821137
16. Puente, C., et al.: Role of universities as drivers of social innovation. Sustainability **13**(24), 13727 (2021). https://doi.org/10.3390/su132413727
17. Caramiello, L.: Frontiere Culturali. Guida, Napoli (2012). ISBN: 9788866661795
18. Mihailidis, P.: Media Literacy and the Emerging Citizen: Youth, Engagement and Participation in Digital Culture. Peter Lang, New York (2014). https://doi.org/10.3726/978-1-4539-1293-5
19. Faccioli, F.: La comunicazione pubblica. Luci e ombre di un'innovazione. Quaderni di sociologia **30**, 16–32 (2022). https://doi.org/10.4000/qds.1234
20. Ogburn, W.: Tecnologia e mutamento sociale. a cura di Iorio G., Armando Editore, Roma (2006). ISBN: 9788860810786
21. APEnet: Verso l'istituzionalizzazione del Public Engagement. In: Report Baro-metro 2022 (2022)
22. Beaulieu, M., Breton, M., Brousselle, A.: Conceptualizing 20 years of engaged scholarship: a scoping review. PLoS ONE **13**(2), e0193201 (2018). https://doi.org/10.1371/journal.pone.0193201
23. Carpentier, N.: Media and Participation: A Site of Ideological-Democratic Struggle. Intellect Ltd., Bristol (2011). https://doi.org/10.26530/OAPEN_606390
24. Liu, B.L., Morales, D., Roser Chinchilla, J.F., Sabzalieva, E., Valentini, A., Vieira do Nascimento, D., Yerovi, C.: Harnessing the Era of Artificial Intelligence in Higher Education. A Primer for Higher Education Stakeholders. UNESCO IESALC, Paris–Caracas, pp. 1–85 (2023). Document code: ED/HE/IESALC/IP/2023/27
25. Eynon, R., Malmberg, L.E.: Understanding the online information-seeking behaviours of young people: the role of networks of support. J. Comput. Assist. Learn. **27**(6), 499–511 (2011). https://doi.org/10.1111/J.1365-2729.2011.00460.x
26. Deng, W.H., Eslami, M., Holstein, K.: Towards "Anytime, Anywhere" Community Learning and Engagement around the Design of Public Sector AI. Carnegie Mellon University (2023). https://doi.org/10.48550/arXiv.2304.00167
27. Zawacki-Richter, O., Marín, V.I., Bond, M., Gouverneur, F.: Systematic review of research on artificial intelligence applications in higher education—where are the educators? Int. J. Educ. Technol. High. Educ. **16**, 39 (2019). https://doi.org/10.1186/s41239-019-0171-0
28. Carroll, N., et al.: Navigating the Utopia and Dystopia Perspectives of AI. Commun. Assoc. Inf. Syst. (CAIS) **55**, 854–874 (2024). https://doi.org/10.17705/1CAIS.05533

29. Chan, C.K.Y., Lee, K.K.W.: The AI generation gap: are Gen Z students more interested in adopting generative AI such as ChatGPT in teaching and learning than their Gen X and Millennial Generation teachers? Centre for the Enhancement of Teaching and Learning. The University of Hong Kong (2023). https://doi.org/10.1186/s40561-023-00269-3
30. Popper, K.: La società aperta e i suoi nemici, vol. 1–2. Armando Editore, Roma (2002). ISBN: 9788883583094
31. Hsu, Y.C. et al.: Empowering Local Communities Using Artificial Intelligence. Delft University of Technology (2022). https://doi.org/10.1016/j.patter.2022.100449
32. Chandra, P.: The Global Impact of AI-Artificial Intelligence: Recent Advances and Future Directions. Department of Professional Security Studies, New Jersey City University, Jersey City, New Jersey, USA (2023). https://doi.org/10.48550/arXiv.2401.12223
33. Castells, M.: Communication Power. Oxford University Press, Oxford (2009). ISBN: 9780199567041
34. Morozov, E.: L'ingenuità della rete. Il lato oscuro della libertà di internet. Codice edizioni (2011). ISBN: 9788875782610
35. Deng, W.H., Lam, M.S., Cabrera, Á.A., Metaxa, D., Eslami, M., Holstein, K.: Supporting User Engagement in Testing, Auditing, and Contesting AI. ACM, Minneapolis (2023). https://doi.org/10.1145/3584931.3611279
36. Giddens, A.: La costituzione della società. Lineamenti di teoria della struttura-zione. Comunità, Milano (1990). ISBN: 9788824504577
37. Maffei, S., Carraro, M.: Dalla comunicazione all'interazione col servizio. Lo user engagement nella comunicazione dei servizi pubblici. MD J **10**(2), 74–85 (2020). ISSN: 2531-9477
38. Università Ca' Foscari Venezia: Artificial Intelligence and Digital Services for Students. Available online at https://www.unive.it/pag/35152. Last accessed 19 Feb 2025
39. Università degli Studi di Milano: Virtual Assistant. In: Student Services. Available online at https://www.unimi.it/it/studiare/servizi-gli-studenti/segreterie-informastudenti/assistente-virtuale. Last accessed 19 Feb 2025
40. Università degli Studi di Napoli Federico II: Public Engagement. In: Third Mission. Available online at https://www.terzamissione.unina.it/activities/public-engagement/. Last accessed 19 Feb 2025
41. Università degli Studi di Palermo: Future Students. Available online at https://www.unipa.it/target/futuristudenti/. Last accessed 19 Feb 2025
42. Università Sapienza di Roma: App Infostud. Available online at https://www.uniroma1.it/it/notizia/app-infostud. Last accessed 19 Feb 2025
43. Visvizi, A.: Artificial intelligence (AI) and sustainable development goals (SDGs): exploring the impact of AI on politics and society. Sustainability **14**, 1730 (2022). https://doi.org/10.3390/su14031730

How Generative AI Enhances Teachers' Creativity in Course Designing and Students' Learning Outcomes

Wei Wang, Yao Tong, and Yenchun Jim Wu

Abstract With the development of Generative AI (GenAI) technologies such as ChatGPT, the landscape of teaching and learning in the educational field is undergoing profound transformations. This study focuses on how GenAI enhances teachers' creativity in course designing and improve students' learning outcomes. Based on the theory of generative learning, this study designed and conducted two experiments to explore the impact of GenAI on course design and its downstream effects on students' learning outcomes. The study found that GenAI significantly enhances the novelty of course designs, regardless of the teachers' levels of experience. Furthermore, course designs involving GenAI effectively improve students' learning effectiveness through cognitive pathways rather than emotional pathways, further validating the potential applications of GenAI in the educational field. The results of this study provide important theoretical foundations for understanding the evolution of teachers' roles in the GenAI era. Additionally, it offers practical guidance on how to leverage GenAI technology to enhance teachers' creativity and improve students' learning outcomes.

Keywords Generative AI · ChatGPT · Teachers' creativity · Course designing · Learning outcomes · SDG

W. Wang · Y. Tong
Huaqiao University, Quanzhou, Fujian, China
e-mail: wwang@hqu.edu.cn

Y. J. Wu (✉)
Graduate Institute of Global Business and Strategy, National Taiwan Normal University, Taipei, Taiwan
e-mail: wuyenchun@gmail.com

A. Visvizi et al. (eds.), *Research and Innovation Forum 2025 Vol2*, Springer Proceedings in Complexity, https://doi.org/10.1007/978-3-032-09062-1_7

1 Introduction

The rapid development of artificial intelligence (AI) has played a significant role in various fields such as intelligent manufacturing, sustainable development, urban governance, and financial risk control, becoming one of the key areas of global technological competition [1–4]. With the widespread application of AI technology, its integration into the field of education is reshaping traditional teaching models and bringing revolutionary changes to personalized learning, intelligent teaching assistance, and educational equity [5, 6]. The application of AI technology may weaken the role of teachers and students' thinking, and this dependency is not conducive to the acceptance of the learning process [7]. On the other hand, some studies have found that the introduction of AI technology can increase the novelty of the classroom and free up teachers' time from routine tasks, allowing them to focus more on knowledge development and improving teaching quality, which is beneficial for course arrangements and student learning [8, 9]. Therefore, we aim to explore the impact of AI integration on teachers' and students' performance under the context of AI-enhanced course design.

The application of AI in teaching involves aspects of efficiency, personalization, and scalability [10]. In terms of course design, AI-generated content (AIGC) can assist in planning the course structure and expanding the scope of knowledge to include novel or interdisciplinary subjects. During the implementation of teaching, generative AI (GenAI) can generate interactive learning tasks to help students understand complex concepts, thereby increasing student engagement [11].

The quality of teaching is closely related to teachers' experience level [12]. Experienced teachers know how to stimulate classroom enthusiasm. In contrast, less experienced teachers tend to be more passive [13]. Therefore, their reliance on AI technology may vary. For instance, teachers with less experience may need it more. Experienced teachers, accustomed to traditional teaching methods, may not be as receptive to new things as younger teachers [14]. We believe that different teachers may have varying attitudes towards the application of AI in teaching design.

As participants in the teaching process, students' perception of the introduction of AI into teaching tasks can reflect its implementation effectiveness [15]. As a novel element, AI to some extent arouses students' interest and enhances classroom interactivity [16]. Moreover, AI possesses extensive knowledge covering various disciplines. Its broad and in-depth knowledge framework provides students with the possibility to understand knowledge better, thereby facilitating their absorption of content and improving the quality of learning [17]. The mechanism by which the integration of AI into teaching affects students' learning outcomes is worth discussing.

Based on the above content, we propose the following three research questions: (1) What are the impacts of AI application in the teaching process on teachers' course design and students' learning outcomes? (2) How do teachers with different levels of experience moderate the effects of AI application on teaching effectiveness? (3) What is the mechanism by which AI-based teaching models influence learning outcomes? To this end, two experiments were conducted to investigate and analyze these questions, examining them from both the teacher and student perspectives in a relatively comprehensive manner. The results of this study not only provide evidence for whether to introduce AI as an auxiliary tool in course education but also enrich the theoretical research in the field of AI education applications.

2 Theoretical Foundations and Hypothesis Development

With the gradual maturation and improvement of AI technology, its ability to understand and generate various types of content, including text, audio, and video, has significantly enhanced. Generative learning theory emphasizes the proactivity of learning, indicating that effective learning relies not only on the passive reception of information but more importantly on how learners actively generate knowledge and construct connections [18, 19]. As recipients of knowledge, students need to rely on strong proactivity to achieve success.

AI has penetrated fields such as product design, achieving significant results in terms of time efficiency and innovation [9, 20]. With continuous technological advancements, the application of GenAI has also gradually extended to the education sector. First, GenAI can rapidly produce a wide range of teaching content, such as case analyses, assignments, and interactive tasks. This significantly reduces the time teachers spend on lesson preparation. By freeing up this time, teachers can focus on how to improve course acceptance rates [21]. Second, the emergence of GenAI has also changed the traditional approach to course design. It can dynamically adjust teaching content according to students' needs and learning progress, thereby enhancing the personalization and interactivity of teaching [22, 23]. Whether it is reducing lesson preparation time or adjusting teaching arrangements, GenAI can improve teachers' work efficiency. This allows teachers to have more time for teaching innovation. Additionally, teachers can use AI to generate ideas or optimize their teaching design, which significantly enhances the innovation and appeal of the course. The innovation of course design directly affects students' interest and engagement in learning [24]. Third, since AI is based on big data and integrates the latest knowledge from interdisciplinary fields, it can create novel teaching content by combining multi-dimensional information. This attracts students' attention and stimulates their desire to explore knowledge [25]. Traditional course design is often limited by teachers' experience and knowledge reserves. In contrast, GenAI, with its extensive knowledge base and powerful generative capabilities, can provide teachers with more diversified and innovative teaching materials. GenAI can also design personalized learning paths and interactive scenarios for students, making the learning process more flexible and engaging. Integrating multidisciplinary knowledge can mitigate the monotony of single-discipline content, leading to more active student participation in the course [26]. Therefore, we propose H1:

H1a: GenAI enhances teachers' creativity in course designing.
H1b: GenAI improves students' learning outcomes.

The role of teachers in classroom teaching cannot be overlooked, especially in the processes of course design and teaching implementation. Teachers' experience and teaching proficiency play a crucial role in teaching effectiveness [27]. Their teaching experience, subject knowledge, and teaching methods directly affect their acceptance and application of GenAI technology [21]. As a new technology, the perceived ease of use and usefulness of AI are influenced by the attitudes of its users. For example, experienced teachers can flexibly utilize teaching materials and effectively integrate them with their in-depth subject knowledge and teaching strategies. Their dependence on GenAI is relatively low, and they may be more cautious about adopting GenAI technology, or they may choose to use it only within the framework of traditional teaching models. In contrast, less experienced teachers may be more receptive to new technologies. They are likely to fully explore the potential of GenAI to assist with their teaching arrangements. This is because they have a stronger ability to adapt to new technologies and a greater spirit of exploration and innovation in teaching practice.

The use of AI by different users yields varying outcomes [14, 28]. Under the same conditions, experienced teachers are more capable of actively fine-tuning AI instructions to better align with their course design and deliver teaching in a manner that is more conducive to student acceptance, thereby achieving a synergistic effect where $1 + 1 > 2$. In contrast, less experienced teachers may have relatively weaker capabilities in this regard. Based on this, we hypothesize that when experienced teachers integrate AI into the teaching process, it positively impacts the innovation of their courses. However, for less experienced teachers, over-reliance on GenAI without proper integration into course design may result in a lack of novelty in course design. Therefore, we propose H2:

> H2: Teachers' experience level positively moderates the impact of GenAI on course design.

Emotion and cognition are two aspects of individual thinking that jointly influence students' learning performance [29]. Positive emotions can enhance students' attention, memory, and problem-solving abilities [30]. Similarly, positive cognitive experiences (such as understanding new knowledge or solving difficult problems) can also trigger positive feedback [31]. Students' learning outcomes are influenced not only by cognitive factors but also by emotional factors. GenAI-generated course design can stimulate students' emotions and cognition through rich teaching content and interactive tasks, thereby enhancing their motivation and effectiveness in learning.

The emergence of AI has brought novelty to the learning process, and novel elements often capture people's attention more effectively. Integrating AI into course design can arouse students' strong interest. Intrinsic motivation is typically associated with an individual's interest in an activity, curiosity, and sense of achievement [32]. Fully mobilizing intrinsic motivation is an important means of improving classroom quality and learning outcomes. This kind of motivation originates from an individual's genuine interest in the subject matter, rather than the desire to obtain good grades or rewards. It is generally considered to have a more positive impact on long-term learning and sustained interest because it enables individuals to remain engaged and committed without the need for external rewards. GenAI can stimulate students' curiosity and desire to explore by generating creative and interactive learning content, thereby enhancing their emotional engagement [33]. This emotional engagement not only increases students' interest in learning but also helps them emotionally identify with the course content, promoting deeper learning.

Overloaded and fragmented knowledge is not conducive to individual integration and absorption. AI, with its powerful autonomous learning capabilities, integrates knowledge across multiple disciplines, emphasizing holism [34]. The integration of teachers and AI, on one hand, enhances the professional competence of teachers, as AI takes on the tedious and time-consuming tasks, allowing teachers to optimize their course design. On the other hand, AI helps students better understand and master knowledge through personalized learning content, instant feedback, and interactive tasks [35]. Personalization and adaptability can effectively reduce students' cognitive load, thereby improving learning outcomes and enabling students to be in a better state when understanding knowledge [36]. Therefore, we propose H3:

H3: GenAI-involved course design enhances students' learning outcomes through both emotional and cognitive pathways.

3 Methods and Materials

To explore the impact of GenAI on teachers' innovation in course design, Study 1 collected data from 60 participants via the Credamo platform, offering monetary rewards as an incentive. The study employed a within-subjects experimental design with two factors: 2 (course design condition: AI-involved vs. AI-not-involved) × 2 (teacher experience level: high vs. low). Participants were asked to evaluate teaching materials under each condition, specifically focusing on the novelty of the materials (measured by items such as "novel," "original," and "rare," where 1 = "strongly disagree" and 7 = "strongly agree") [37]. Subsequently, the composite measure of material novelty was calculated as the average score of the three items.

To assess the impact of GenAI on student learning outcomes, Study 2 collected data from 80 participants via the Credamo platform, offering monetary rewards as an incentive. The study designed two types of course content: courses designed with GenAI involvement and without GenAI involvement. After viewing the teaching content, participants were required to complete a learning outcomes questionnaire. We measured learning outcomes using a perceived usefulness scale (items included "appropriate," "feasible," and "publishable," where 1 = "strongly disagree" and 7 = "strongly agree") [37]. Additionally, participants reported their emotional and cognitive experiences during the learning process, which were assessed using the Positive and Negative Affect Schedule (PANAS-Pos) and the Academic Self-Efficacy Scale (ASES) [38, 39]. These scales were adapted to fit the specific context of this study.

4 Findings

Independent samples t-tests revealed significant differences in the nature of course design, with higher ratings for courses designed with GenAI involvement (M_{GenAI} = 5.622, SD = 0.976) compared to those designed without GenAI (M_{HUMAN} = 3.592, SD = 1.420; $t = -12.907, p = 0.000$) (Fig. 1). This indicates that GenAI involvement indeed enhances the novelty of teachers' courseware design, supporting H1a. Further, to test the moderating effect of teacher experience level, we conducted a 2 (course design condition) × 2 (teacher experience level) factorial ANOVA. Results showed a significant main effect of course design condition (F(1, 59) = 11.976, $p = 0.001$, $\eta^2 = 0.169$) (Fig. 2). However, the interaction between course design condition and teacher experience level was not significant (F(1, 59) = 0.417, $p = 0.521$, $\eta^2 = 0.007$). Thus, the results do not support H2.

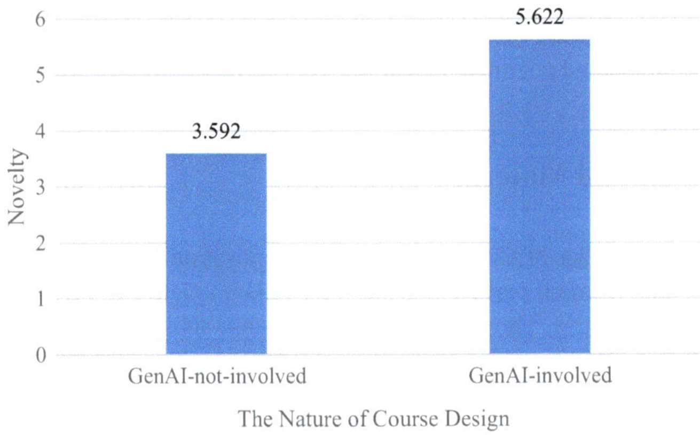

Fig. 1 The impact of GenAI on teachers' innovation in course design. *Source* Authors

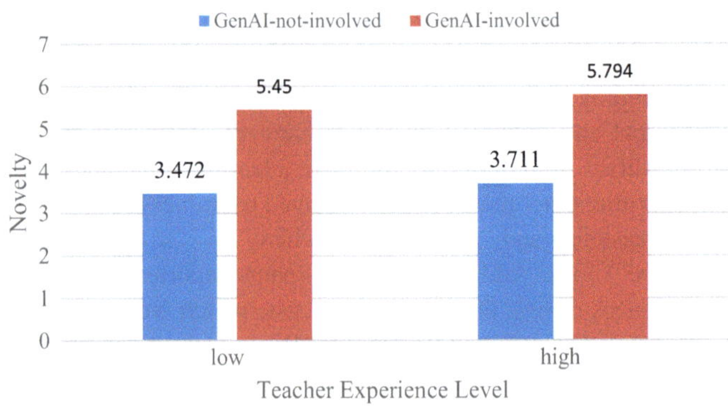

Fig. 2 The moderating effect of teacher experience level. *Source* Authors

In terms of student learning outcomes, independent samples t-tests showed significant differences in the nature of course design (with or without GenAI involvement) ($M_{GenAI} = 5.792, SD = 0.756; M_{HUMAN} = 5.388, SD = 0.983; t = -2.915, p = 0.004$) (Fig. 3). This indicates that GenAI involvement in course design indeed enhances student learning outcomes, supporting H1b. Further, to examine the mediating role of emotion and cognition in the main effect, we conducted mediation analysis using Model 4. Figure 4 showed that the direct effect of course design nature on student learning outcomes was not significant ($\beta = 0.153$, SE = 0.093, LLCI = − 0.030, ULCI = 0.336). Meanwhile, the indirect effect of course design nature on student learning through cognitive level was significant ($\beta = 0.230$, SE = 0.105, LLCI = 0.038, ULCI = 0.450), while the indirect effect through emotional level was not significant ($\beta = 0.021$, SE = 0.039, LLCI = − 0.019, ULCI = 0.134). The results partially support H4.

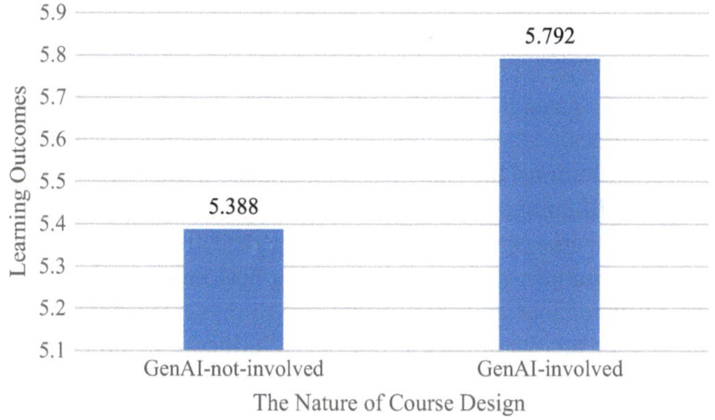

Fig. 3 The impact of GenAI on student learning outcomes. *Source* Authors

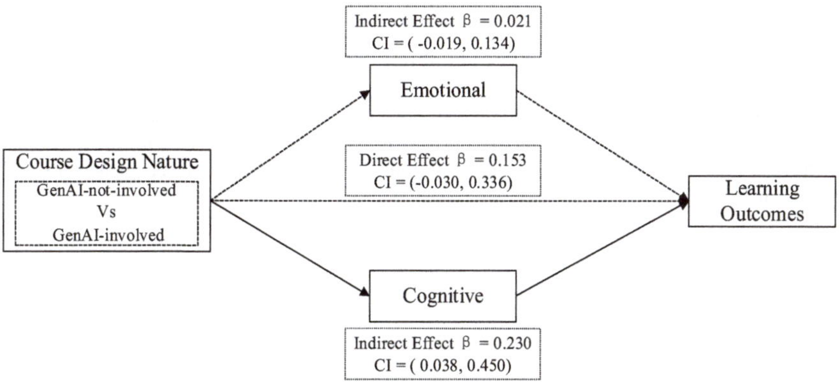

Fig. 4 The mediating role of emotion and cognition in the main effect. *Source* Authors

5 Discussion and Conclusions

The study underscores the importance of learners actively generating knowledge and constructing connections as emphasized by generative learning theory, offering a new theoretical perspective on the integration of educational technology and teaching practice [18]. First, by introducing GenAI, the study extends the theory. GenAI not only provides diverse teaching content and interactive tasks but also dynamically adjusts content to meet students' needs, enhancing their learning experience [23]. Second, the study examines the evolving role of teachers in the GenAI era. With GenAI, they become course optimizers and learning facilitators [21]. The findings also show that teachers of varying experience levels can utilize GenAI to optimize course design. Finally, the study provides a theoretical foundation for educational innovation. By demonstrating GenAI's significant effects on course novelty and student learning, it supports the application of AI in education, promoting personalization and interactivity [25].

The study offers concrete guidance for applying GenAI in education. First, GenAI allows teachers to focus more on teaching innovation. Second, the diverse content and interactive tasks generated by GenAI effectively increase student engagement and learning outcomes.

Despite its contributions, the study has limitations. Future research could explore GenAI's effects in different teaching scenarios, such as online education [15]. Additionally, ethical concerns and the risk of teacher deskilling due to GenAI should be addressed. For example, personalized teaching may involve student data, posing privacy risks. Over-reliance on GenAI could also diminish teachers' ability to independently design and innovate courses [7]. Finally, further research should investigate GenAI's applications across various disciplines to confirm its universal applicability and effectiveness in education.

This study explored the impact of GenAI on the novelty of course design and student learning outcomes through two experiments. The results indicate that GenAI significantly enhances the novelty of course design, regardless of teachers' experience levels. Moreover, GenAI effectively improves student learning outcomes through cognitive pathways rather than emotional ones.

Acknowledgements This work is partially supported by the National Natural Science Foundation of China Grant (72472053, 72072062), and the 2025 Postgraduate Tutor Team Construction Project of Huaqiao University (Generative AI enabled high-quality Chinese business management research tutor team).

References

1. Ramírez-Gordillo, T., Mora, H., Maciá-Lillo, A., Amador, S., Gil, D.: Human-centric solutions and AI in the smart city context: the industry 5.0 perspective. In: Visvizi, A., Troisi, O., Corvello,

V. (eds.) The International Research and Innovation Forum. Springer, pp. 193–203, (2023). https://doi.org/10.1007/978-3-031-44721-1_16
2. Bosco, G., Riccardi, V., Sciarrone, A., D'Amore, R., Visvizi, A.: AI-driven innovation in smart city governance: achieving human-centric and sustainable outcomes. Transform. Government: People, Process and Policy (2024). https://doi.org/10.1108/TG-04-2024-0096
3. Visvizi, A.: Artificial intelligence (AI) and sustainable development goals (SDGS): exploring the impact of AI on politics and society. Sustainability **14**(3), 1730 (2022). https://doi.org/10.3390/su14031730
4. Visvizi, A., Kozlowski, K., Calzada, I., Troisi, O.: Multidisciplinary movements in AI and generative AI: Society, business, education (2025)
5. Strielkowski, W., Grebennikova, V., Lisovskiy, A., Rakhimova, G., Vasileva, T.: AI-driven adaptive learning for sustainable educational transformation. Sustain. Dev. **33**, 1921–1947 (2024). https://doi.org/10.1002/sd.3221
6. Loia, F., Capolupo, N., Gravili, G., Adinolfi, P.: Bridging sustainable higher education and metaverse during the post-pandemic era: Highlights from a multiple case study. In: Visvizi, A., Troisi, O., Corvello, V. (eds.) The International Research and Innovation Forum. Springer, pp. 363–371 (2023). https://doi.org/10.1007/978-3-031-44721-1_27
7. Lin, H., Chen, Q.: Artificial intelligence (AI)-integrated educational applications and college students' creativity and academic emotions: students and teachers' perceptions and attitudes. BMC Psychol **12**(1), 487 (2024). https://doi.org/10.1186/s40359-024-01979-0
8. Zhang, J., Zhang, Z.: AI in teacher education: unlocking new dimensions in teaching support, inclusive learning, and digital literacy. J. Comput. Assist. Learn. **40**(4), 1871–1885 (2024). https://doi.org/10.1111/jcal.12988
9. Mariani, M.M., Machado, I., Magrelli, V., Dwivedi, Y.K.: Artificial intelligence in innovation research: a systematic review, conceptual framework, and future research directions. Technovation **122**, 102623 (2023). https://doi.org/10.1016/j.technovation.2022.102623
10. Copur-Gencturk, Y., Li, J., Atabas, S.: Improving teaching at scale: Can AI be incorporated into professional development to create interactive, personalized learning for teachers? Am. Educ. Res. J. **61**(4), 767–802 (2024). https://doi.org/10.3102/00028312241248514
11. Wang, Y., Cao, Y., Gong, S., Wang, Z., Li, N., Ai, L.: Interaction and learning engagement in online learning: the mediating roles of online learning self-efficacy and academic emotions. Learn. Individ. Differ. **94**, 102128 (2022). https://doi.org/10.1016/j.lindif.2022.102128
12. Van Katwijk, L., Jansen, E., Van Veen, K.: Pre-service teacher research: a way to future-proof teachers? Eur. J. Teach. Educ. **46**(3), 435–455 (2023). https://doi.org/10.1080/02619768.2021.1928070
13. Lazarides, R., Fauth, B., Gaspard, H., Göllner, R.: Teacher self-efficacy and enthusiasm: relations to changes in student-perceived teaching quality at the beginning of secondary education. Learn. Instr. **73**, 101435 (2021). https://doi.org/10.1016/j.learninstruc.2020.101435
14. Wang, X., Li, L., Tan, S.C., Yang, L., Lei, J.: Preparing for AI-enhanced education: conceptualizing and empirically examining teachers' ai readiness. Comput. Hum. Behav. **146**, 107798 (2023). https://doi.org/10.1016/j.chb.2023.107798
15. Wang, W., Liu, H., Wu, Y.J.: Effects of sentiment discreteness on MOOCs' disconfirmation: text analytics in online reviews. Interact. Learn. Environ. 1–18 (2024). https://doi.org/10.1080/10494820.2024.2391050
16. Huang, H.-W., Teng, D.C.-E., Tiangco, J.A.N.Z.: The impact of ai chatbot-supported guided discovery learning on pre-service teachers' learning performance and motivation. J. Sci. Educ. Technol. 1–15 (2024). https://doi.org/10.1007/s10956-024-10179-9
17. Gao, Z., Cheah, J.-H., Lim, X.-J., Luo, X.: Enhancing academic performance of business students using generative AI: an interactive-constructive-active-passive (ICAP) self-determination perspective. Int. J. Manage. Educ. **22**(2), 100958 (2024). https://doi.org/10.1016/j.ijme.2024.100958
18. Fiorella, L., Mayer, R.E.: Eight ways to promote generative learning. Educ. Psychol. Rev. **28**(4), 717–741 (2016). https://doi.org/10.1007/s10648-015-9348-9

19. Wang, X., Mayer, R.E., Zhou, P., Lin, L.: Benefits of interactive graphic organizers in online learning: evidence for generative learning theory. J. Educ. Psychol. **113**(5), 1024 (2021). https://doi.org/10.1037/edu0000606
20. Rammer, C., Fernández, G.P., Czarnitzki, D.: Artificial intelligence and industrial innovation: evidence from German firm-level data. Res. Policy **51**(7), 104555 (2022). https://doi.org/10.1016/j.respol.2022.104555
21. Huang, K.-L., Liu, Y.-C., Dong, M.-Q.: Incorporating AIGC into design ideation: a study on self-efficacy and learning experience acceptance under higher-order thinking. Think. Ski. Creat. **52**, 101508 (2024). https://doi.org/10.1016/j.tsc.2024.101508
22. Guo, J., Ma, Y., Li, T., Noetel, M., Liao, K., Greiff, S.: Harnessing artificial intelligence in generative content for enhancing motivation in learning. Learn. Individ. Differ. **116**, 102547 (2024). https://doi.org/10.1016/j.lindif.2024.102547
23. Siddiquee, T.A.R., Salman, M., Kunwar, R.K., Abdulla, A., Najid, S., Abdullah, F., Sanusi, A.N.Z., Hasan, M.K.: Architecture education: rubrics in google classroom as a tool of improving the assessment and learning. In: Visvizi, A., Troisi, O., Corvello, V. (eds.) The International Research and Innovation Forum. Springer, pp. 373–386 (2023). https://doi.org/10.1007/978-3-031-44721-1_28
24. El-Sabagh, H.A.: Adaptive e-learning environment based on learning styles and its impact on development students' engagement. Int. J. Educ. Technol. High. Educ. **18**(1), 53 (2021). https://doi.org/10.1186/s41239-021-00289-4
25. Liu, C.-C., Liao, M.-G., Chang, C.-H., Lin, H.-M.: An analysis of children'interaction with an AI chatbot and its impact on their interest in reading. Comput. Educ. **189**, 104576 (2022). https://doi.org/10.1016/j.compedu.2022.104576
26. Mamlook, R., Garadah, T., Riaz, S., Salama, A.-A.M.J., Alkhazali, T.M.N., Ismail, A.M.A., Tabassum, M.: Intelligent information and communication technology innovation management in higher education. In: Visvizi, A., Troisi, O., Saeedi, K. (eds.) Research and Innovation Forum 2021: Managing Continuity, Innovation, and Change in the Post-Covid World: Technology, Politics and Society. Springer, pp. 155–163 (2021). https://doi.org/10.1007/978-3-030-84311-3_15
27. Faez, F., Karas, M., Uchihara, T.: Connecting language proficiency to teaching ability: a meta-analysis. Lang. Teach. Res. **25**(5), 754–777 (2021). https://doi.org/10.1177/1362168819868667
28. Ojan, M.-P., Lara-Navarra, P., Sánchez-Navarro, J.: Profiling soft skills in the labour market: the case of communication. In: Visvizi, A., Troisi, O., Corvello, V. (eds.) The International Research and Innovation Forum. Springer, pp. 427–434 (2023). https://doi.org/10.1007/978-3-031-44721-1_32
29. Mangaroska, K., Sharma, K., Gašević, D., Giannakos, M.: Exploring students' cognitive and affective states during problem solving through multimodal data: lessons learned from a programming activity. J. Comput. Assist. Learn. **38**(1), 40–59 (2022). https://doi.org/10.1111/jcal.12590
30. Juárez Varón, D., Bellido-García, I., Gupta, B.B.: Analysis of stress, attention, interest, and engagement in onsite and online higher education: a neurotechnological study. Comunicar **31**(76):21–34 (2023). https://doi.org/10.3916/C76-2023-02
31. Schneider, S., Beege, M., Nebel, S., Schnaubert, L., Rey, G.D.: The cognitive-affective-social theory of learning in digital environments (castle). Educ. Psychol. Rev. **34**(1), 1–38 (2022). https://doi.org/10.1007/s10648-021-09626-5
32. Poupard, M., Larrue, F., Sauzéon, H., Tricot, A.: A systematic review of immersive technologies for education: learning performance, cognitive load and intrinsic motivation. Br. J. Edu. Technol. **56**(1), 5–41 (2025). https://doi.org/10.1111/bjet.13503
33. Chen, X., Hu, Z., Wang, C.: Empowering education development through aigc: a systematic literature review. Educ. Inf. Technol. **29**(13), 17485–17537 (2024). https://doi.org/10.1007/s10639-024-12549-7
34. Zeng, J., Jia, X.: Systems theory-driven framework for AI integration into the holistic material basis research of traditional chinese medicine. Engineering **40**, 28–50 (2024). https://doi.org/10.1016/j.eng.2024.04.009

35. Salman, M., Samadi, Z.B., Saquib, M., Yusoff, Z.M.: A critical review on the various issues and challenges of virtual pedagogy in architecture education. In: Visvizi, A., Troisi, O., Corvello, V. (eds.) The International Research and Innovation Forum. Springer, pp. 387–406 (2023). https://doi.org/10.1007/978-3-031-44721-1_29
36. Feng, L.: Investigating the effects of artificial intelligence-assisted language learning strategies on cognitive load and learning outcomes: a comparative study. J. Educ. Comput. Res. **62**(8), 1961–1994 (2025). https://doi.org/10.1177/07356331241268349
37. Doshi, A.R., Hauser, O.P.: Generative AI enhances individual creativity but reduces the collective diversity of novel content. Sci. Adv. **10**(28):eadn5290 (2024). https://doi.org/10.1126/sciadv.adn5290
38. Watson, D., Clark, L.A., Tellegen, A.: Development and validation of brief measures of positive and negative affect: the PANAS scales. J. Pers. Soc. Psychol. **54**(6), 1063 (1988). https://doi.org/10.1037/0022-3514.54.6.1063
39. Feldman, D.B., Kubota, M.: Hope, self-efficacy, optimism, and academic achievement: distinguishing constructs and levels of specificity in predicting college grade-point average. Learn. Individ. Differ. **37**, 210–216 (2015). https://doi.org/10.1016/j.lindif.2014.11.022

Detecting the Digital Ghostwriter in the Classroom: Accuracy of Pedagogy Students in Identifying AI-Generated Essays

Jana Šindlerová, Jana Vejvodová, and Lucie Rohlíková

Abstract The rapid development of generative artificial intelligence (AI) presents new challenges and opportunities in education, particularly in assessing students' writing. This study examines the ability of pedagogy students to distinguish between human-written and AI-generated essays. Through an experimental approach, participants evaluated a set of texts, identifying key linguistic and structural features that signaled AI involvement. The findings provide insights into students' accuracy in recognizing AI-generated content and the implications for academic integrity, assessment methods, and the role of AI literacy in teacher training. The study highlights the need for adapting pedagogical strategies to the evolving landscape of digital writing technologies.

Keywords AI literacy · Teacher training · AI-generated content · Student's writing

1 Introduction

Large language models can produce texts that closely resemble those written by humans. In education, this raises concerns about the authenticity of student work and the ability of teachers to respond appropriately. If a teacher fails to recognize an AI-generated essay, it may lead to misjudging a student's skills, unfair advantages, and a weakened classroom climate.

J. Šindlerová
Faculty of Arts, Charles University, Prague, Czech Republic
e-mail: jana.sindlerova@ff.cuni.cz

J. Vejvodová · L. Rohlíková (✉)
Faculty of Education, University of West Bohemia, Plzeň, Czech Republic
e-mail: lrohlik@kvd.zcu.cz

J. Vejvodová
e-mail: vejvod@kcj.zcu.cz

A. Visvizi et al. (eds.), *Research and Innovation Forum 2025 Vol2*, Springer Proceedings in Complexity, https://doi.org/10.1007/978-3-032-09062-1_8

Essays serve not only as grading tools but also as diagnostic instruments for evaluating students' linguistic competencies and critical thinking. When artificial intelligence replaces student writing, teachers lose insight into student abilities and opportunities for skill development. Recognizing such cases enables teachers to adapt assignments, emphasize writing processes, and address ethical dimensions of artificial intelligence use.

This study investigates how accurately master's students in Czech language teacher education can distinguish between AI-generated and student-written essays. Related research (e.g., [1]) found that even experienced teachers often fail to detect AI-generated content. Although some showed slightly better results, many overestimated their detection skills.

Previous studies have examined human ability to recognize AI-generated texts in fields such as journalism [2], scientific writing [3], and literature [4, 5]. However, little research has focused on educational settings, particularly involving pre-service teachers and real classroom genres like essays.

Our study addresses this gap by examining detection accuracy and interpretive strategies among Czech language teaching students.

2 Methodology

2.1 Research Aims

The present study interprets the results of research focused on determining the ability of master's students in Czech language teacher education to identify essays written by real students and texts generated by AI tools. We formulated the following research questions:

- To what extent are pre-service Czech language teachers able to identify AI-generated texts among texts written by students?
- How confident are pre-service teachers when determining the origin of a text?
- Which aspects of the texts lead them to identify a text as AI-generated or student-written?

We offer the following hypotheses:

1. Pre-service teachers will generally not be able to predict their performance correctly. They will likely overestimate their own abilities.
2. Pre-service teachers with extensive personal experience with generative large language models will recognize the AI-generated texts with both high precision and recall. Inexperienced pedagogy students will score lower.
3. Students with previous teaching experience will recognize the AI-generated texts with higher accuracy than inexperienced students.
4. Prompting large language models to write as a high school student of a certain age will result in texts that are less detectable than those with neutral prompts.

5. ChatGPT 4o will produce high quality, less recognizable texts than the ChatGPT 3.5 version or Microsoft Copilot.

2.2 Methods and Data

A total of 60 university students enrolled in a master's program in Czech language teacher education from four Czech universities participated in the study. All participants were first- or second-year master's students. Only seven were men; the rest were women. Within the experiment, they were asked to evaluate 14 texts with respect to their authorship. Out of the 14 texts, 8 were written by real high school students, in the pre-terminal year of study (approximately 17–18 years of age), 6 were produced by prompting a large language model.

The high school students were given the following instruction: Write an opinion essay on the following topic: "What is, in your opinion, the most important prerequisite of a happy life." The writing conditions were strictly controlled to ensure none of the students uses large language models. Out of the resulting texts, we have chosen 8 most suitable for the experiment—the choice included strongly subjective opinionated pieces, as well as rather general and vague ones, also, both essays with perfect grammar and spelling, and quite imperfect essays were included.

Six AI-generated essays were produced using ChatGPT (GPT-3.5 and GPT-4o) and Microsoft Copilot—tools commonly available to students. These included texts from both the paid (GPT-4o) and free (GPT-3.5) versions of ChatGPT, as well as from Microsoft Copilot available through a school license. Each of the large language models were then prompted twice—first with a general prompt similar to the one the real students got: "Write an opinion essay on the following topic: "What is, in your opinion, the most important prerequisite of a happy life." This reflects a common way of student prompting—simply copying the teacher's assignment. Second, we added a simple specification of age to the same prompt: "Write as a 17-year-old high school student." This reflects a more sophisticated prompting which is likely to produce a text more similar to the real student texts.

Data collection was conducted via an online Google Form. Participants were asked to indicate whether they believed each text had been written by a real student or by artificial intelligence. They were also asked to explain the reasoning behind their decisions.

Within the questionnaire, the participants were asked to give important personal information (their teaching experience and experience with generating texts), to guess how successful in the task they will likely be, read the texts, decide on the authorship and provide clues that led them to the specific decision.

Neither of the university students had extensive practical experience with teaching writing. Also, they were not given any clue regarding the number of AI-generated texts in the sample, so that they could not attempt guessing.

We then measured the precision and recall for each of the students [6], and also the ratio of right and wrong assessments for each of the texts. The recall, in this case, is the ratio of the number of correctly recognized AI-generated texts to the number of all AI-generated texts in the data. A perfect recall score tells us that there are no false negatives, i.e., the pre-service teacher did not miss any ghost-written texts. The lower the recall score ranges, the more AI-generated texts escaped the teacher's notice. A perfect precision score tells us there were no false positives, i.e., the pre-service teacher did not falsely accuse any student of ghost-writing. The lower the precision score gets, the more human texts were labeled as AI-generated.

Based on the number of false positives and false negatives in the assessment, we have defined the following range of successfulness.

- We assess the students as **very successful** in the task, in case there were 0–1 FNs and 0 FPs at most.
- We assess the students as **rather successful**, in case there were 2 FNs and 1 FP. at most.
- We assess the students as **rather unsuccessful**, in case there were 3–4 FNs and 2 FPs at most.
- We assess the students as **unsuccessful**, in case there were at least 5–6 FNs or 3 and more FPs.

Note that we draw the line between successful and unsuccessful performance at a point where half of the AI-generated texts (3/6) remains unrecognized and/or a quarter of real student texts (2/8) is falsely reported as AI-generated. In our opinion, it is much more problematic when a teacher falsely accuses an individual of AI-plagiarism, or digital ghost-writing, putting at stake the very basis of the mutual trust, then when they do not recognize the AI-plagiarism and let it eventually escape without consequences.

3 Results

3.1 Students

The average recall of the group was 0.575. The average precision was 0.592. This basically means that the task of identifying an AI-generated text is difficult and the pedagogy students are generally unsuccessful in it. In this respect, our findings are not surprising and correlate with the findings of [1]. Nevertheless, 15 people (25% of the test group) reached at least "rather successful" category and 5 of them were even very successful (4 of them identifying every text correctly, and one with one false negative). It might be interesting then to compare their original estimate of their own abilities to their actual performance.

In the beginning, the students expressed generally either mild confidence or mild disbelief in their own abilities to recognize an AI-generated text. Only one person expressed a strong belief in their abilities, interestingly, the performance of this person showed perfect recall score, but low precision, i.e., a tendency to excessively label human-generated texts as AI-generated (see Table 1).

If we consider broader categories of successfulness or unsuccessfulness, we can say that 33 people from the test group predicted their performance more or less correctly, while 18 people (30% of the students) overestimated their performance and 9 people (15%) underestimated it. Note that overestimating one's abilities with respect to the identification of AI plagiarism may lead to serious problems in the assessment process and mutual trust in the classroom.

Also, out of the 24 people who believed they will succeed in the task, only one fourth really did. One fourth of the people who did not believe in their abilities did better than they expected.

We hypothesized that students with extensive experience with prompting and text generation will generally score better than those without. Surprisingly, this did not prove correct, see Table 2.

Nor does the teaching experience seem to play a role. The students did not have an extensive experience with teaching writing, nevertheless, we tried to split them into four groups according to the experience stated in the questionnaire: 2–5 years of teaching—experienced; 0–2 years of teaching—slightly experienced; supervized teaching within the university studies (few hours)—beginners; no teaching experience at all—inexperienced (see Table 3).

Most surprisingly, those students that do not have almost any practical experience with teaching, with the exception of supervised teaching that is part of their university curriculum, seem to be significantly more successful than any of the remaining groups.

Table 1 Participant confidence versus actual performance

Actual performance	Previous estimate rather unsuccessful	Previous estimate rather successful or very successful
Very successful	2	3
Rather successful	7	3
Rather unsuccessful	8	5
Unsuccessful	19	13

Table 2 Prior experience with LLMs versus predicted AI detection ability

Experience with LLMs	Very successful	Rather successful	Rather unsuccessful	Unsuccessful
Often	2	0	2	3
Sometimes	1	5	4	11
Have tried	1	4	6	13
No experience	1	1	1	5

Table 3 Prior teaching experience versus predicted AI detection ability

Teaching experience	Very successful	Rather successful	Rather unsuccessful	Unsuccessful
2–5 years	1	0	2	6
0–2 years	1	1	4	5
Few hours	2	7	4	8
No experience	1	2	3	13

Table 4 Correct identification of AI-generated texts by students

	General prompt (%)	Prompt with age (%)
Copilot	70	53.3
GPT 3.5	73.3	45
GPT 4o	60	38.3

It seems that the ability to recognize generated text lies somewhere else—for example, there might be a correlation with the individual reading comprehension or critical thinking skills. This may be one of the questions for future work.

Summary

The findings indicate that pre-service teachers generally struggled to accurately identify AI-generated texts. Notably, prior experience with large language models or teaching did not significantly enhance performance. Additionally, a considerable number of participants misjudged their own abilities in this task.

3.2 *Texts*

Table 4 shows percentage of students who correctly identified the text as AI-generated.

The hypothesis that incorporating the desired age and status of the writing agent in the prompt will make the task more difficult, i.e., the result will resemble the real student text more, has proved correct. The pre-service teachers were significantly more successful with the generally prompted texts than with those with the age specification.

The hypothesis that the paid version of GPT will yield better results than GPT 3.5 or Microsoft Copilot also holds.

3.2.1 Identification of AI-Generated Texts

For the correctly identified AI-generated texts, students provided the following reasons for their decisions.

Low degree of subjectivity and personal experience

The text synthesizes general ideas that feel universal and detached. The arguments are not sufficiently specific and often present general truths rather than a personal approach.

Overly smooth and balanced structure, formulaic expression

Overall, the text appears polished and well-crafted, but also unnatural.

Repetitive themes

There is frequent emphasis on non-material values, with repeated use of expressions such as *balance, gratitude, external factors, spiritual development, personal growth*—suggesting the use of widely available sources and templates.

Repetition of words

Students noted excessive use of verbs that express thinking or feeling—so-called "mental ownership" verbs that reflect subjective modality. For some respondents, however, the overuse of these verbs suggested student authorship, as such expressions are often emphasized in classroom instruction on writing reflective essays. If students submit a text with a "low index of reflection," they are given a lower grade. As a result, students often **overuse these formal features**, which can make their writing sound mechanical or artificial.

Overly sophisticated wording

Students pointed to phrases they felt a high school student would not use in a reflective essay, such as: "a deeply personal matter, the illusion of happiness defined by external factors, the meaning of life brings motivation to face challenges…".

Seeming depth of content, but actual lack of meaning

Although the text appears linguistically rich, in the end it doesn't offer new or interesting ideas. It relies on general phrases and formulates them in a way that seems deep but lacks real content value.

Excessive formal correctness

"The text feels too perfect." Students also noted that the **frequent use of em dashes** is not typical of authentic student writing."

Formal errors

In Text 1, the verb "dosáhnout" (to achieve) appeared with the accusative instead of the genitive case. Some students saw this as a clue pointing to AI authorship, while others considered it typical of human error. Similarly, nonstandard word order was interpreted both as a sign of AI generation and as natural for student writing.

3.2.2 Identification of Student-Written Texts

The most common reasons respondents cited for identifying a text as student-written were as follows:

- Subjectivity and Personal Experience
- Spelling, Grammatical, and Lexical Errors; Stylistic Imperfections
- Authenticity and Spontaneity; Use of Informal Language
- Irregular Structure and Thought Progression

3.2.3 Interpreting the Same Clues Differently

The same textual element was interpreted differently by various respondents, particularly when the text referenced specific media, cultural, scientific, or social contexts:

Literary Reference

- Reasons for identifying the text as AI-generated: "a reference to Anne of Green Gables"
- Reasons for identifying the text as student-written: "I think AI would write more formally and wouldn't reference a book character."

Reference to the Magazine 'Modern Mother'

- Reasons for identifying the text as AI-generated: "It seems unlikely that third-year high school students would read the magazine 'Modern Mother'."
- Reasons for identifying the text as student-written: "Choice of unique/unusual themes—'Modern Mother'."

References to Hollywood and Sri Lanka in the Text

- Reasons for identifying the text as AI-generated: "The author uses broader knowledge, such as the fact that a Sri Lankan woman must pick 13 kg of petals per day, so I think the text was generated by AI."
- Reasons for identifying the text as student-written: "Comparison of a Sri Lankan woman to a billionaire on a yacht."

Mention of Tying Shoelaces

- Reasons for identifying the text as AI-generated: "Who would help us tie our shoelaces? Students wouldn't express themselves this way."
- Reasons for identifying the text as student-written: "There are also original questions that point more to student expression: 'Who would help us tie our shoelaces?' Artificial intelligence would formulate more neutral sentences."

Mention of Maslow's Hierarchy of Needs

- Reasons for identifying the text as AI-generated:

  - "This text feels artificial to me; who would seek happiness in Maslow's hierarchy???"
- Reasons for identifying the text as student-written:
  - "AI probably wouldn't bring up Maslow's hierarchy."

Summary

AI-generated texts created with prompts specifying the author's age were harder to identify, confirming that such prompt engineering increases authenticity. GPT-4o outputs were the most difficult to detect. Respondents often relied on features such as generality, overly polished style, or complex vocabulary to flag texts as AI-generated. However, identical textual elements (e.g., cultural references, modal verbs) were interpreted in opposing ways. These ambiguities underline the challenges of AI detection and the limits of human judgment.

4 Discussion

The results of our study confirm the complexity of the task pre-service teachers face when attempting to identify AI-generated texts. With average precision and recall scores below 0.6, the group was largely unsuccessful, which aligns with [1] and supports the hypothesis that even linguistically trained readers often fail to detect AI authorship reliably.

Interestingly, the findings challenge some of our initial assumptions. Contrary to expectations, neither prior teaching experience nor frequent use of large language models led to significantly better performance. The most successful subgroup had only limited supervised teaching practice, suggesting that detection ability may depend more on analytical aptitude and sensitivity to stylistic nuance than on teaching experience or AI familiarity.

Nearly one-third of participants overestimated their detection abilities, a notable cognitive bias with potential pedagogical consequences. Overconfidence may lead to misjudging student work or making unfounded accusations. Conversely, about 15% underestimated their skills and performed better than expected, suggesting that cautious self-assessment could support the development of reliable evaluation strategies.

Prompt formulation also significantly influenced recognizability. AI-generated texts mimicking the style of a 17-year-old student were harder to detect, supporting the notion that simple prompt engineering can enhance believability. GPT-4o outputs proved the most difficult to identify, reinforcing that more advanced models produce highly natural-sounding texts. The hypothesis that GPT-4o would outperform GPT-3.5 and Microsoft Copilot in evading detection was confirmed.

Certain textual features, such as cultural references, were inconsistently interpreted by respondents, underscoring the subjective nature of detection and the limitations of intuition-based judgment.

These findings suggest a pressing need to include critical AI literacy and forensic stylistics in teacher education. Teachers must be equipped not only to detect dishonesty but also to preserve classroom trust—false accusations can cause more harm than undetected misuse.

While this study did not test instructional interventions or assessment formats, the results highlight the importance of developing awareness and critical reflection in pre-service teachers. Given the observed overconfidence and interpretive ambiguity, curricula should include training in stylistic analysis, forensic comparison, and authorship ethics. Assessment design may also benefit from a shift toward process-based evaluation and structured student reflection. Future research could assess the impact of such strategies and examine detection competence across countries or over time through longitudinal tracking. Cross-national comparative studies may reveal cultural or systemic differences in how AI-generated texts are perceived and evaluated, while longitudinal research could capture how detection skills evolve with growing exposure to generative tools.

5 Conclusions

Our study shows that identifying AI-generated texts among student essays remains a significant challenge even for academically trained individuals preparing for a teaching career. The findings highlight the limitations of intuitive detection and reveal the risk of both false positives and false negatives in assessment contexts. Neither teaching experience nor familiarity with large language models proved to be decisive predictors of performance, suggesting that detection may depend on more subtle cognitive or personality-based traits.

Given the growing availability and sophistication of AI tools, teacher education must urgently address the dual challenge of equipping future educators with strategies to recognize AI-generated texts and encouraging pedagogical practices that reduce the incentive for their misuse. This includes redefining writing assignments to emphasize process over product, integrating discussions on authorship ethics, and fostering reflective use of AI technologies.

Future research should focus on identifying the cognitive and stylistic indicators that successful individuals rely on when evaluating text authenticity and on developing training tools that can enhance these competencies. Moreover, the ethical dimension of detection—especially the potential harm of false accusations—must remain central to any institutional response to AI-authored student work.

When comparing individual texts, it is evident that the formulation of the prompt significantly affects recognizability. AI-generated texts that mimicked the style of a 17-year-old student were more difficult to detect, supporting the idea that minimal prompt engineering can significantly increase textual believability. Furthermore, our data confirm that GPT 4o outputs were the most difficult to identify, reinforcing the notion that more advanced models can produce highly human-like outputs with fewer detectable anomalies. The hypothesis that ChatGPT 4o would produce higher-quality, less detectable texts than ChatGPT 3.5 or Microsoft Copilot was also confirmed.

The analysis also highlights that certain textual features, like references to specific cultural or media contexts, were interpreted differently by respondents.

These findings suggest a pressing need to include critical AI literacy and forensic stylistics in teacher education programs. The role of the teacher is not only to detect dishonesty but also to maintain classroom trust. Erroneous accusations can cause significant damage to student–teacher relationships, potentially more than undetected cheating. In conclusion, it is essential to emphasize that developing competences in the field of Information and Communication Technology (ICT) and Artificial Intelligence (AI) represents a critical imperative for all stakeholders within the educational system. This digital literacy—among both students and educators—is a fundamental prerequisite for successful digital transformation and for fully leveraging the potential of "smart education" through an ecosystem approach [7]. The systematic cultivation of these skills is crucial for addressing issues such as the digital divide and gender gaps, and for fostering students' entrepreneurial orientation within higher education environments [8]. Acquired ICT and AI skills enable an effective transition towards the "smart value co-education" model [9], where the active involvement of both parties leads to the dynamic value cocreation within the educational process [10]. Ensuring these competences is therefore vital for creating truly modern, adaptable, and inclusive education for the 21st century.

Acknowledgements This work was supported by the project "Human-centred AI for a Sustainable and Adaptive Society" (reg. no.: CZ.02.01.01/00/23_025/0008691), co-funded by the European Union.

This publication was supported by the Cooperatio Program provided by Charles University, research area Linguistics, implemented at the Faculty of Arts of Charles University.

References

1. Fleckenstein, J., Meyer, J., Jansen, T., Keller, S.D., Köller, O. Möller, J.: Do teachers spot AI? Evaluating the detectability of AI-generated texts among student essays. Comput. Educ.: Artif. Intell. **6**(June 2024), 100–209 (2024). https://doi.org/10.1016/j.caeai.2024.100209
2. Graefe, A., Haim, M., Haarmann, B., Brosius, H.B.: Readers' perception of computer-generated news: credibility, expertise, and readability. Journalism **19**(5), 595–610 (2018). https://doi.org/10.1177/146488491664126
3. Gao, C.A., Howard, F.M., Markov, N.S., Dyer, E.C., Ramesh, S., Luo, Y., Pearson, A.T.: Comparing scientific abstracts generated by ChatGPT to real abstracts with detectors and blinded human reviewers. NPJ Digit. Med. **6**(1), 75 (2023). https://doi.org/10.1038/s41746-023-00819-6
4. Gunser, V.E., Gottschling, S., Brucker, B., Richter, S., Çakir, D., Gerjets, P.: The pure poet: how good is the subjective credibility and stylistic quality of literary short texts written with an artificial intelligence tool as compared to texts written by human authors?. In: Proceedings of the Annual Meeting of the Cognitive Science Society, vol. 44, no. 44 (2022)

5. Köbis, N., Mossink, L.D.: Artificial intelligence versus Maya Angelou: experimental evidence that people cannot differentiate AI-generated from human-written poetry. Comput. Hum. Behav. **114**, 106553 (2021). https://doi.org/10.1016/j.chb.2020.106553
6. Powers, D.M.W.: Evaluation: from precision, recall and F-measure to ROC, informedness, markedness and correlation. J. Mach. Learn. Technol. **2**(1), 37–63 (2011)
7. Visvizi, A., Troisi, O., Grimaldi, M.: Exploring the drivers for digital transformation in smart education: an ecosystems approach. In: ITM Web of Conferences, vol. 51. EDP Sciences, p. 01003 (2023)
8. Barra, C., Grimaldi, M., Muazzam, A., Troisi, O., Visvizi, A.: Digital divide, gender gap, and entrepreneurial orientation: how to foster technology adoption among Pakistani higher education students? Socioecon. Plann. Sci. **93**, 101904 (2024)
9. Botti, A., Grimaldi, M., Tommasetti, A., Troisi, O., Vesci, M.: Modeling and measuring the consumer activities associated with value cocreation: an exploratory test in the context of education. Serv. Sci. **9**(1), 63–73 (2017)
10. Loia, V., Maione, G., Tommasetti, A., Torre, C., Troisi, O., Botti, A.: Toward smart value co-education. In: Smart Education and e-Learning 2016. Springer International Publishing, pp. 61–71. (2016)

Regulatory Frameworks and Policy Making for Competitiveness, Growth and Well-Being

China, Twin Transition and Patents: The Case of Electric Vehicles (EVs)

Anna Visvizi, Krzysztof Karwowski, and Vilma Çekani

Abstract In a context defined by shifting patterns of global competition, technological rivalry, and the resurgence of power diplomacy, by referencing the case of China's approach to electromobility, this paper examines the evolving landscape of green innovation seen as both a mechanism for sustainable development and a tool of industrial policy. Using the electric vehicle (EV) sector as a case study, and patent data from WIPO's PATENTSCOPE database as a key indicator of innovation activity, the paper addresses two main research questions: (1) What is the geographical distribution of EV-related patents? and (2) How does the evolution of EV patent ownership influence the dynamics of the global EV market? The analysis reveals China's growing dominance across core EV technologies, especially in batteries, platforms, and, to a lesser extent, charging systems. This suggests a deliberate and strategic use of green innovation as a means of technological catch-up and market consolidation. The finding confirm the existence of a major shift in R&D activities in the EV sector globally, largely to the detriment of the traditionally established players, e.g. European countries, and to the advantage of primarily China, Korea, the United States, and others.

Keywords China · Global competition · Competitiveness · Electric vehicles (EV) · Patents · PATENSCOPE · SDG-7 · SDG-13

A. Visvizi (✉) · K. Karwowski
SGH Warsaw School of Economics, Warsaw, Poland
e-mail: avisvi@sgh.waw.pl

A. Visvizi
Effat University, Jeddah, Saudi Arabia

V. Çekani
Epoka University, Tirana, Albania

A. Visvizi et al. (eds.), *Research and Innovation Forum 2025 Vol2*, Springer Proceedings in Complexity, https://doi.org/10.1007/978-3-032-09062-1_9

1 Introduction

In context of changing patterns of competition worldwide, including the quest for new drivers of economic growth, the consolidation of influence of the already influential players on the global market, and the re-turn of power diplomacy [1–4], innovation-driven competitiveness has become the key issue of global policy debates. These debates, and resulting policy considerations and strategies, including—increasingly so—deployment of protectionist policies and tools, unfold against the backdrop of another debate, i.e. digital and green transition (aka twin transition). Whereas the debate on the origins, rationale, logic, and mechanisms underpinning twin transition represents a vibrant field of research in itself [5–7], in the context of this paper, its role in promoting innovation in very specific domains of economy is stressed. Admittedly, as research suggests, while the drive toward digital and green transition, may have been employed in the European Union (EU), for instance, as a means of reconciling otherwise conflicting outcomes of the push for more growth and the need to protect the environment [8, 9], it may have been employed by other players to serve different objectives [10, 11]. The case of China is very telling in this respect. Traditionally criticized for exploitative environmental policies and environmental degradation, over the past years, by focusing on "green innovation" China succeeded in recasting itself not only as an environmentally responsible a player, but also as one of the leading actors globally in green-technology dependent products [12, 13]. Again, research addressing selected aspects of this issue, including domestic and international factors behind the, let's call it "quiet green transformation" or "green turn", emerges gradually [10, 11, 14, 15]. That being said, several questions pertaining to the strategy underpinning China's green turn still need to be addressed. By referencing the case of electromobility [16], i.e. including electric vehicles, and especially the unfolding map of patents globally, this paper seeks to address this issue. That being said, the really interesting question is what lessons can be drawn from an analysis of China's patent activity in the EVs' domain. Given the frequently contested role of China in the global order, recognizing a degree of ignorance of laymen and misinformation, considering the debate on global competition, trade wars, access to raw materials and technology, the objective of this paper is to navigate the complex set of issues. To this end, two questions are addressed:

RQ1: What is the geographical distribution of EV related patents?
RQ2: How does the evolution of EV patent ownership influence the dynamics of the global EV market?

To address these questions, the remainder of this paper is structured as follows. Section 2 offers an insight into selected aspects of the technology development-patent application-strategy nexus in the field of EV manufacturing. Section 3 elaborates on methods and materials. In Sect. 4 the findings are presented and discussed. Discussion (Sect. 5) and conclusions (Sect. 6) follow.

2 The Technology Development-Patent Application-Strategy Nexus in the EV Industry

Electromobility is considered a key component of the twin transition, in that it aligns with attempts at decarbonization of transportation, and necessitates digital innovation in the manufacturing process [10]. The electric vehicle (EV) industry stands at the center of the drive toward less pollution as well as toward meeting emissions' standards. Clearly, as a relatively new development on the market, the prospect of benefiting from the first mover advantage in the EV sector, seemed not only very attractive, but also—sort of—guaranteed to historically established car producers from the United States (US) and the EU (see Table 1 for an insight). And yet, today, the global EV market is dominated by Chinese producers [16], whereby the key performance indicators (KPIs) are not solely a function of production capacity and demand, but primarily of control over knowledge assets, particularly patents, as well as access to rare earths. For the purpose of the analysis here, rare earth and other factors of production, less patents, will be considered as a constant. The focus will be directed at patents. The latter serve as key indicators of a country's innovation capacity, technology development potential, strategic planning, economic diplomacy, and many other.

Patents are the key measure of innovation and market potential in the rapidly evolving EV sector, where companies are engaged in fierce competition not only for technological supremacy but also for strategic positioning in the global market place [17, 18]. Although initially dominated by patent applications from the US and the EU, global patent activity has undergone dynamic changes [19–22] over the past decade. With regard to global innovation trends in the EV industry, research suggests that patent activity, including patent application submissions and registrations have been

Table 1 The comparison of the largest car manufacturing countries (in millions of cars) in 2003 and 2023

Year 2003			Year 2023		
Country	Total production (in millions of units)	Global %	Country	Total production (in millions of units)	Global %
World	60.63	100	World	93.55	100
USA	12.11	19.97	China	25.75	27.53
Japan	10.29	16.97	USA	10.89	11.64
Germany	5.51	9.09	Japan	9.68	10.35
China	4.44	7.32	Germany	4.66	4.98
France	3.62	5.97	India	4.52	4.83
S. Korea	3.18	5.24	Mexico	4.01	4.29
Spain	3.03	5.00	S. Korea	3.95	4.22

Source Authors, based on International Organization of Motor Vehicle Manufacturers (OICA, Organisation Internationale des Constructeurs d'Automobiles [36, 37])

dominated by players from outside the EU [23–25]. The work of the European Patent Office (EPO), initially established in 1973, essentially consistent with increasing the ease of access to European applicants, and ensuring international recognition of patents registered with EPO [26, 27] were intended to facilitate the EU-based inventors to apply for patents and secure their rights to them. Nevertheless, the ease of access that the EPO process and procedures offer, have served applicants from outside the EU as well, thus enabling them to register and protect increasingly sophisticated (industrial) solutions [28]. China, and so Chinese applicants, belong to the most active ones, especially in the domain of EV industry and related sectors [29].

An analysis of patent data, available at WIPO's PATENTSCOPE reveals that the EV industry is characterized by a very dynamic landscape of innovation globally. Significant growth in patent activity related to green technologies indicates a focus on carbon reduction, where new solutions in the fields of green mobility and EVs are the more common. While Asia leads this innovation trend, the United States excels in the quality of registered patents [30, 31].

One of the most strategic components of EVs is the battery, usually lithium-ion and lithium-polymer. Its quality, including the travel distance it secures, and time to recharge it needs, are the key factors that define a battery's and EV's utility [32, 33]. Battery-related patents dominate the landscape of EV technologies, suggesting a strong link between storage efficiency and market competitiveness [34, 35]. Apart from batteries, charging systems are equally important for the EVs' manufacturing and marketing. For years, long charging times were seen as a disadvantage, especially when compared to combustion vehicles that are ready to drive immediately after refueling. Today, fast charging technology changes the status quo, i.e. EVs are now on par with internal combustion engines in terms of time needed to secure their readiness to move. The third issue that weighs on EVs strategic advantage is the EV platform itself. Here, notions such as ease of production, scalability and usability are of key significance. Notably, EV platforms are more than the traditional vehicle frames. EV platforms are a complex and integrated system that combines the battery pack, usually integrated with the drive axle, powertrain, energy management and cooling systems, chassis and suspension, along with an integrated electronics and communication modules [30, 36].

A basic insight into the technical aspects of EV's construction and operation is necessary if the implications of the evolution of the manufacturing process and change in global car market share are to be comprehended. Data presented in Table 1, reveals a substantial change in the overall volume of cars produced over the years 2003–2023. Specifically, whereas in 2003 the US with 12.11 million units produced dominated the global car market, in 2023 with a production output at 10.89 million units, it ranked second, i.e. after China. Notably, China's output in this domain increased from 4.4 million units in 2003 to 25.75 million units in 2023. Two issues need to be highlighted here. First, the overall output of the car industry increased by more than 50% (from 60.63 million to 93.55 million) over the period 2003–2023. Several factors contribute to these developments, including growing demand (driven by the rise in the middle class in several emerging markets), improvements in the manufacturing process (driven by global supply chain consolidation, lean manufacturing processes, automation and robotization), and post 2008 + economic recovery and so on [36, 37].

These processes have been catalyzed by public policies designed to stimulate demand for new environmentally friendlier vehicles. China is a piece and parcel of these developments, whereby both domestic and external factors play a role in Chinese government's pursuit to boost the development of Chinese car industry [38–40]. Examples would include policies supporting growth of car manufacturing, securing access to raw materials, technology and talent [39, 40], followed by a gradual increase in production capacity (including car subcomponents and finished cars), and aggressive penetration of foreign markets. Domestically, the Chinese government would employ such instruments as subsidies, tax incentives, and horizontal strategies aiming at the development of sustainable road transport [41–44]. In brief, the EV industry and the EV market have changed over these past few years [45], also in terms of the key brands and manufacturers, most of them from China. For instance, Chinese producers, with brands otherwise considered exotic, e.g. BYD, Geely, and Chery Auto, succeeded in consolidating their position in markets earlier dominated by European manufacturers. In addition, new brands have been created, such as the Chinese Nio, Avatr or BAIC (commonly known as Beijing). Other Asian producers, led by Korean chaebols, such as Hyundai and its subsidiaries, e.g. KIA and Genesis [46]; brands from India, e.g. Ather, Ola Electric, Ultraviolette; from Japan, e.g. ASPARK; and an unusual Vietnamese-Singapore cooperative, Vinfast that gained popularity due to effective marketing strategies.

3 Methods and Materials

To address the research questions outlined above (RQ1-RQ2), a mixed methods approach was employed, including desk research and literature review, to set the conceptual framework for the discussion in this study. Subsequently, primary data was collected to add substance to claims and assumptions raised earlier. For the sake of data collection, the PATENTSCOPE tool provided by the World Intellectual Property Organization (WIPO) was used. Given the features of the PATENTSCOPE database, i.e. it offers access to published international patent applications, in full-text, already on the day of publication, thus it is representative; it is published by an international organization specializing in intellectual property rights, thus is it objective, comprehensive, and transparent; it has been widely used in research [47, 48] on diverse aspects of patent activity in international and regional contexts. PATENSCOPE is a comprehensive, multilingual, and internationally standardized repository of patent application documents from participating national and regional offices. Based on the provisions of the Patent Cooperation Treaty (PCT) and the mechanisms of cooperation entailed therein, PATENTSCOPE provides access to internationally recognized patents. Importantly, it integrates regional patent institutions, i.e. the European Patent Office (EPO). As a result, the PATENTSCOPE data covers

123.5 million patent documents, including 5.2 million published international PCT patent applications.

As for the actual use of the database, the original plan involved scraping patents containing in their titles and abstracts keywords, such as “electric vehicle” and “electric vehicles”. Both variants were used due to inconsistencies in terminology and the need to capture all relevant, both in singular, i.e. vehicle, and plural, i.e. vehicles, cases. Nevertheless, the initial plan failed. Accordingly, given the vastness of the database, to arrive at a valid and manageable sample to be investigates, in line with best practices [49], the records were scraped and examined in line with the steps and precepts of standard (though, minor variations were necessary) structured literature review (SLR) protocol (PRISMA). The steps included: formulation of the research question; team building; structuring the details of the protocol (keywords/strings; steps); conducting the search; extracting the data; based on predefined inclusion and exclusion criteria (see Table 2), selecting the actual sample of patent applications; presentation of the results; discussion. This approach secured that the search terms were clearly defined and aligned with RQ1 and RQ2; that the criteria of inclusion and exclusion were clearly defined; and that the replicability of the process was secured. The search covered the January 1, 2015–January 18, 2025. The source material was extracted on January 18, 2025.

Accordingly, the keywords included: “electric vehicl*” “electric vehicl* + batter*”, “electric vehicl* + charging systems”, and “electric vehicl* + platforms”, all in connection to EVs. The search included both titles of patent applications as well summaries of the applications’ content. The keywords were subsequently associated with patent codes to ensure more precise data identification and extraction, i.e. the search focused on patent codes, rather than on keywords. As patent codes are linked information on year of registration, and the nationality of the patent applicant, basing the search on the keyword-patent code nexus, the search allowed to limit the initial sample of 123,500,000–485,487 records. This data was downloaded as an Excel file.

Based on this set of data, patents were analyzed based on the following criteria: the most patent application active countries, i.e. China, Germany, Japan, Korea, and the

Table 2 PRISMA protocol-based PATENSCOPE search

Inclusion criteria	Exclusion criteria
Globally recognized patents, including the European Patent protocol	Locally recognized patents
Globally recognized and registered patents	Applications filled but still processed
Globally recognized patents focused on such technical features of EVs as batteries, charging systems, and platforms	Globally recognized patents focused on other than technical features of EVs, e.g. design and decorations, software
Patents registered over the period 2015–2025 (Jan 18, 2025)	Patents registered prior to 2015 and as of January 18, 2025
Language in which the patent application was submitted: English	Language in which the patent application was submitted: other than English

Source Authors

United States; patents registered collectively as "European Patents" (EP) were also included. The following exclusion criteria were applied: patents not recognized as global, as well as those unrelated to the focus areas identified in the search: batteries, charging systems, and platforms, have been excluded from the analysis. Inclusion criteria were defined as: patent applications where the preponderant part of the patent description is related to technology, and not for instance design or decorative features of an EV have been excluded from the analysis. As a result of applying the exclusion criteria, the size of the sample declined to 14,821. This sample was filtered manually to identify only those patent applications that are centered on technical (batteries, charging systems, platforms) aspects of EVs. As a result, the sample shrunk to 9822 records. Finally, only patent applications submitted and patents registered over the period 2015–2024 were included in the final analysis. This resulted in the final sample size stabilizing at 6 920 records. The process of data filtering has been presented in Fig. 1.

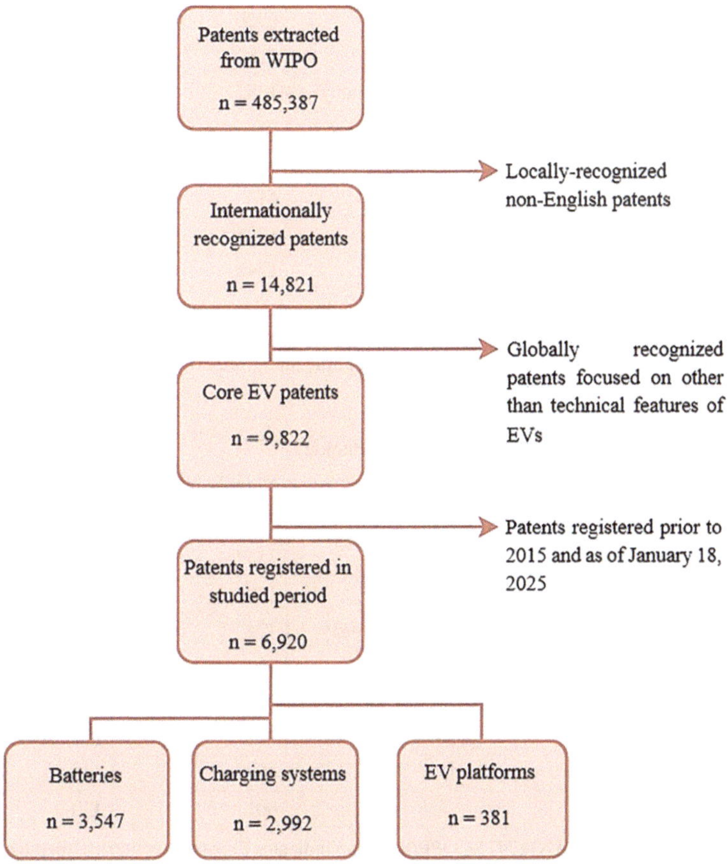

Fig. 1 The PRISMA process applied to the PATENTSCOPE-based data search. *Source* Authors

4 Results

The results of data collection are presented in Tables 2, 3, 4, 5 and 6. Each of these tables demonstrates a specific aspect of patent activity in the domain of EV. Table 3 offers a general insight into the evolution of the field. Tables 4, 5 and 6 feature, respectively patent activity in domain such as battery-related technology (Table 4), charging systems (Table 5), and EV platform-related technology (Table 6). They illustrate the distribution of patents published over the past decade, divided by the years 2015 and 2025. To highlight the dynamics of respective developments, the past five years, i.e. 2020–2025, have been placed in spotlight too.

Table 3 Distribution of core EV patents

Country of origin	Number of patents	
	2015–Jan 2025	2020–Jan 2025
All	6920	4278
China	1782	1065
Remaining European countries*, excluding Germany	1282	747
Germany	95	71
Japan	837	416
Korea	613	438
USA	1065	688
Other**	1246	853

Source Authors, based on PATENTSCOPE/WIPO, * indicates European countries, members of the European Patent Organization (EPO), https://www.epo.org/en/about-us/foundation/member-states; ** indicates submissions from countries, with insignificant number of patent registrations

Table 4 Distribution of patents for batteries' technology

Country of origin	Number of patents	
	2015–Jan 2025	2020–Jan 2025
All	3547	2266
China	1085	690
Remaining European countries*, excluding Germany	474	308
Germany	26	18
Japan	607	288
Korea	267	201
USA	487	331
Other**	601	430

Source Authors, based on PATENTSCOPE/WIPO, * indicates European countries, members of the European Patent Organization (EPO), https://www.epo.org/en/about-us/foundation/member-states; ** indicates submissions from countries, with insignificant number of patent registrations

Table 5 Distribution of patents for EV's 'charging systems'

Country of origin	Number of patents	
	2015–Jan 2025	2020–Jan 2025
All	2992	1956
China	619	359
Remaining European countries*, excluding Germany	698	436
Germany	62	53
Japan	217	128
Korea	328	233
USA	499	338
Other**	569	409

Source Authors, based on PATENTSCOPE/WIPO, * indicates European countries, members of the European Patent Organization (EPO), https://www.epo.org/en/about-us/foundation/member-states; ** indicates submissions from countries, with insignificant number of patent registrations

Table 6 Distribution of patents for the EV platforms

Country of origin	Number of patents	
	2015–Jan 2025	2020–Jan 2025
All	381	56
China	78	16
Remaining European countries*, excluding Germany	110	3
Germany	7	0
Japan	13	0
Korea	18	4
USA	79	19
Other*	76	14

Source Authors, based on PATENTSCOPE/WIPO, * indicates European countries, members of the European Patent Organization (EPO), https://www.epo.org/en/about-us/foundation/member-states; ** indicates submissions from countries, with insignificant number of patent registrations

As for the distribution of core EV patents (Table 3), over the past decade, China has solidified its position as a major player in global innovation, particularly in terms of patent filings. In the broader EV category, China accounted for approximately 25.8% of global patents between 2015 and 2025, i.e. 1782 patents out of a total of 6920. This made China the single largest contributor during the full 10-year period, outpacing traditional leaders such as the USA, Japan, and Korea. However, a closer look at the more recent timeframe, i.e. from 2020 to January 2025, reveals a nuanced shift. China still led in total patent contributions over the entire period, but its output decelerated, with 1065 patents filed. While this placed China ahead of most other countries, China's rate of growth slowed down.

As for patents for batteries' technology (Table 4), between 2015 and 2025, China established itself as the leading contributor to battery technology patents related to electric vehicles (EVs), reflecting its strategic investment in core components of EV innovation. In the broader battery-related patent category, China accounted for 1085 out of 3547 global filings, i.e. approximately 30.6%, and retained a dominant position in the more recent 2020–2025 period with 690 out of 2266 filings (about 30.4% out of total) (Table 4). This highlights China's sustained commitment to research and development (R&D) activities related to battery technology development. This trend confirms China's leadership in battery innovation, positioning it not only as a dominant EV manufacturer but also as a principal force in the development of strategic technologies underpinning the global transition to electric mobility. A case can be made that innovation in the domain of EV batteries constitutes a critical driver of China's EV industrial strategy.

As for patents for charging systems, between 2015 and 2025, China played a prominent role in the global innovation landscape for electric vehicle (EV) charging systems, although its dominance in this domain was more balanced in comparison to battery technologies. In the broader category of EV charging system patents, China accounted for 619 out of 2.992 filings (approximately 20.7%) between 2015 and 2025 (Table 5). Overall, while China continues to lead or remain highly competitive across EV-related technologies, the data on charging systems suggest a more distributed innovation ecosystem, with Europe and the United States emerging as especially strong players in this critical segment of EV development.

As for patents related to EV platforms, between 2015 and 2025, China emerged as one of the two leading contributors to innovation in this domain, nearly matching the United States in terms of patent activity. China filed 78 out of the total 381 global patents in this category, accounting for 20.5% of the total (Table 6). China has been a leading force in EV platform innovation. Meanwhile, the United States is increasingly asserting itself in this segment, and the European Patent Office holds a strong, though uneven, position—particularly in the narrower category. These trends point to a maturing field where global leadership is still taking shape.

5 Discussion

The analysis of patent activity in the EV sector between 2015 and January 2025 (Tables 3, 4, 5 and 6) reveals interesting, at times counterintuitive developments in the global innovation landscape. While China's overall dominance across EV-related patent categories was expected, given its aggressive national policies and industrial strategies, a closer look at the distribution of activity over time and by technology domain offers a more nuanced picture. Specifically, despite accounting for the highest number of EV-related patents (25.8% of the global total) over the full decade, China's contribution in the 2020–2025 period fell to 24.89%. From a different angle, the performance of EPO member states, including Germany, is worth noting. In that collectively, these countries outperformed Japan and Korea in nearly every category. Germany is clear leader among this group of countries and so data presented in Tables 3, 4, 5 and 6 accounts for it separately. The EPO members' (thus including Germany) lead in EV charging systems' technology is noticeable. With a total of 760 patents registered over 2015–2025, these countries outpaced even China, with a total of 619 patents. This may suggest a strategic specialization of the EPO members. That is, they may lag in battery and platform technologies, but gradually are building their leadership in infrastructural technologies critical to large-scale EV deployment. The US has shown a strong positive patent activity over 2015–2025. It maintained its position in general and established a part with China in the domain of EV platforms (79 vs. 78 patents).

The case of Germany is an interesting one. Despite its status as a traditional automotive powerhouse, German patent activity remained limited across all segments, withan overall number of patent registrations in the EV sector at 95 over 2015–2025. Only 7 patents related to EV platform technology were identified. This seems to confirm a slower initial response to the nascent EV market as well as a more risk-averse approach to innovation and product development strategy of German car producers. The latter may be a function of senior managers' perceptions of the market and conceptions of consumers' expectations [49]. The impact of the political cycle and government's incentives (or not) for EV production/consumption play a role too [50].

As argued earlier, innovation in battery technology is a key component of building competitive advantage in the EV sector. Here, China's (with over 30% of global filings in this sector) dominance remains undisputed. Nevertheless, as the number of patent registrations attributed to the US and the EPO members, Germany included, increases, the landscapes of technology and competition become more diversified. In the domain of EV platform technologies, only 381 patents were filed globally over the past 10 years, of which only 56 in the past five years. This trend may suggest a strategic shift of car producers toward integration and system-level innovation rather than R&D associated with platform development.

In sum, this analysis illustrates that while China retains a commanding lead in EV-related patents overall, global competition intensifies. The role of the EPO members, although often underestimated, is striking in the charging systems' technology. It may thus signals a potential strategic pivot toward support infrastructure and regulatory-driven innovation, both areas critical for mass EV adoption.

6 Conclusions

The starting point for the discussion in this paper was the realization that the today's politics and economy, and so also the performance of key actors on the global scene, can only be understood if the following five factors are included in the analysis. These include, first, a complex interplay of economic rivalry among established powers, countries consolidating their power, and countries aspiring to power. Regional variability exists, whereas and rivalry, or claims to leadership, play a role also in regional contexts. Second, advances in technology, while of great potential in view of manufacturing, new products, quality of life improvements etc. exist. Technology, though, may also be a source of disruption across levels, including questions of equitable access, equality, inclusion/exclusion etc. These are followed by considerations of strategy and policy related to technology transfer, including extraction of resources, flow of resources, extraction/production infrastructure, and standards. These issues, finally, are reflective of the re-emerging asymmetries of power around the world. Third, the past decade has seen a resurgence of power politics. At its core is a shift from a relatively cooperative, liberal global order, based on principles of multilateralism, and echoing values and principles of the Transatlantic partnership, to a fragmented and competitive landscape. Fourth, given the external (economic rivalry) and domestic pressures ((youth) unemployment, economic uncertainty, skills' misalignment, the crumbling of social insurance and healthcare systems), a return to protectionist policies is not a taboo anymore. Here, an increasing variety of tools and measures are employed by national governments to protect, favor, or advance domestic industries, frequently under the disguise of other policies and policy imperatives. Twin transition is a good case in point. Fifth, because ideas matter, and discourses are constitutive of social reality, the depletion of the quality of the political discourse at the global, regional, national levels, is a factor that plays a role in contemporary politics and economy. By zooming in the analysis to the case of China, the objective of this paper was to explore what lessons might be drawn from an analysis of China's patent activity in the EVs' domain. Two RQs were addressed: What is the geographical distribution of EV related patents? And How does the evolution of EV patent ownership influence the dynamics of the global EV market?

The findings of this paper suggest that innovation in the EV sector has become both a strategic and politically charged domain of global competition. China's dominant position in patent filings across core EV technologies, including batteries, charging systems, and EV platforms, underscores the instrumental role the state can play in catalyzing (green) innovation [51, 52] Through a deliberate combination of industrial planning, state-backed incentives, and targeted investment in strategic technologies, China has recast itself, or at least its image, from an environmental laggard to a frontrunner in clean mobility. This strategic repositioning is not incidental. It is a function of a broader framework of state-driven economic modernization in China.

Green innovation, as evidenced in the case of electromobility, is not merely a response to environmental imperatives or global sustainability goals. Instead, it functions as a dual-purpose mechanism: one that responds to growing international pressure for climate action, while simultaneously serving as a protective and competitive industrial tool. The focus on patent development in China's EV sector demonstrates that green innovation is also a means to consolidate technological sovereignty, secure future market share, and strengthen geopolitical leverage; especially in view of the re-emergence of power diplomacy and the intensification of techno-economic competition among global powers [1–3].

From this perspective, green innovation becomes a camouflaged instrument of economic security. The Chinese government's policies, ranging from subsidies, tax relief, and R&D funding, to strategic support for key industrial players, underscore the state's central role not only in promoting green transitions but also in orchestrating innovation-led industrial development (cf. [51]). The patent data examined in this paper reflect this trajectory. That is, China leads in battery and EV platform patents and maintains a strong, though more balanced, position in EV charging technologies. This suggests both technological capacity and strategic foresight, i.e. China has effectively used the momentum of the twin transition to consolidate its role as a global innovation powerhouse.

In contrast, the patenting activity in the US and the EU reflects more decentralized, market-driven innovation systems. While the US retains a strong position in terms of patent quality and Europe shows leadership in specific niches, e.g. charging systems, neither has matched the scale and coherence of China's state-led innovation model. This divergence is particularly significant in the context of the twin transition, i.e. the push for digital and green transformation, which increasingly shapes national and supranational policy agendas. In this context, green innovation has evolved from being an environmental necessity into a core component of geopolitical strategy and economic resilience.

In conclusion, the findings of this paper suggest that EVs' production and green innovation are interrelated today. The quest to establish a position on the market though is contingent not only on the ability to successfully apply for patents and register them, but also—quite strikingly—on managerial decisions taken at a given company level. The case of Germany seems to suggest that. In all cases examined here, including primarily China, but then also the US, the EPO members, the pervasive role of regulation, and resulting policy tools, has become apparent. This highlight a shift toward industrial policy, competition and free trade. Future research should necessarily address these issues in-depth.

References

1. Wu, C.X.: A Bargaining theory of US–China economic rivalry: differentiating the trade and technology wars. Chin. J. Int. Polit. **17**(4), 323–345. https://doi.org/10.1093/cjip/poae017Nye
2. Pape, R.A.: Why economic sanctions do not work. Int. Secur. **22**(2), 90–136 (1997). https://muse.jhu.edu/article/446841
3. Muñoz de Bustillo Llorente, R.: A critical review of the digital and green twin transitions: implications, synergies and trade-offs. In: JRC Working Papers Series on Labour, Education and Technology, No. 2024/07, European Commission, Joint Research Centre (JRC), Seville (2024). https://www.econstor.eu/handle/10419/311129
4. Nye, J.S. Jr.: Soft power and public diplomacy revisited. Hague J. Dipl. **14**(1–2), 7–20 (2019). https://doi.org/10.1163/1871191X-14101013
5. Müller, M., Lang, S. Stöber, L.-F.: Twin transition—hidden links between the green and digital transition. J. Innov. Econ. Manage. **45**(3), 57–94 (2024). https://shs.cairn.info/journal-of-innovation-economics-2024-3-page-57?lang=en
6. Diodato, D., Huergo, E., Moncada-Paternò-Castello, P., Rentocchini, F., Timmermans, B.: Introduction to the special issue on "the twin (digital and green) transition: handling the economic and social challenges." Ind. Innov. **30**(7), 755–765 (2023). https://doi.org/10.1080/13662716.2023.2254272

7. Kovacic, Z., García Casañas, C., Argüelles, L., Yáñez Serrano, P., Ribera-Fumaz, R., Prause, L., March, H.: The twin green and digital transition: high-level policy or science fiction? Environ. Plan. E: Nat. Space **7**(6), 2251–2278 (2024). https://doi.org/10.1177/25148486241258046(Originalworkpublished2024)
8. Gao, X.: The EU's twin transitions towards sustainability and digital leadership: a coherent or fragmented policy field? Reg. Stud. **59**(1), 2360053 (2025). https://doi.org/10.1080/00343404.2024.2360053
9. Wang, J., Visvizi, A., Nan, F., Meng, F.: Placing China's green technology innovation in a context. In: Visvizi, A., Troisi, O., Corvello, V. (eds.) Research and Innovation Forum 2023. RIIFORUM 2023. Springer Proceedings in Complexity. Springer, Cham (2024a). https://doi.org/10.1007/978-3-031-44721-1_36
10. Wang, J., Wu, B., Zhou, X., Qu, A.: China and its market-oriented green technology innovation system: the rationale, the mechanisms, and the policymaking process. In: Visvizi, A., Troisi, O., Corvello, V. (eds.) Research and Innovation Forum 2023. RIIFORUM 2023. Springer Proceedings in Complexity. Springer, Cham (2024b). https://doi.org/10.1007/978-3-031-44721-1_38
11. Brueck, C., Losacker, S., Liefner, I.: China's digital and green (twin) transition: insights from national and regional innovation policies. Regional Stud. **59**(1) (2024). https://doi.org/10.1080/00343404.2024.2384411
12. Ahmad, N., Youjin, L., Žiković, S., Belyaeva, Z.: The effects of technological innovation on sustainable development and environmental degradation: evidence from China. Technol. Soc. **72**, 102184 (2023). https://doi.org/10.1016/j.techsoc.2022.102184
13. Brueck, C.: The role of foreign MNEs in China's twin transition: a study on the organization of green and digital innovation processes. Compet. Rev. 29 October 2024 **34**(5), 879–895 (2024). https://doi.org/10.1108/CR-08-2023-0207
14. Zhai, X., An,Y., Shi, X., Liu, X.: Measurement of green transition and its driving factors: evidence from China. J. Clean. Prod. **335**, 130292 (2022). https://doi.org/10.1016/j.jclepro.2021.130292
15. Altenburg, T., Corrocher, N., Malerba, F.: China's leapfrogging in electromobility. A story of green transformation driving catch-up and competitive advantage. Technol. Forecast. Soc. Change **183**, 121914 (2022). https://doi.org/10.1016/j.techfore.2022.121914
16. Hertel, D., Bräunig, G., Thürer, M.: Towards a green electromobility transition: a systematic review of the state of the art on electric vehicle battery systems disassembly. J. Manuf. Syst. **74**(2024), 387–396 (2024). https://doi.org/10.1016/j.jmsy.2024.03.013
17. Yang, Z., Huang, H., Lin, L.: Sustainable electric vehicle batteries for a sustainable world: perspectives on battery cathodes, environment, supply chain, manufacturing, life cycle, and policy. Adv. Energy Mater. **2022**(12), 2200383 (2022). https://doi.org/10.1002/aenm.202200383
18. Wingender, P., Yao, J., Zymek, R., Carton, B., Cerdeiro, D., Weber, A.: Europe's shift to EVs amid intensifying global competition. In: IMF Working Papers 2024, vol. 218 (2024). Accessed 1 Aug 2025. https://doi.org/10.5089/9798400291371.001
19. Wong, C-Y., Yeung, H-W, Huang, S., Song, J., Lee, K.: Geopolitics and the changing landscape of global value chains and competition in the global semiconductor industry: rivalry and catch-up in chip manufacturing in East Asia. Technol. Forecast. Soc. Change **209**, 123749 (2024). https://doi.org/10.1016/j.techfore.2024.123749
20. Marszałek, A.: Innowacje w sektorze biotechnologii jako katalizator procesu transferu wiedzy i technologii. Kwartalnik Nauk O Przedsiębiorstwie **65**(3), 48–61 (2022). https://doi.org/10.33119/knop.2022.65.3.3
21. Di, Y., Zhou, Y., Zhang, L., Indraprahasta, G.S. Cao, J.: Spatial pattern and evolution of global innovation network from 2000 to 2019: global patent dataset perspective. Complexity **2022**, 5912696, 11 (2022). https://doi.org/10.1155/2022/5912696
22. Yuan, X., Li, X.: Mapping the technology diffusion of battery electric vehicle based on patent analysis: a perspective of global innovation systems. Energy **222**(2021), 119897 (2022). https://doi.org/10.1016/j.energy.2021.119897

23. Schmitt, G., Scott, J., Davis, A., Utz, T.: Patents and progress; intellectual property showing the future of electric vehicles. World Electr. Veh. J. **8**(3), 635–645 (2016). https://doi.org/10.3390/wevj8030635
24. Golembiewski, B., vom Stein, N., Sick, N., Wiemhöfer, H.D.: Identifying trends in battery technologies with regard to electric mobility: evidence from patenting activities along and across the battery value chain. J. Clean. Prod. **87**(2015), 800–810 (2015). https://doi.org/10.1016/j.jclepro.2014.10.034
25. Plomer A.: The EPO as patent law-maker in Europe. Eur. Law J. 2019 **25**, 57–74 (2019). https://doi.org/10.1111/eulj.12304
26. Plomer, A.: The unified patent court and the transformation of the European patent system. IIC **51**, 791–796 (2020). https://doi.org/10.1007/s40319-020-00963-6
27. Chávez, A., Lara, A.: The diversity of agents and patent thicket evolution in electric vehicles. Int. J. Automot. Technol. Manage. **20**(1), 76–107 (2020). https://doi.org/10.1504/IJATM.2020.105309
28. Wu, L., Xu, M.: Research on cooperative innovation network structure and evolution characteristics of electric vehicle industry. Sustainability **14**(10), 6048 (2022). https://doi.org/10.3390/su14106048
29. Li, X., Yuan, X.: Tracing the technology transfer of battery electric vehicles in China: a patent citation organization network analysis. Energy **239**, 122265 (2022). https://doi.org/10.1016/j.energy.2021.122265
30. Consolati, D., Marmaglio, P., Canziani, L., Tiboni, M., Amici, C.: Electric actuation of transport vehicles: overview of technical characteristics and propulsion solutions through a systematic patent analysis. Actuators **13**(1), 15 (2024). https://doi.org/10.3390/act13010015
31. Sun, B., Ju, Z.: Research on the promotion of new energy vehicles based on multi-source heterogeneous data: consumer and manufacturer perspectives. Environ. Sci. Pollut. Res. **30**, 28863–28873 (2023). https://doi.org/10.1007/s11356-022-24304-x
32. Damjanov, K., Crouch, D.: Powering wicked futures: EV batteries as vital matters of care. Continuum **39**, 692–703 (2025). https://doi.org/10.1080/10304312.2025.2532128
33. Ma, S.-C., Xu, J.-H., Fan, Y.: Characteristics and key trends of global electric vehicle technology development: a multi-method patent analysis. J. Clean. Prod. **338**, 130502 (2022). https://doi.org/10.1016/j.jclepro.2022.130502
34. Tian, Y., Zhang, Q.: Development overview of electric vehicle industry based on patent data analysis. In: Proceedings SPIE 12510, International Conference on Statistics, Data Science, and Computational Intelligence (CSDSCI 2022), vol. 125100T (13 Jan 2023) (2022). https://doi.org/10.1117/12.2656790
35. Liu, Z., Liu, X., Zhao, F.: Research on NEV platform development strategies for automotive companies. World Elect. Veh. J. **12**(4), 201 (2021). https://doi.org/10.3390/wevj12040201
36. OICA: International Organization of Motor Vehicle Manufacturers, Statistics | (2003). www.oica.net. Retrieved 23 Dec 2024, from https://www.oica.net/category/production-statistics/2003-statistics/
37. Christiansen, B.: China's economic growth and innovation in globalization. In: Visvizi, A., Lytras, M., Zhang, X., Zhao, J. (eds.) (2019) Foreign Business in China and Opportunities for Technological Innovation and Sustainable Economics. IGI Global (2019). https://doi.org/10.4018/978-1-5225-8980-8
38. Karwowski, K., Visvizi, A., Troisi, O.: Explaining China's pivots and priorities through the aerospace industry development strategy. In: Visvizi, A., Troisi, O., Corvello, V. (eds.) Research and Innovation Forum 2023. Springer International Publishing, pp. 543–558 (2024). https://doi.org/10.1007/978-3-031-44721-1_41
39. Karwowski, K., Visvizi, A.: Knowledge and technology transfer in China. The case of the Chinese HEIs sector. In: Visvizi, A., Troisi, O., Corvello, V., Grimaldi, M. (eds.) Research and Innovation Forum 2024. RIIFORUM 2024. Springer Proceedings in Complexity. Springer, Cham (2026). https://doi.org/10.1007/978-3-031-78623-5_61
40. State Council of China: The Notice of the State Council on the Publication of "Made in China 2025". Retrieved on 30 July 2025 from English (2015). www.gov.cn/policies/latest_releases/2015/05/19/content_281475110703534.htm

41. Huang, K.G.-L., Huang, C., Shen, H., Mao, H.: Assessing the value of China's patented inventions. Technol. Forecast. Soc. Change **170**, 120868 (2021). https://doi.org/10.1016/j.techfore.2021.120868
42. Zhang, X., Luo, W., Xiang, D.: Strategic emerging industries and innovation: evidence from China. Int. Rev. Econ. Financ. **98**, 103858 (2025). https://doi.org/10.1016/j.iref.2025.103858
43. South China Morning Post: China is recruiting embassy staff with a focus on electric vehicles, source says 17 Dec 2024 (2024). https://www.scmp.com/news/china/politics/article/3291139/china-recruiting-embassy-staff-focus-electric-vehicles-source-says
44. Sparke, M.: Defining geoeconomics amid shifts in global hegemony: critical geographies of new international conjunctures. Environ Plan A **57**(1), 103–107 (2024). https://doi.org/10.1177/0308518X241265293
45. Kim, B.: Cultural forces and historical events to shape organizing principles in Korea: an exploratory perspective. Manag. Decis. **48**(5), 826–841 (2010). https://doi.org/10.1108/00251741011043948
46. Cernaro, A., Minutoli, F.: The innovation imagined for the window: international patents in comparison. In: Corrao, R., Campisi, T., Colajanni, S., Saeli, M., Vinci, C. (eds.) Proceedings of the 11th International Conference of Ar. Tec. (Scientific Society of Architectural Engineering). Colloqui.AT.e 2024. Lecture Notes in Civil Engineering, vol. 612. Springer, Cham (2025). https://doi.org/10.1007/978-3-031-71867-0_14
47. Atiencia. D, Pincay, J.: Classifying patents in Spanish within the subclasses of the standardized international patent classification system: a case study. In: 2024 Tenth International Conference on eDemocracy & eGovernment (ICEDEG), Lucerne, Switzerland, pp. 1–6 (2024) https://doi.org/10.1109/ICEDEG61611.2024.10702062
48. Chuprat, S., Novianto, E.H.D., Matsuura, Y. et al.: A closer look on patent analytics through systematic literature review. Manage. Rev. Q. (2024). https://doi.org/10.1007/s11301-024-00452-x
49. Loder, J., Rinscheid, A., Wüstenhagen, R.: Why do (some) German car manufacturers go electric? The role of dynamic capabilities and cognitive frames. Bus. Strateg. Environ. **33**(2), 1129–1143 (2024). https://doi.org/10.1002/bse.3538
50. Kędzierski, M.: Unplugged. The Uncertain Future of Electromobility in Germany, OSW Commentary, 2024–03–13, Centre for Eastern Studies (OSW) (2024). https://www.osw.waw.pl/en/publikacje/osw-commentary/2024-03-13/unplugged-uncertain-future-electromobility-germany. Accessed 15 July 2025
51. Meckling, J., Nahm, J.: When do states disrupt industries? Electric cars and the politics of innovation. Rev. Int. Polit. Econ. **25**(4), 505–529 (2018). https://doi.org/10.1080/09692290.2018.1434810
52. Cao, C., Su, Y., Zheng, Q.: Impact of policy incentives on technological innovation and diffusion within the new-energy vehicle industry: an ecosystem approach. Technol. Anal. Strat. Manage. 1–17 (2024). https://doi.org/10.1080/09537325.2024.2306648

Industrial Symbiosis Networks in Italian Industrial Parks: The Case of Fosso Imperatore

Carlotta D'Alessandro, Giuseppe Ioppolo, Grazia Calabrò, and Giuseppe Caristi

Abstract Industrial parks (IPs) represent critical nodes in sustainable industrial development, with Industrial Symbiosis (IS) emerging as a promising approach for enhancing resource efficiency and environmental performance. This paper examines the Fosso Imperatore IP in Southern Italy to address the geographical gap in IS studies focused on this region. Using a computational approach combining Python-based waste flow estimation with network analysis, the study quantifies potential material exchanges and identifies barriers to implementation. Results reveal the technical potential for symbiotic relationships with 319.74 tons of metal, 1,775.02 tons of plastic, 186.55 tons of paper, and 547.06 tons of organic waste that could potentially be exchanged within the park. However, network analysis shows that despite favorable network density (0.32) and compatible material flows, companies currently process 100% of waste externally, primarily due to economic incentives from external buyers. The research identifies key barriers to IS implementation including economic considerations, information asymmetry, and limited internal processing infrastructure. Digital technologies including IoT-enabled waste monitoring, blockchain-based transaction systems, and AI-powered matching algorithms present promising solutions to overcome these barriers. However, technological innovation must be coupled with economic incentives and institutional support to transform technical potential into implemented symbiotic relationships. Indeed, geographical proximity and technical compatibility alone are insufficient drivers for symbiosis development without addressing underlying economic and social barriers.

C. D'Alessandro (✉) · G. Ioppolo · G. Calabrò · G. Caristi
Università Degli Studi Di Messina, Messina, Italy
e-mail: carlotta.dalessandro@unime.it

G. Ioppolo
e-mail: giuseppe.ioppolo@unime.it

G. Calabrò
e-mail: grazia.calabro@unime.it

G. Caristi
e-mail: giuseppe.caristi@unime.it

A. Visvizi et al. (eds.), *Research and Innovation Forum 2025 Vol2*, Springer Proceedings in Complexity, https://doi.org/10.1007/978-3-032-09062-1_10

Keywords Industrial parks · Industrial symbiosis · Network analysis · Digital Innovation · Sustainability · SDG 9 (Industry · Innovation · Infrastructure)

1 Introduction

Climate change, driven by greenhouse gas emissions, is causing global temperatures to rise, leading to severe weather events and ecosystem disruptions [1]. Additionally, the unsustainable consumption of natural resources is depleting vital reserves of water, minerals, and forests [2]. To tackle these challenges the concept of sustainable development (SD) has evolved from an aspirational goal to an urgent imperative. Indeed, SD represents the delicate balance between meeting present needs without compromising future generations' ability to meet their own [3]. This global sustainability challenge has profound implications for industrial activities, which have traditionally been resource-intensive and environmentally disruptive. A manifestation of such resource-consuming industrial activities is represented by Industrial Parks (IPs).

IPs are large land areas developed and planned to accommodate multiple businesses with shared infrastructure [4] and represent both a challenge and an opportunity in this context. Also, their geographical proximity and shared resources [5] create possibilities for sustainability innovation, considering their potential environmental footprints.

Contemporary IPs are increasingly adopting Industrial Symbiosis (IS) as a core strategy to minimize environmental impact, foster a circular economy, and enhance overall sustainability [6]. IS has emerged as a transformative approach within these industrial ecosystems, as it reimagines traditional industrial relationships by fostering collaborative networks where one company's waste becomes another's resource. This approach creates circular systems where materials, energy, water, and by-products flow between entities that have historically operated independently [7].

Beyond environmental benefits, IS represents a compelling business opportunity that drives eco-innovation while enhancing competitiveness [8].

Digital technologies (DTs) can significantly enhance IS implementation in IPs by mapping resource flows, identifying synergies, and facilitating transactions between companies with complementary material needs [9]. Despite physical proximity reducing logistical barriers, information gaps often prevent optimal resource matching; digital platforms address this by visualizing waste streams, algorithmically identifying exchange opportunities, and streamlining collaboration, ultimately transforming traditional IPs into connected innovation ecosystems [10].

While not all IPs implement IS principles, they provide potential platforms where such practices could be developed. Some IPs may evolve into Eco-Industrial Parks (EIPs) through the intentional implementation of IS strategies, representing a more advanced stage of sustainable industrial development [11].

Despite growing interest in IS implementation across industrial ecosystems, significant geographical disparities exist in research coverage, with Southern Italian regions remaining notably underexplored. This geographical imbalance limits the understanding of how regional contexts influence the potential for IS and which specific barriers might affect implementation in diverse socioeconomic environments. Additionally, while DTs show promise for enabling IS, their application in addressing region-specific barriers requires further investigation. Therefore, an examination of these issues will be presented in the literature review section.

With this in mind, this research addresses two primary questions: (1) What is the technical potential for implementing IS within Southern Italian IPs? (2) How can DTs overcome the specific barriers to IS implementation in these regional contexts?

This paper focuses on the Fosso Imperatore IP as a case study of Southern Italian industrial development, with attention to the analysis of material waste flows and potential symbiotic relationships. Moreover, the study examines how DTs can address implementation barriers within the Fosso Imperatore context and proposes policy recommendations to support IS development in Southern Italian IPs.

Understanding these regional differences and opportunities is essential for supporting Italy's transition toward more sustainable industrial development. Furthermore, expanding studies to Southern Italy could enrich the current literature of IPs in the Italian context.

The remainder of this paper is structured as follows: Sect. 2 continues on IS literature with focus on Italian IPs; Sect. 3 details the methodological approach; Sect. 4 presents the findings; and Sect. 5 discusses the identified implementation barriers and concludes with policy implications.

2 Literature Review

Building on the IS concepts introduced above, this section examines their specific application in Italian industrial contexts. The implementation of IS principles varies significantly across different geographical and industrial settings, influenced by regional economic conditions, governance structures, and technological capabilities. Indeed, many studies have investigated IP development across different regions in Italy, examining their characteristics, management models, and potential for sustainability improvements (Table 1).

Table 1 Summary of Italian IPs case studies

Park's name	Region	References
Bomporto IP	Emilia-Romagna	[12]
Raibano IP	Emilia-Romagna	[13]
Macrolotto IP	Tuscany	[14]
Ponterosso IP	Friuli-Venezia Giulia	[14]
Padova IP	Veneto	[14]
Bussi chemical site	Abruzzo	[15]

Italian IPs exhibit different models ranging from single-sector specialization to multi-sector diversification. Management approaches vary between bottom-up initiatives led by companies seeking improved performance through resource sharing and top-down approaches driven by local governments or research institutions. Additionally, they differ in size (22–150 ha), industrial composition, and advancement in implementing sustainable practices and resource optimization strategies.

Despite the interest on IS in Italian IPs, a geographical imbalance exists. Most studies have focused on Northern and Central Italian IPs, creating a knowledge gap regarding Southern Italian industrial ecosystems. Southern Italian IPs typically feature different industrial compositions, smaller firm sizes, distinct resource availability patterns, and unique socioeconomic contexts. These regional characteristics necessitate tailored approaches to IS that may diverge from models successful in Northern regions.

This knowledge gap is particularly significant given how these regional economic and environmental factors likely influence industrial development trajectories in Southern Italy. The potential of DTs to address region-specific IS implementation barriers remains underexplored, despite their demonstrated capacity to facilitate complex industrial ecosystem relationships. Understanding how DTs can enable IS within Southern Italian IPs (considering their specific constraints and opportunities) represents a valuable research direction with practical implications for sustainable industrial development across Italy's diverse regions.

Furthermore, this work conceptually aligns with Ciasullo and colleagues' exploration of socio-ecological systems [16]. Their framework considers how industrial environments adapt to technological changes while maintaining a balance between economic, social, and environmental dimensions. The present study extends this ecological systems thinking specifically to industrial contexts in Southern Italy, demonstrating how DTs might facilitate more sustainable resource exchanges within regionally-specific industrial ecosystems. This intersection of digital innovation, regional industrial contexts, and ecological systems thinking represents a fruitful area for expanding the scope of discussions within the Resource Innovation and Industry Forum.

3 Methods and Materials

3.1 Case Study Approach and Description

This research employed a case study approach to examine the Fosso Imperatore IP, addressing the identified research gap concerning Southern Italian IPs. A case study analysis is a research method that involves an examination of a specific subject, such as an individual, group, organization, or event, within its real-life context [17]. This methodological choice aligns with previous research on IPs, where case studies have proven effective for capturing the economic, social, and environmental factors within specific regional contexts [14]. This case study serves as a microcosm of the broader challenges and opportunities facing IS implementation in Southern Italy, which exhibits significant differences from the more extensively studied Northern regions.

The Fosso Imperatore IP in the Agro Nocerino Sarnese area of Campania is an industrial center in Southern Italy, managed by the Co.I.F.IM. consortium. The park spans 144,846 m^2 and is home to 28 companies employing approximately 1000 workers. These companies collectively generate an annual turnover of over 300 million euros [18].

Strategically located with direct access to infrastructure networks and close to the Caserta-Roma highway, the park offers excellent accessibility for industrial traffic. As illustrated in Fig. 1, Fosso Imperatore accommodates a diverse industrial composition spanning multiple sectors, including metal manufacturing, plastic processing, machinery production, printing, food industry, and various supporting services. This sectoral diversity positions the park as a vital economic driver for the Agro Nocerino Sarnese region while providing opportunities for potential IS relationships.

The Co.I.F.IM. Consortium's mission statement highlights the potential for by-products and energy exchange by focusing on raw materials and product exchange databases. The consortium is committed to knowledge and technology exchange, offering shared IT services and market information systems. Networking is actively promoted through initiatives to enhance integration and collaboration among members. Such initiatives involve the promotion of innovative products and limited space for exhibitions. The consortium also involves local stakeholders by being open to joining other business aggregations and playing a significant role in regional economic development.

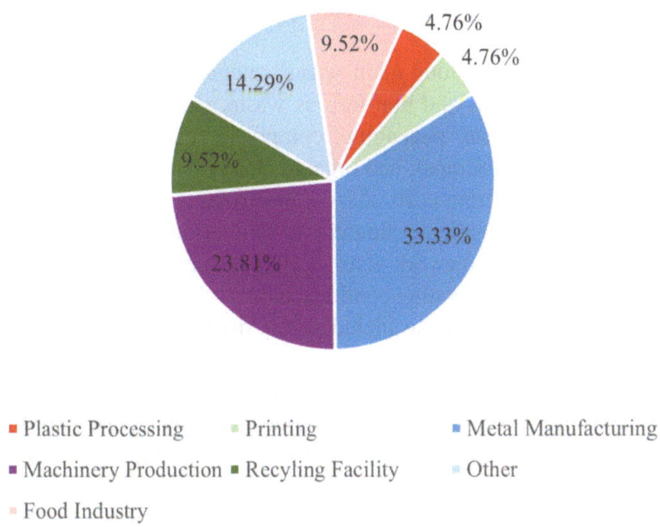

Fig. 1 Industrial composition of Fosso Imperatore IP. *Source* Authors

3.2 Data Estimation and Analysis

The process of data estimation commenced with the identification of each company operating in the park and their classification according to their respective NACE industrial sectors. Due to very limited company collaboration, direct measurement of resource flows proved feasible only for a few firms. This necessitated the development of a data-driven estimation methodology based on publicly available data, to estimate industrial waste generation and identify suitable material flows for IS within the IP.

Data was collected from multiple sources: company data including business names, ATECO and NACE codes, employee counts, and turnover figures from public registries; regional waste generation statistics from ISPRA categorized by economic activity and material type [19]; regional economic indicators from ISTAT providing turnover and employment figures by sector [20]; and direct survey data from the subset of companies that provided waste information. The data preparation phase required standardization efforts to ensure compatibility across different classification systems, mapping NACE and ATECO codes to consistent formats and categorizing companies into appropriate industrial sectors aligned with ISTAT economic sections.

A custom Python-based script was developed to estimate waste flows within the IP. The waste estimation phase employs an employee-based approach founded on the principle that workforce size correlates with waste generation volume. For each economic sector, waste generation multipliers are calculated as regional waste per employee, then applied to individual companies. The model incorporates constraints to ensure realistic estimates, including upper bounds on per-employee waste generation based on industry benchmarks, sector-specific waste composition profiles derived from regional patterns, and minimum thresholds for material flows to focus on significant streams. Primary waste data from surveys replaces estimates where available, improving overall accuracy while maintaining methodological consistency.

The analysis focuses on four main waste categories with high potential for IS: metal (including ferrous and non-ferrous waste, shavings, packaging); plastic (including production waste, packaging, composite materials); paper (including packaging and production scraps); and organic material (including food processing waste and biological materials). To address uncertainties, the model incorporates sensitivity analysis functionality, evaluating how variations in key parameters affect the overall results. This framework tests multiplier variations ranging from 50 to 150% of baseline values, generating confidence intervals for total waste estimates. It is important to note that the analysis included only firms generating waste potentially useful for IS.

The computational model generates standardized input for Sankey diagram visualization, which effectively represents resource flows in IS contexts by visually encoding both flow direction and magnitude [21]. The process for Sankey diagram generation involves aggregation of flows by source, waste type, and destination; filtering to include only significant flows above one ton annually.

Network analysis has been widely recognized as an effective approach for mapping and analyzing resource exchanges in industrial ecosystems [22]. In a network analysis, companies are represented as nodes and potential material flows as directed edges, creating a visual and quantitative representation of symbiotic opportunities.

Results are visualized using a customized Mermaid-based flowchart system that employs color-coding for different industrial sectors, node sizing based on company scale, edge thickness representing material flow volume, and clear typology differentiating waste generators from recyclers. This visual representation serves both analytical purposes and as a communication tool for engaging stakeholders.

4 Findings

4.1 Material Flow and Network Analysis

Figure 2 presents the material flows stemming from the most waste-generating companies. Total estimated recyclable waste generation amounted to 1,044.6 tons annually, distributed across four main categories: metal (319.74 tons), plastic (1,775.02 tons), paper (186.55 tons), and organic waste (547.06 tons). Most contributions come from manufacturing sectors, particularly metal fabrication and plastic processing companies. Metal industry companies like GAMBARDELLA SRL, STARPUR SRL, OM.P.M. SRL, and TECNOLAT SRL collectively generated significant portions (19.4%) of the park's metal waste. The plastic processing sector was dominated by NEW DIMENSION PLASTIC with 1,754.62 tons of plastic waste, representing most of all plastic waste in the park. Organic waste generation was concentrated in the food industry, with ITALORTO and IMEPA SRL accounting for the total of organic waste flows. This concentration of specific waste streams within industrial sectors creates theoretically favorable conditions for IS by providing consistent, high-volume material flows that could support economically viable exchange relationships. Despite these potential synergies remain largely unexploited, companies predominantly prefer to sell their waste to external buyers.

Nonetheless, the network analysis (Fig. 3) identified potential material exchange relationships within the IP. These connections represent theoretically viable flows based on material compatibility and geographical proximity, though they are not currently implemented. Network density, measuring the ratio of actual connections to possible connections, was calculated at 0.32, indicating a moderately connected industrial ecosystem in theory. This value exceeds the network density threshold of 0.2 [23], suggesting that Fosso Imperatore has sufficient connectivity to support symbiotic relationships if companies chose to implement them.

Centrality analysis identified key potential actors within the network. NEW DIMENSION PLASTIC emerged as the node with highest degree centrality (connected to 7 other companies), positioning it as a critical waste generator within

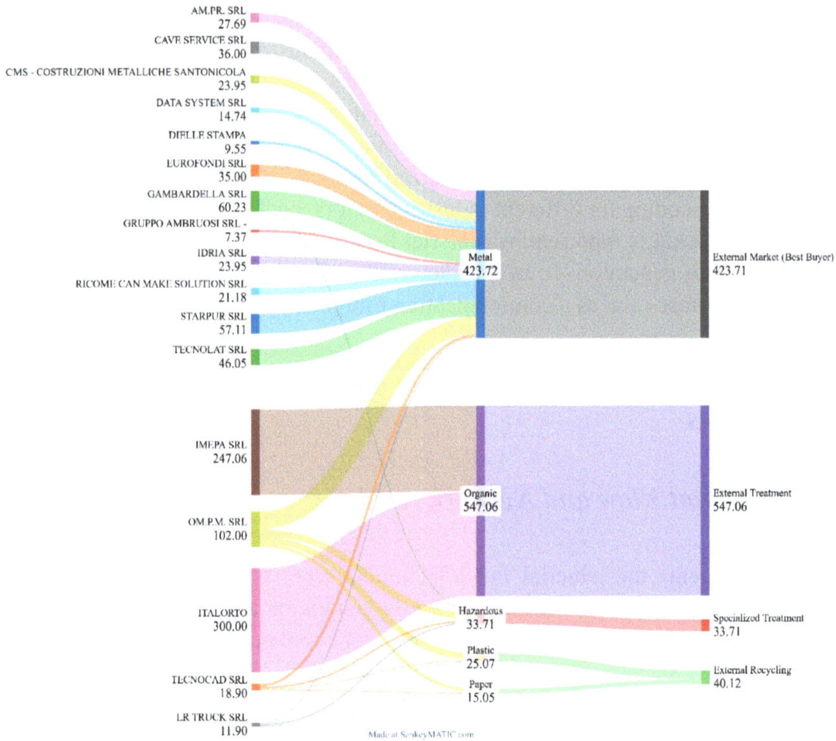

Fig. 2 Sankey diagram visualizing the material flows of the most relevant waste generating firms in Fosso Imperatore IP. *Source* Authors

the potential network. The analysis also highlighted potential for a biogas facility that could serve as an intermediary in more complex exchange relationships, particularly for organic waste flows.

Community detection algorithms identified three distinct potential clusters within the network: a metal exchange community that could develop around companies like GAMBARDELLA SRL and CAVE SERVICE SRL; a plastic processing community that could be anchored by NEW DIMENSION PLASTIC; and a potential organic waste-to-energy community that could connect food producers with energy generation facilities.

4.2 Digital Innovation Opportunities for Enhanced Symbiosis

The findings address the first research question by quantifying substantial material flows that could theoretically form symbiotic relationships within Fosso Imperatore. Regarding the second research question on how DTs can overcome implementation

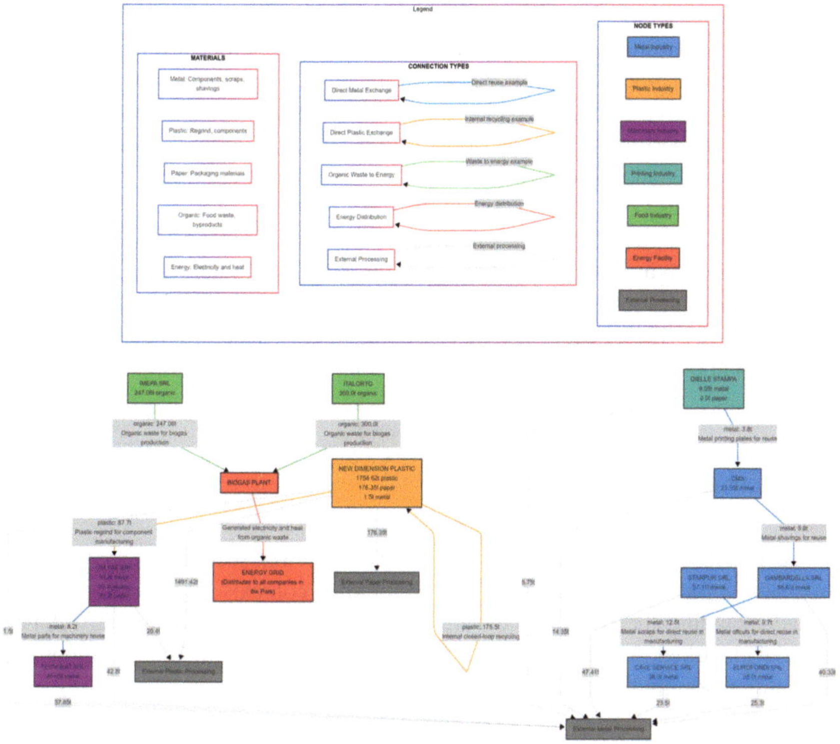

Fig. 3 Network analysis of most relevant waste-generation firms of Fosso Imperatore IP showing the potential for IS. *Source* Authors

barriers, the following digital solutions directly address the economic, informational, and infrastructural challenges identified above.

The computational methods developed in this research highlight the transformative potential DTs in overcoming barriers to IS implementation at Fosso Imperatore. Advanced digital platforms [24] represent a promising approach to address the information asymmetry and economic barriers identified.

Such platforms could incorporate real-time waste stream monitoring through Internet of Things (IoT) sensors [25], enabling companies to track material flows and identify exchange opportunities with unprecedented precision. Blockchain technology offers additional potential by creating immutable records of waste transactions, enhancing trust between potential symbiosis partners while ensuring regulatory compliance [26].

The integration of artificial intelligence (AI) algorithms could further enhance symbiosis development through automated matching of waste generators with potential receivers based on material properties, volume, and timing considerations [27].

Digital twins of industrial processes, as proposed by [28], could enable companies to simulate and optimize potential symbiotic relationships before physical implementation, reducing perceived risk. For Fosso Imperatore specifically, where economic considerations drive companies to sell waste externally, AI-powered economic analysis tools could perform comprehensive cost–benefit analyses that account for transportation savings, supply security benefits, and long-term resilience improvements of local exchanges [29].

These technologies align with the Industry 4.0 paradigm that increasingly shapes manufacturing environments, creating synergies between digitalization and circular economy principles [30]. Notably, digital platforms can increase IS implementation rates by 40–60% [31], suggesting significant potential for digital transformation to catalyze IS at Fosso Imperatore despite current barriers.

5 Discussions and Conclusions

This section discusses the findings presented in the previous section, interpreting their implications for IS development in Southern Italian contexts. The analysis focuses on understanding the barriers preventing IS implementation despite the technical potential identified. A conclusive paragraph contains policy implications, limitations, and future research directions.

5.1 Barriers and Digital Solutions to Industrial Symbiosis Implementation

Currently, 100% of materials are processed externally, including 319.74 tons of metal, 1,775.02 tons of plastic, 186.55 tons of paper, and 547.06 tons of organic waste that could potentially be redirected to internal exchanges or processing facilities. Based on typical economic benefits documented in similar EIPs [32], capturing even 50% of these external flows could generate estimated annual savings through reduced transportation and disposal costs, while creating additional revenue streams through resource valorization.

Nonetheless, companies prioritize immediate financial returns from selling waste to external buyers over the potential long-term benefits of symbiotic relationships. This economic barrier aligns with literature [33], as economic feasibility as a critical determinant of IS adoption.

Another barrier is the imbalance in material generation, with plastic waste concentrated in a single company (NEW DIMENSION PLASTIC) accounting for approximately 70% of all waste in the park. This concentration might create challenges for establishing reciprocal relationships.

The limited internal processing capacity for all waste streams represents another significant barrier. Unlike some EIPs that have developed dedicated infrastructure for material processing, Fosso Imperatore lacks these facilities, forcing companies to seek external options for waste management. Although there is one recycling facility within the park (ECOSIDER), companies still predominantly choose external buyers offering better prices.

Information and trust barriers likely also play a role in limiting symbiosis development. The difficulty in obtaining direct waste data from companies during this research suggests information sharing practices are limited. Indeed, information availability and trust are critical factors in IS development [34].

The barriers identified in this study (economic prioritization of short-term gains, material flow imbalances, infrastructure limitations, and information asymmetry) represent systemic challenges requiring targeted interventions. While the DTs presented in the findings offer technical capabilities to address these barriers, their successful implementation requires understanding the specific connections between challenges and solutions. For instance, economic barriers necessitate not just technological tools but shifts in decision-making frameworks that incorporate long-term resilience and sustainability metrics. Similarly, information barriers reflect not merely technical data limitations but deeper trust and coordination issues within industrial networks. The technical solutions identified must therefore be embedded within broader institutional and policy contexts, particularly in Southern Italian industrial environments where collaborative traditions may differ from Northern regions where most IS studies have focused. This underscores the importance of regionally-tailored policy approaches discussed in the following section.

5.2 Conclusions and Policy Implications

This study of the Fosso Imperatore IP highlights the untapped IS potential in Southern Italy, with important policy implications at multiple levels. IP management should implement economic incentives that make internal material exchanges competitive with external sales, potentially through collective waste management contracts or shared processing infrastructure development. Regional authorities should consider tax advantages and funds for companies engaging in symbiotic exchanges. National policies should address the North–South sustainability gap by prioritizing digital innovation grants for Southern IPs to implement the technologies discussed. Indeed, technical potential alone is insufficient without addressing underlying economic and informational barriers. The study does, however, face limitations including reliance on estimated rather than measured waste flows, limited exploration of social dynamics influencing company decision-making, and the inherent constraints of a single-case approach. Future research should expand to comparative analyses across multiple Southern Italian IPs to validate the generalizability of these findings, implement and evaluate pilot digital platforms in these contexts, and develop economic models that quantify the specific threshold conditions under which symbiotic exchanges become financially advantageous compared to external sales. Addressing these research needs while implementing the suggested policy recommendations, could transform IPs, like Fosso Imperatore, from sites of missed opportunity to showcases of circular economy principles, contributing to regional economic development and environmental sustainability in Southern Italy.

This transformation would not only contribute to regional economic development and environmental sustainability in Southern Italy but also provide valuable insights for similar industrial contexts worldwide, where geographical proximity alone has proven insufficient to catalyze industrial symbiosis relationships. The integration of digital technologies with appropriately tailored economic incentives represents a promising pathway to unlock the circular economy potential of underutilized industrial ecosystems.

Acknowledgements Thanks to the Project "CE_IPsa: A development model for Circular Economy between Industrial Parks and surroundings areas" CUP J53D23016630001 - codice identificativo P20227NFA4.

References

1. Rusoke, T.: Global warming and climate change: projections and implications. In: Handbook of Energy and Environment in the 21st Century. CRC Press, pp. 335–356
2. Lu, C., Wang, K.: Natural resource conservation outpaces and climate change: roles of reforestation, mineral extraction, and natural resources depletion. Resour. Policy **86**, 104159 (2023)
3. Liwarska-Bizukojc, E., Bizukojc, M., Marcinkowski, A., Doniec, A.: The conceptual model of an eco-industrial park based upon ecological relationships. J. Clean. Prod. **17**(8), 732–741 (2009)
4. Shevchuk, N., Tulchynska, S., Severyn-Mrachkovska, L., Pidlisna, O., Kryshtopa, I.: Conceptual principles of the transformation of industrial parks into eco-industrial ones in the conditions of sustainable development. IJCSNS Int. J. Comput. Sci. Netw. Secur. **21**(12), 349–355 (2021)
5. Côté, R.P., Cohen-Rosenthal, E.: Designing eco-industrial parks: a synthesis of some experiences. J. Clean. Prod. **6**(3–4), 181–188 (1998)
6. Chen, X., Dong, M., Zhang, L., Luan, X., Cui, X., Cui, Z.: Comprehensive evaluation of environmental and economic benefits of industrial symbiosis in industrial parks. J. Clean. Prod. **354**, 131635 (2022)
7. Chertow, M.R.: Industrial symbiosis: literature and taxonomy. Annu. Rev. Energy Env. **25**(1), 313–337 (2000)
8. Lombardi, D.R., Laybourn, P.: Redefining industrial symbiosis: crossing academic–practitioner boundaries. J. Ind. Ecol. **16**(1), 28–37 (2012)
9. Krom, P., Piscicelli, L., Frenken, K.: Digital platforms for industrial symbiosis. J. Innov. Econ. Manage. I124-XXVI (2022)
10. Makropoulos, C., Kritikos, N.A., Pantazis, C.: Matchmaking for industrial symbiosis: a digital tool for the identification, quantification and optimisation of symbiotic potential in industrial ecosystems. Front. Chem. Eng. **6**, 1363888 (2024)
11. Galvan-Cara, A.L., Graells, M., Espuña, A.: Application of industrial symbiosis principles to the management of utility networks. Appl. Energy **305**, 117734 (2022)

12. Aguinaldo, M.E.C., Daddi, T., Hamza, M., Gasbarro, F.: Climate change perspectives and adaptation strategies of business enterprises: a case study from Italy. Int. J. Sust. Dev. World **26**(2), 129–140 (2019)
13. Conticelli, E., Tondelli, S.: Application of strategic environmental assessment to eco-industrial parks: Raibano case in Italy. J. Urban Plan. Dev. **139**(3), 185–196 (2013)
14. Daddi, T., Tessitore, S., Testa, F.: Industrial ecology and eco-industrial development: case studies from Italy. Prog. Ind. Ecol. Int. J. **9**(3), 217–233 (2015)
15. Taddeo, R., Simboli, A., Morgante, A.: Implementing eco-industrial parks in existing clusters. Findings from a historical Italian chemical site. J. Clean. Prod. **33**, 22–29 (2012)
16. Ciasullo, M.V., La Sala, A., Ferrara, M.: Normalizing digitization into a new 5.0 equilibrium. Insights from socio-ecological systems. In: The International Research and Innovation Forum. Springer International Publishing, Cham, pp. 107–115 (2023)
17. Takahashi, A.R.W., Araujo, L.: Case study research: opening up research opportunities. RAUSP Manage. J. **55**, 100–111 (2020)
18. CO.I.F.IM. (Consorzio Imprenditori Fosso Imperatore): (2025). Accessible at http://www.coifim.it/Siamo.html
19. ISPRA—Istituto Superiore per la Protezione e la Ricerca Ambientale: Rapporto Rifiuti Speciali Edizione 2024 (Rapporti 402/2024). ISPRA, Roma (2024). ISBN 978-88-448-1219-5
20. ISTAT—Istituto Nazionale di Statistica: Risultati economici delle imprese e delle multinazionali a livello territoriale, anno 2022 (2022). Available at https://www.istat.it/tavole-di-dati/risultati-economici-delle-imprese-e-delle-multinazionali-a-livello-territoriale-anno-2022
21. Schmidt, M.: The Sankey diagram in energy and material flow management: part II: methodology and current applications. J. Ind. Ecol. **12**(2), 173–185 (2008)
22. Domenech, T., Davies, M.: Structure and morphology of industrial symbiosis networks: the case of Kalundborg. Procedia Soc. Behav. Sci. **10**, 79–89 (2011)
23. Ashton, W.: Understanding the organization of industrial ecosystems: a social network approach. J. Ind. Ecol. **12**(1), 34–51 (2008)
24. Krom, P., Piscicelli, L., Frenken, K.: Digital platforms for industrial symbiosis. J. Innov. Econ. Manage. **39**(3), 215–240 (2022)
25. Song, B., Yeo, Z., Kohls, P., Herrmann, C.: Industrial symbiosis: exploring big-data approach for waste stream discovery. Procedia CIRP **61**, 353–358 (2017)
26. Esmaeilian, B., Sarkis, J., Lewis, K., Behdad, S.: Blockchain for the future of sustainable supply chain management in Industry 4.0. Resour. Conserv. Recycl. **163**, 105064 (2020)
27. Fraccascia, L., Yazan, D.M.: The role of online information-sharing platforms on the performance of industrial symbiosis networks. Resour. Conserv. Recycl. **136**, 473–485 (2018)
28. Low, J.S.C., Tjandra, T.B., Lu, W.F., Lee, H.M.: Adaptation of digital twin technology to industrial symbiosis: a framework and case study of adaptable industrial internet of things. Procedia CIRP **80**, 655–660 (2018)
29. Tseng, M.L., Tan, R.R., Chiu, A.S., Chien, C.F., Kuo, T.C.: Circular economy meets industry 4.0: can big data drive industrial symbiosis? Resour. Conserv. Recycl. **131**, 146–147 (2018)
30. Lopes de Sousa Jabbour, A.B., Jabbour, C.J.C., Godinho Filho, M., Roubaud, D.: Industry 4.0 and the circular economy: a proposed research agenda and original roadmap for sustainable operations. Ann. Oper. Res. **270**(1), 273–286 (2018)
31. van Capelleveen, G., Amrit, C., Yazan, D.M.: A quantitative study on the effectiveness of big data to predict industrial symbiosis activities. J. Ind. Inf. Integr. **24**, 100228 (2021)
32. Jacobsen, N.B.: Industrial symbiosis in Kalundborg, Denmark: a quantitative assessment of economic and environmental aspects. J. Ind. Ecol. **10**(1–2), 239–255 (2006)
33. Neves, A., Godina, R.G. Azevedo, S., Pimentel, C., CO Matias, J.: The potential of industrial symbiosis: case analysis and main drivers and barriers to its implementation. Sustainability **11**(24), 7095 (2019)
34. Mouzakitis, Y., Aminalragia-Giamini, R., Adamides, E.D.: From the treatment of Olive Mills wastewater to its valorisation: towards a bio-economic industrial symbiosis. In: Sustainable Design and Manufacturing 2017: Selected Papers on Sustainable Design and Manufacturing 4. Springer International Publishing, pp. 267–276 (2017)

Measuring Innovation in EU Research Infrastructures through Public–Private Co-Publications: Evidence from ACTRIS

Marcella De Martino, Giuseppe Giordano, Simone Gagliardi, Giuseppe Gargano, Rosa Maria Petracca Altieri, Giulia Saponaro, and Carmela Cornacchia

Abstract Against the backdrop of the European Union's (EU) initiative consistent with the establishment of so-called Research Infrastructures (RI), by focusing on the case of Aerosol, Clouds and Trace Gases Research Infrastructure (ACTRIS), this paper examines how public–private partnerships (PPPs) contribute to innovation and research excellence. Through the lens of ACTRIS' main output, i.e. publications, effectively public–private co-publications (PPCs), the objective is to explore to what extent European RIs contribute not only to research excellence, but most importantly to an uptake of research findings by the private sector. The analysis of 176 PPCs, drawn from peer-reviewed literature indexed in Web of Science (WoS), suggests that PPCs show higher citation impact, stronger international co-authorship, and greater thematic alignment with applied research, as compared to general scientific publication. However, the analysis also shows a temporal gap between academic advancements and their industrial uptake, consistent with knowledge transfer dynamics. PPCs represent useful indicators of knowledge transfer and innovation. Still, additional

M. De Martino (✉) · S. Gagliardi · G. Gargano · R. M. Petracca Altieri · C. Cornacchia
CNR-IMAA (Italy), Tito, Italy
e-mail: marcella.demartino@cnr.it

S. Gagliardi
e-mail: simone.gagliardi@cnr.it

G. Gargano
e-mail: giuseppe.gargano@cnr.it

R. M. Petracca Altieri
e-mail: rosamaria.petraccaaltieri@cnr.it

C. Cornacchia
e-mail: carmela.cornacchia@cnr.it

G. Giordano
University of Salerno (Italy), Fisciano (Salerno), Italy
e-mail: ggiordano@unisa.it

G. Saponaro
ACTRIS-ERIC, Helsinki, Finland
e-mail: giulia.saponaro@actris.eu

A. Visvizi et al. (eds.), *Research and Innovation Forum 2025 Vol2*, Springer Proceedings in Complexity, https://doi.org/10.1007/978-3-032-09062-1_11

indicators, such as co-patents, joint research and development (R&D) projects, and informal collaboration networks, represent additional indicators necessary to offer a comprehensive insight into the RIs' innovation potential.

Keywords Research infrastructure · Public · Private Partnership (PPP) · Innovation · Public · Private Co-Publication (PPC) · Bibliometric analysis · SDG 9—industry · Innovation and infrastructure · SDG 13—climate action

1 Introduction

Collaboration between research and industry is widely recognized as a cornerstone of innovation and sustainable growth [1]. These collaborations facilitate the co-creation of knowledge, accelerate the translation of research into innovative products and services, and contribute to the competitiveness of national and regional economies [2]. Among the different collaborative models, public–private partnerships (PPPs) have gained prominence as effective mechanisms that foster innovation. Traditionally emerged as mechanisms to sustain infrastructure development in sectors such as transportation, healthcare, and education by leveraging private capital [3], PPPs have evolved significantly to support research and innovation [4–6] and citizen participation as a further pillar of sustainable development framework [7].

In the research and innovation domain, innovation-oriented PPPs are designed to align public and private resources with clearly defined societal missions—such as climate neutrality, digital sovereignty, and health resilience—through long-term, co-governed collaboration [8]. Unlike traditional PPPs, which focus on infrastructure delivery, risk-sharing and cost -efficiency objectives [3], these next-generation partnerships are aimed at fostering collaborative innovation, accelerating knowledge diffusion, and contributing to broader societal impact [9]. These partnerships are central to the European Union's mission-driven policy agenda under Horizon Europe (2021–2027), as they enable the alignment of scientific excellence with industrial application and public policy objectives, ultimately fostering innovative solutions that drive systemic change across economic, societal, and environmental domains [10].

Research infrastructures (RIs) play a crucial role in advancing knowledge and fostering innovation through industry collaboration to address global challenges. As publicly funded and often multinational entities, RIs provide the scientific backbone for addressing complex societal challenges by offering advanced facilities, curated data, technological services, and highly skilled human capital [11]. RIs are hubs of knowledge and innovation, enabling interdisciplinary collaboration across borders and sectors, functioning as platforms for technology validation and innovation co-development [12], contributing to industry competitiveness and sustainable growth.

However, there is still limited research on how RIs collaborate with industry and the extent to which these collaborations contribute to innovation outcomes and broader societal impact [13]. To address this gap, the study aims to explore the impact of innovation-oriented PPPs within the context of European RIs, focusing on the case of ACTRIS (Aerosol, Clouds and Trace Gases Research Infrastructure)—a pan-European, distributed research infrastructure operating in the environmental domain.

In pursuing this objective, the study is guided by the following research questions: RQ1: How are innovation-oriented PPPs defined and implemented within research infrastructures, RQ2: How do these collaborations enable joint knowledge production, resource sharing, and innovation-oriented outcomes?

The study adopts a bibliometric approach, using public–private co-publications (PPCs) as a proxy for measuring knowledge co-creation and innovation outcomes within ACTRIS's collaborative ecosystem. By analysing scientific publications from 2016 to 2024, the study examines the citation impact and thematic evolution of PPCs to trace the development of science–industry collaboration. It identifies key research domains and industrial engagement areas, assessing the role of PPCs as indicators of both scientific output and innovation outcomes.

The paper is structured as follows: Sect. 2 reviews PPPs in the context of research infrastructures, addressing PPCs subsequently. Section 3 introduces the case study, ACTRIS, while Sect. 4 describes the methodology and data sources used in the analysis. Section 5 presents the results of the bibliometric analysis, examining the volume and temporal trends PPCs, their citation impact, and the thematic structure of collaborative research. Finally, Sect. 6 concludes with recommendations and directions for future research.

2 Innovation-oriented Public and Private Partnerships in Research Infrastructures

As EU innovation policy shifted towards knowledge co-creation and societal impact, a new generation of PPPs emerged, emphasizing adaptive, trust-based, and co-creative collaboration among public research organizations, private firms, and civil society [9, 14]. These innovation-oriented PPPs are transdisciplinary and cross-sectoral, bringing together public and private stakeholders with complementary expertise, resources, and interests [10], and require new governance models that prioritize relational dynamics and iterative learning over transactional efficiency [14]. An example of this evolution is found in the public–private licensing agreements, which effectively channel public-sector innovation into economic development, particularly in high-tech sector [15]. Furthermore, these partnerships have evolved to include open, collaborative models that accelerate pre-competitive research and foster innovation, such as Open Science Partnerships (OSPs) [9].

European RIs increasingly engage with industry and non-academic stakeholders through open access models, innovation co-development activities, and technology transfer mechanisms [12]. Collaborative models between RIs and industry may align with the characteristics of innovation-oriented PPPs [16] and OSPs [9]. These encompass a spectrum of formal collaborative arrangements, including joint research and development (R&D) activities, technology procurement partnerships, spin-offs, and consultancy agreements [17, 18]. These collaborations often occur across various stages of the innovation process—from early exploratory research to the validation of new technologies and their deployment in the market—enabling both public and private actors to mobilize complementary expertise and resources [13].

In public procurement, pre-commercial procurement (PCP) and public procurement of innovation (PPI) are formal collaborative agreements between RIs and industrial partners aimed to stimulate innovation by enabling the RI to act as an early adopter of new technologies, thereby reducing market risks for companies involved in high-risk, pre-commercial R&D activities [12]. These co-innovation frameworks rely on formal incentives and relational dynamics to generate high-impact, mission-oriented outcomes [14]. For example, in large-scale facilities such as CERN, partnerships in technological procurement enable companies to co-develop advanced instrumentation and benefit from knowledge spillovers [19].

RIs also collaborate with industry by providing access to their facilities and services for various innovation-related activities, including product testing, prototype validation, calibration, and data-intensive research [12]. The European Charter for Access to Research Infrastructures [20] defines access modalities—physical, remote virtual and hybrid, and priority-driven—following open science principles and criteria of research excellence. The EU Transnational Access (TNA) scheme provides financial support to cover travel, accommodation, and access costs, thereby fostering co-development of technologies, technical and training resources and mutual knowledge exchange. RIs adhere to open and FAIR (findable, accessible, interoperable, reusable) principles, ensuring responsible and transparent data governance while supporting innovation uptake in the private sector [11]. However, the balance between openness and ethical responsibility represents an important challenge in the governance of collaborative research [21].

In particular, the misalignment in timeframes and incentives, the presence of divergent cultures, and the challenges in intellectual property management represent important barriers to RI and industry collaboration [22]. Industry often prioritizes short-term, market-driven innovation, while RIs are typically embedded in longer-term scientific missions. Moreover, concerns about proprietary knowledge and commercialization rights can constrain openness and mutual trust. In contrast, enabling conditions include open science and data sharing policy, trust-based inter-organisational relationships and shared intellectual property rights (IPR) frameworks. These conditions are strengthened by the integration of RIs into innovation ecosystems, and by the active role of research performing organizations (RPOs) in bridging scientific advancements with market and policy needs [23].

The outputs of such collaborations range from co-publications and joint patents to technology development, industrial PhD training programmes, and standardised methodologies [24]. These outputs are tangible measures of knowledge co-creation and innovation [25]. Beyond these measurable indicators, such collaborations contribute to public health, environmental quality, and to broader societal well-being and resilience, particularly in the case of environmental RIs.

2.1 Public–Private Co-Publications (PPCs) as an Indicator of Knowledge Transfer and Innovation

Public–private co-publications (PPCs) are typically conceptualized as academic publications co-authored by researchers affiliated with universities or public research institutes and private-sector companies [26]. PPCs have increasingly attracted attention as an indicator of knowledge transfer, innovation, and collaborative research activity between public and private undertakings [27–29]. They underscore how innovation-oriented PPPs can serve as effective vehicles of shared scientific advancement and technology transfer mechanisms [24, 30]. Such co-authorships are facilitated by open access policies, strengthening the European research and innovation ecosystem [31]. PPCs provide insights into research cooperation beyond knowledge exchange, illustrating tangible collaboration and mutual trust between the public and private sectors [26]. Similarly, PPCs directly reflect the complementary integration of distinct expertise and resources, enabling more impactful and innovative scientific outcomes than purely academic publications [27, 29].

PPCs measure innovation by showcasing direct industry engagement in scientific research and how industry-relevant problems are addressed through academic expertise [24, 28]. Qualitative case studies support the causal link between PPCs and commercialization outcomes (e.g., patents and spin-offs), showing that co-authored research often leads to technological applications that are later patented or commercialized [28, 29]. Furthermore, cross-boundary collaboration has a significant positive impact on generating disruptive and novel research within large-scale RIs [32].

Empirical studies have methodologically relied upon bibliometric data sources, most commonly the Web of Science (WoS), to identify and analyse co-publication patterns [33] and network dynamics [30]. Nonetheless, to comprehensively assess knowledge transfer pathways, bibliometric analysis should be complemented with additional innovation data sources, such as co-patents, start-up activity, and subsidized collaborative R&D projects [29, 30].

PPCs generally exhibited higher citation impacts than purely academic research, particularly when combined with international collaboration [27]. Geographic proximity also plays a critical role in shaping PPC intensity. Studies highlight that regional innovation ecosystems, characterized by strong institutional linkages and policy support, foster more frequent and effective university–industry collaborations [24, 26, 28, 30].

3 The Case Study: ACTRIS

The Aerosol, Clouds and Trace Gases Research Infrastructure (ACTRIS) is a pan-European research infrastructure dedicated to understanding atmospheric processes, producing high-quality data and information on short-lived atmospheric constituents and their interactions and impact on climate, air quality, and health. Established to provide high-quality data, services, and technological advancements, ACTRIS enables cutting-edge atmospheric research while fostering innovation. ACTRIS has evolved over more than 15 years of continuous development, supported by Member States and the European Commission. ACTRIS, initially developed through a series of European projects, was included in the European Strategy Forum on Research Infrastructures (ESFRI) roadmap in 2016 and granted landmark status in 2021. In 2023, it became a European Research Infrastructure Consortium (ERIC), a legal framework that facilitates sustainable, collaborative research infrastructure across Europe. The ACTRIS ERIC was established with 17 founding members and observer countries [34].

The increasing demand for reliable atmospheric data and advanced environmental monitoring technologies has positioned ACTRIS as a critical player in scientific and industrial domains. ACTRIS provides access to state-of-the-art observational platforms and exploratory facilities for studies in natural and controlled environments to support atmospheric research, from air pollution monitoring to climate change studies. These resources enhance fundamental scientific discoveries and the development of market-ready technologies, particularly for small and medium-sized enterprises (SMEs) engaged in producing atmospheric monitoring instruments. Through several TNA opportunities under EU-funded projects [35], ACTRIS enables researchers and industry stakeholders to access its facilities, fostering interdisciplinary collaboration and joint technology development. However, as highlighted in several reports [36], TNA statistics provide only a partial picture of private sector access; in reality, many companies have engaged with ACTRIS facilities outside the TNA framework, particularly through informal research collaborations. As ACTRIS continues to expand its network and partnerships, its role in sustainable innovation becomes increasingly significant. By integrating scientific excellence in the atmospheric domain with industrial engagement, ACTRIS serves as a “knowledge hub” that bridges the gap between academia, policy, and industry to foster a more sustainable and knowledge-driven society.

4 Methodology and Data Source

This study adopts a bibliometric analysis approach to analyse PPCs between 2016–2024, corresponding to ACTRIS' entry into the ESFRI roadmap. This milestone officially recognised ACTRIS' strategic importance as a distributed European research infrastructure, thereby significantly enhancing its visibility and appeal to the industrial sector.

In particular, the study examines trends in publications, and citation impact to trace the evolution of scientific productivity. Then, topic trends and thematic evolution are analysed through keyword co-occurrence and clustering, showing research domains most strongly associated with industrial collaboration. Advanced bibliometric tools, specifically the Bibliometrix package in R-Studio [37] were employed.

Data collection was rigorously performed using Clarivate Analytics' Web of Science (WoS) database, which indexes peer-reviewed scientific literature. Hence, there are 1,381 publications from 2016 to 2024 related to all ACTRIS scientific publications, among which 176 are identified as PPCs. PPCs were identified in the WoS database based on author affiliations. A publication qualifies as a PPC if it includes at least one author from a private-sector and one from a public research institution, such as a university or public research centre.

5 Results

The results presented in the study offer a comparison between all ACTRIS scientific publications (1,381) and the subset classified as PPCs (176). Table 1 offers an overview of key metrics.

Overall, PPCs demonstrate a higher scientific impact compared to the full corpus of publications, with an average of 29.88 citations per publication versus 15.94. PPCs generate outputs with greater visibility, potentially due to their applied orientation, industry relevance, and innovation potential. Regarding authorship, PPCs involve more co-authors (19.9 versus 9.95), reflecting their collaborative, multidisciplinary nature that integrates academic and industry expertise. Furthermore, international collaboration is also more prevalent in PPCs (87.5% versus 70.02%), underscoring

Table 1 ACTRIS scientific outputs and public–private co-publications (PPCs) in 2016–2024

Main info	Scientific outputs	PPC
Number of publications	1381	176
Average citations per publication	15.94	29.88
Co-authors per publication	9.95	19.9
International co-authorship %	70.02	87.5
Single-authored publications	11	1

A comparison of key metrics

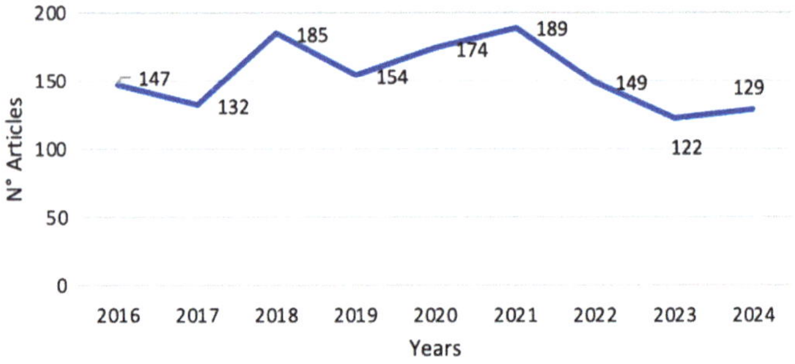

Fig. 1 Annual scientific production is considered from the entire dataset 2016–2024. *Source* Authors

global orientation and the role of the TNA scheme in facilitating industry collaboration. Single-authored PPCs are nearly absent, further highlighting their inherently collaborative nature.

Figure 1 shows the annual scientific production related to ACTRIS from 2016 to 2024, characterised by an overall fluctuating trend, potentially influenced by various factors, including project funding cycles, completion of collaborative research phases, or external economic and policy factors.

Although a separate figure is not provided for PPCs due to sample size limitations, a descriptive analysis reveals a similar pattern, with a notable peak in 2020 (28 of 174 publications). This increase coincides with intensified industry–academic collaboration during the COVID-19 pandemic, aimed at developing innovative technologies and monitoring systems. For example, ACTRIS atmospheric simulation chambers were used to test the filtering efficiencies of protective face masks before they entered the market. Additionally, ACTRIS mobilised its community's expertise to quantify the impacts of stay-at-home policies on atmospheric composition changes over Europe using ACTRIS-related observations. It also launched a special issue involving two of the major atmospheric peer-reviewed journals, namely Atmospheric Chemistry and Physics, and Atmospheric Measurement Techniques.

Corresponding author's country analyses (Figs. 2 and 3) also show differences in collaborative patterns. PPCs predominantly involve multi-country collaborations (MCP), suggesting a strong international dimension of industry-academia partnerships, and reflecting the transnational and strategic role of ACTRIS in facilitating international cooperation. Additionally, the participation of countries not officially part of ACTRIS ERIC—including Belarus, the UK, and Greece—highlights its role in addressing worldwide environmental challenges and facilitating knowledge transfer across borders beyond its formal membership.

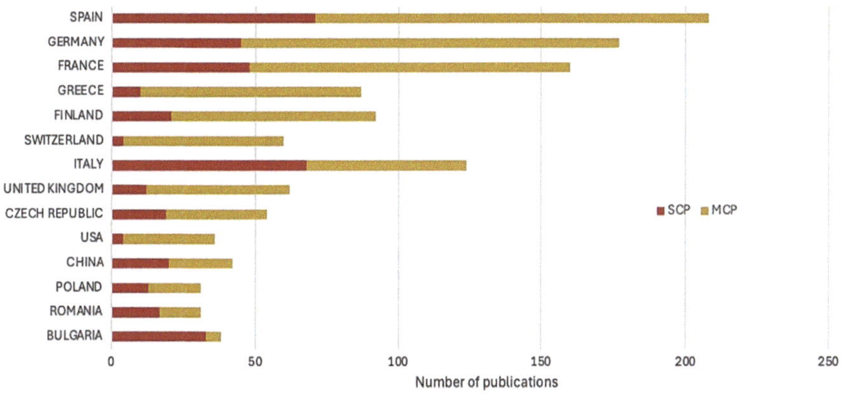

Fig. 2 Corresponding author's countries: all publications. SCP = Single country publications; MCP = Multiple countries publications. *Source* Authors

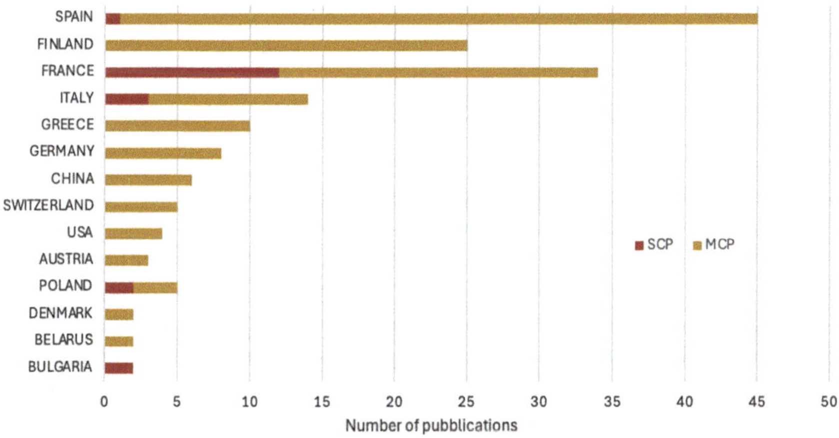

Fig. 3 Corresponding author's countries: PPCs. SCP = Single country publications; MCP = Multiple countries publications. *Source* Authors

5.1 Thematic Trends and Factor Analysis

The temporal evolution and thematic dynamics of research topics within the ACTRIS and PPCs are shown in Figs. 4 and 5, respectively. Concerning the whole corpus of scientific publications (Fig. 4), the early emergence (2016–2019) of core scientific topics such as "aerosol," "optical properties," "retrieval," "nucleation," and "Saharan dust," reflects strong academic focus on atmospheric measurement and modeling (Fig. 4). Later, these topics became increasingly relevant in PPCs between 2019 and 2022 (Fig. 5). In line with this pattern, themes like "quality control," "microphysical

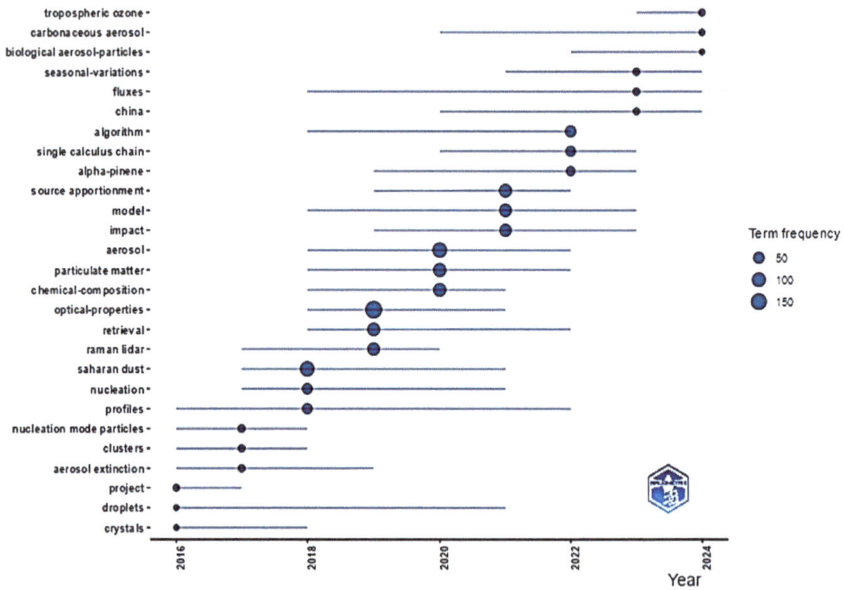

Fig. 4 Topic trends: all publications. *Source* Authors

properties," and "network," which appear in the broader scientific literature in earlier years, have become more prevalent in PPCs only in recent years.

This temporal lag is likely due to the inherent knowledge transfer process from academia to industry. Scientific research typically involves exploring and developing new methodologies, theoretical frameworks, or advanced measurement technologies. Initially, these innovations undergo rigorous academic validation, reflected in their early prominence in purely academic publications. Subsequently, once consolidated and validated, they progressively become attractive to industry partners seeking mature, robust, and market-ready solutions.

In 2022–2024, the interest of the scientific community broadens to socially relevant issues, including "carbonaceous aerosol," "biological aerosol particles," "tropospheric ozone," and "seasonal variations," indicating a move toward comprehensive environmental and climate-health assessments (Fig. 4). These topics will likely underpin future market-oriented solutions, particularly in sectors addressing environmental monitorinig, public health, and sustainable urban development.

Factor analysis provides an additional layer to identify clusters of thematic content in PPCs. This method, implemented in Biblioshiny [37], involves keyword co-occurrence analysis to reduce the complexity of the dataset into meaningful thematic groups, visualized in a two-dimensional factor map divided into four quadrants. Figure 6 presents the results of Multiple Correspondence Analysis (MCA), a technique used to explore relationships between categorical variables. The analysis has been applied to the Keywords plus [38] extracted from our collection.

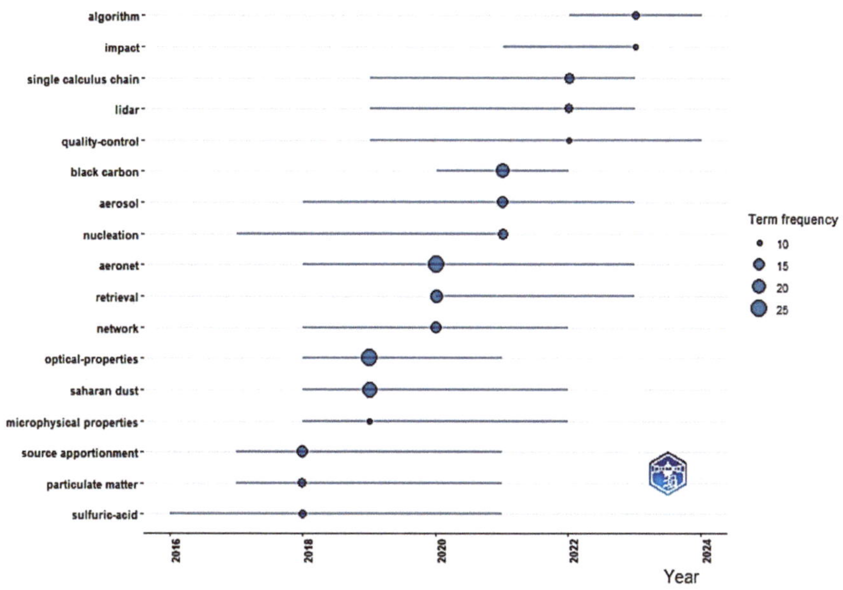

Fig. 5 Topic trends: PPCs. *Source* Authors

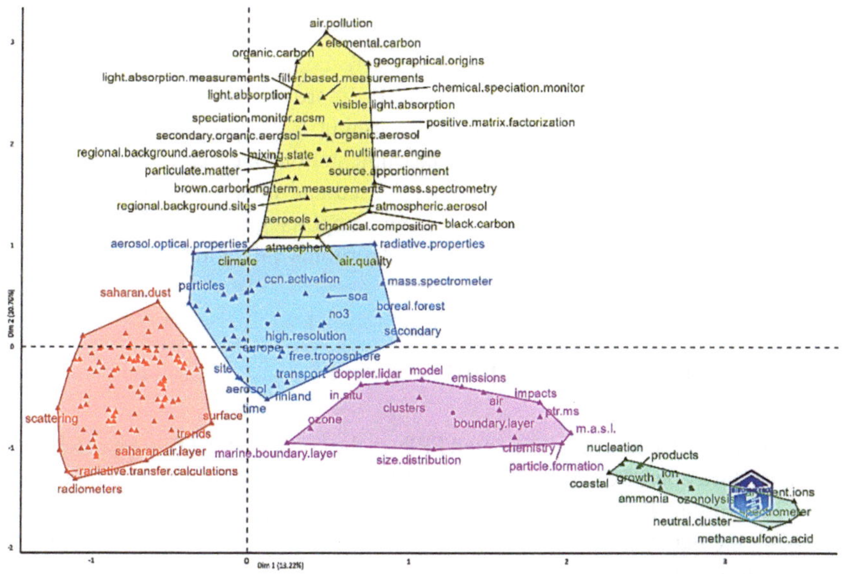

Fig. 6 Multiple correspondence analysis. Dim. 1: 13.22%—Dim. 2: 10,76%. *Source* Authors

The first two dimensions (factorial axes) derived from the MCA indicate how much diversity each keyword contributes to the overall data structure. A high positive or negative value in the two dimensions (Dim1 or Dim2) suggests that the keyword is strongly associated with a particular theme. Each keyword has been grouped into a specific cluster based on its MCA position. The MCA and the derived clustering successfully differentiate the key research themes of ACTRIS as reported in [22]. Clustering can be considered a valuable descriptive tool for potential collaboration areas with industrial applications.

Cluster 1 (Blue) is associated with optical properties and remote sensing of aerosols, emphasizing keywords such as lidar, aerosol optical depth (AOD), inversion, and calibration. Research in this area underpins the development and optimization of ground-based observation networks used for aerosol monitoring.

Cluster 2 (Yellow) focuses on chemical composition, air quality, and pollution sources, with representative terms including black carbon, particulate matter, VOCs, and source apportionment. This research directly informs applications in pollution control, climate mitigation, and the assessment of air pollution impacts on human health.

Cluster 3 (Violet) addresses climate interactions, long-term trends, and regional impacts, involving climate variability, radiative transfer, and tropospheric ozone. These studies support global climate monitoring and contribute to pollution and climate risk mitigation strategies.

Cluster 4 (Light red) explores aerosol–cloud interactions and atmospheric modelling, highlighting processes such as hygroscopic growth, parameterization, and radiative effects. This cluster is critical for improving the accuracy of climate models and mechanisms relevant to global warming.

Cluster 5 (Light green) pertains to instrumentation, measurement techniques, and field campaigns, characterized by field-based and in situ measurements, lidar systems, and sun photometers. This cluster supports ground-based observation networks and their use for satellite retrieval algorithms validation.

6 Discussion and Conclusions

The study explored innovation-oriented PPPs in the context of the European RIs with a specific focus on PPCs as an indicator of knowledge co-production and collaborative innovation. By analysing ACTRIS' scientific outputs and PPCs from 2016 to 2024, the study contributes to the growing literature on research–industry collaboration in large-scale RIs [13].

The empirical findings highlight distinct collaborative patterns of PPCs in comparison to scientific publications. Specifically, PPCs are characterized by a higher average number of co-authors, with a greater prevalence of international partnerships, and impact in terms of citation, underscoring the innovation potential of transnational RI–industry collaboration. This pattern aligns with the literature, which suggests that PPC is increasingly embedded within global scientific networks [24, 30] and contributes significantly to generating innovation outcomes [30, 32]

The analysis also identifies a temporal lag in adopting emerging scientific topics within industry-driven research, a pattern commonly observed in knowledge transfer process. Factor analysis provides additional insights into the thematic clustering within PPCs, highlighting how industry priorities can shape research topics and identifying areas with the highest potential for innovation and collaboration. These findings highlight the importance of sustained industry-academia dialogues to ensure that scientific advancements are effectively translated into practical, market-ready solutions [25].

Nevertheless, the study presents some limitations in using PPCs, as they may not fully capture informal collaborations—such as early-stage co-design, consultancy, or informal exchanges—which can be equally critical to the innovation process [24, 33]. Furthermore, the observed temporal lag suggests that co-publications often emerge only after foundational research has been established, thereby overlooking the upstream dynamics of knowledge co-creation—such as joint problem definition, early-stage experimentation, and exploratory design.

Future research should address these limitations by adopting a more comprehensive and multi-dimensional approach.

Future analyses should incorporate complementary innovation indicators such as co-patenting activity, collaborative project participation, technology licensing agreements and spinoff creation. Furthermore, combining quantitative and qualitative methods—such as interviews with industry partners, project-level case studies, and social network analysis—could provide deeper insights into the motivations, barriers, and success factors underpinning innovation-oriented partnerships. Such a mixed approach would enhance the measurement of innovation-oriented PPPs and improve the effectiveness of policies aimed at maximizing their societal and technological impact.

Acknowledgments The author(s) wish to acknowledge the ACTRIS (Aerosol, Clouds and Trace Gases Research Infrastructure) for its essential role in providing information on access to high-quality data, facilities and technical support that were integral to the execution of this research.

Funding This paper has been funded by EU—Next Generation EU Mission 4, Component 2—CUP B53C22002150006—Project IR0000032—ITINERIS—Italian Integrated Environmental Research Infrastructures System.

References

1. European Commission: Directorate-General for Research and Innovation. Why Investing in Research and Innovation Matters for a Competitive, Green, and Fair Europe—A Rationale for Public and Private Action (2024a). https://doi.org/10.2777/01237
2. Rossoni, A.L., de Vasconcellos, E.P.G., de Castilho Rossoni, R.L.: Barriers and facilitators of university-industry collaboration for research, development and innovation: a systematic review. Manage. Rev. Q **74**, 1841–1877 (2024). https://doi.org/10.1007/s11301-023-00349-1

3. Hodge, G.A., Greve, C.: Public–private partnerships: an international performance review. Publ. Adm. Rev. **67**(3), 545–558 (2007). https://doi.org/10.1111/j.1540-6210.2007.00736.x
4. OECD: Commercialising public research: new trends and strategies. OECD Publ. (2013). https://doi.org/10.1787/9789264193321-en
5. Koppenjan, J., Klijn, E.-H., Verweij, S., Duijn, M., van Meerkerk, I., Metselaar, S., Warsen, R.: The performance of public–private partnerships: an evaluation of 15 years DBFM in dutch infrastructure governance. Publ. Perform. Manage. Rev. **45**(5), 998–1028 (2022). https://doi.org/10.1080/15309576.2022.2062399
6. Lattemann, C., Stieglitz, S., Kupke, S., Schneider, A.-M.: Impact of PPPs to broadband diffusion in Europe. Transf Gov. People Process. Policy **3**(4), 355–374 (2009). https://doi.org/10.1108/17506160910997874
7. Kibria, M.G., Hong, P.: E-government in Asian countries: a conceptual framework for sustainable development. Transf. Gov. People Process. Policy **18**(4), 616–637 (2024). https://doi.org/10.1108/TG-01-2023-0003
8. Mazzucato, M.: Mission-Oriented Research & Innovation in the European Union: A Problem-Solving Approach to Innovation-Led Growth. Brussels Eur. Comm. (2018). https://op.europa.eu/en/publication-detail/-/publication/5b2811d1-16be-11e8-9253-01aa75ed71a1/language-en
9. Norn, M.T., Pujol Priego, L., Ramos-Vielba, I., Ryan, T.K., Conradsen, M.L., Durcan, T.M., Hulcoop, D.G., Edwards, A., Müller, S.: Archetypes of open science partnerships: connecting aims and means in open biomedical research collaborations. Humanit. Soc. Sci. Commun. **11**(1), 1184 (2024). https://doi.org/10.1057/s41599-024-03682-2
10. European Commission: Directorate-General for Research and Innovation, Performance of European Partnerships. Biennial monitoring report 2024 on partnerships in Horizon Europe, Publications Office of the European Union (2024b). https://doi.org/10.2777/991766
11. ESFRI: White Paper "Making Science Happen. A New Ambition for Research Infrastructures in the European Research Area (2020). https://www.esfri.eu/sites/default/files/White_paper_ESFRI-final.pdf
12. ESFRI: Innovation-Oriented Cooperation of Research Infrastructures, ESFRI Scripta, Vol. 3, Working Group (2018). https://www.esfri.eu/sites/default/files/u4/ESFRI_SCRIPTA_VOL3_INNO_double_page.pdf
13. D'Ippolito, B., Rüling, C.C.: Research collaboration in large-scale research infrastructures: collaboration types and policy implications. Res. Policy **48**, 1282–1296 (2019). https://doi.org/10.1016/j.respol.2019.01.011
14. Callens, C., Verhoest, K., Boon, J.: Combined effects of procurement and collaboration on innovation in public-private partnerships: a qualitative comparative analysis of 24 infrastructure projects. Publ. Manage. Rev. **24**(6), 860–881 (2022). https://doi.org/10.1080/14719037.2020.1867228
15. Raile, E.D., Austin, E.K., Wallner, M.P., Peterson, J., Lewandowski, B., Kapps, D., Sellegren, B., Hutton, J.: Technology license agreements as public–private partnerships for economic development: evaluations using surveys, input–output modeling, and regression analysis. J. Innov. Entrepreneurship **14**(14), 1 (2025). https://doi.org/10.1186/s13731-025-00467-y
16. OECD: Fostering research infrastructure ecosystems for addressing complex scientific and societal challenges. Policy Paper n 183 (2025). https://www.oecd.org/en/publications/fostering-research-infrastructure-ecosystems-for-addressing-complex-scientific-and-societal-challenges_34797721-en.html
17. Perkmann, M., Walsh, K.: University–industry relationships and open innovation: towards a research agenda. Int. J. Manage. Rev. **9**(4), 259–280 (2007). https://doi.org/10.1111/j.1468-2370.2007.00225.x
18. Chabbi, A., Loescher, H.W., Dillon, M.S.: Integrating environmental science and the economy: innovative partnerships between the private sector and research infrastructures. Front. Environ. Sci. **5**, 49 (2017). https://doi.org/10.3389/fenvs.2017.00049
19. Castelnovo, P., Florio, M., Forte, S., Rossi, L., Sirtori, E.: The economic impact of technological procurement for large-scale research infrastructures: evidence from the large hadron collider at CERN. Res. Policy **47**, 1853–1867 (2018). https://doi.org/10.1016/j.respol.2018.06.018

20. European Commission: Directorate-General for Research and Innovation. European Charter for Access to Research Infrastructures—Principles and Guidelines for Access and Related Services, Publications Office of the European Union (2024c). https://doi.org/10.2777/8299402
21. Spinello, A.O., Giglitto, D., Lockley, E.: Management of Open Access Research Infrastructures in Large EU Projects: the "CultureLabs" Case. CNR-IRCrES Work. Pap. 9/2021 (2021). https://doi.org/10.23760/2421-7158.2021.009
22. Scarrà, D., Piccaluga, A.: The impact of technology transfer and knowledge spillover from big science: a literature review. Technovation **116**, 102165 (2022). https://doi.org/10.1016/j.technovation.2020.102165
23. Rådberg, K.K., Löfsten, H.: The entrepreneurial university and development of large-scale research infrastructure: exploring the emerging university function of collaboration and leadership. J. Technol. Transf. **49**, 334–366 (2024). https://doi.org/10.1007/s10961-023-10033-x
24. Abramo, G., D'Angelo, C.A.: Drivers of academic engagement in public–private research collaboration: an empirical study. J. Technol. Transf. **47**, 1861–1884 (2022). https://doi.org/10.1007/s10961-021-09884-z
25. Perkmann, M., Tartari, V., McKelvey, M., Autio, E., Broström, A., D'Este, P., Sobrero, M.: Academic engagement and commercialisation: a review of the literature on university–industry relations. Res. Policy **42**(2), 423–442 (2013). https://doi.org/10.1016/j.respol.2012.09.007
26. Tijssen, R.J.W.: Co-authored research publications and strategic analysis of public–private collaboration. Res. Eval. **21**(3), 204–215 (2012). https://doi.org/10.1093/reseval/rvs013
27. Bloch, C., Ryan, T.K., Andersen, J.P., Schneider, J.W.: Public-private collaboration and scientific impact: an analysis based on Danish publication data for 1995–2013. J. Inform. **13**, 593–604 (2019). https://doi.org/10.1016/j.joi.2019.03.003
28. Fritsch, M., Titze, M., Piontek, M.: Identifying cooperation for innovation—a comparison of data sources. Ind. Innov. **27**(6), 630–659 (2020). https://doi.org/10.1080/13662716.2019.1650253
29. Wong, P.K., Singh, A.: Do co-publications with industry lead to higher levels of university technology commercialization activity? Scientometrics **97**(2), 245–265 (2013). https://doi.org/10.1007/s11192-013-1029-1
30. Beaudry, C., Allaoui, S.: Impact of public and private research funding on scientific production: the case of nanotechnology. Res. Policy **41**(9), 1589–1606 (2012). https://doi.org/10.1016/j.respol.2012.03.022
31. European Commission: Directorate-General for Research and Innovation. Science, Research and Innovation Performance of the EU—2024—A Competitive Europe for a Sustainable Future (2024d). https://op.europa.eu/en/publication-detail/-/publication/c683268c-3cdc-11ef-ab8f-01aa75ed71a1/language-en
32. Zhang, M., Wang, L., Li, Z.: The impact of team compositions on disruptive and novel research in large-scale research infrastructures. Scientometrics (2025). https://doi.org/10.1007/s11192-025-05319-w
33. De Stefano, D., Fuccella, V., Vitale, M.P., Zaccarin, S.: Quality issues in co-authorship data of a national scientific community. Netw. Sci. (2023). https://doi.org/10.1017/nws.2022.40
34. Laj, P., et al.: Aerosol, Clouds and Trace Gases Research Infrastructure (ACTRIS): the European Research Infrastructure Supporting Atmospheric Science. Bull. Am. Meteorol. Soc. **105**, E1098–E1136 (2024). https://doi.org/10.1175/BAMS-D-23-0064.1
35. ATMO-ACCESS (Access to Atmospheric Research Facilities) Funded in the frame of the programme H2020-EU.1.4.1.2—Grant Agreement n. 101008004—(1 April 2021–31 March 2025); ACTRIS-IMP (Implementation project). Funded in the frame of the programme H2020-EU.1.4.1.1.—Grant Agreement 871115—(1 January 2020–31 December 2023); ACTRIS-2 (Aerosol, Clouds, and Trace gases Research Infrastructure), Funded in the frame of the programme EC H2020, Grant Agreement n. 654109 (1 May 2015–1 May 2019); ACTRIS (Aerosol, Clouds, and Trace gases Research Infrastructure network), Funded in the frame of the programme EC FP7, Grant Agreement n. 262254—(1 April 2011–31 March 2015)

36. Gagliardi, S., Gargano, G., Petracca Altieri, R.M., Cornacchia, C., Dubost, A., Olivieri, M: Milestone 9.2: Identification of collaboration models between ACTRIS and the private sector. (2021). https://actris.eu/sites/default/files/Documents/ACTRIS%20IMP/Milestones/ACTRIS_IMP_WP9_MS52_Identification%20of%20collaboration%20models%20between%20ACTRIS%20and%20the%20private%20sector.pdf
37. Aria, M., Cuccurullo, C.: Bibliometrix: an R-tool for comprehensive science mapping analysis. J. Inform. **11**, 959–975 (2017). https://doi.org/10.1016/j.joi.2017.08.007
38. Garfield, E., Sher, I.H.: Keywords Plus™—algorithmic derivative indexing. J. Am. Soc. Inform. Sci. **44**(5), 298–299 (1993). https://doi.org/10.1002/(SICI)1097-4571(199306)44:5%3c298::AID-ASI5%3e3.0.CO;2-A

The Impact of Disclosure Quality, ESG Targets and Compensation Schemes on Sustainability Performance: Evidence from Italy

Roberto De Luca, Ornella Malandrino, Vincenzo Piluso, and Valentina Cicalese

Abstract The purpose of the present research is to evaluate the determinants of firms' sustainability performance, focusing on the level of disclosure, management commitment, and compensation schemes. Regarding the first facet, the aim of the paper is to understand whether company reports are really informative and are a good predictor of real sustainability actions. The other main factors analyzed refer to the presence of specific ESG targets in the reports and to management compensation schemes, to assess if the presence of sustainability incentives genuinely affect sustainability policies. A time-series analysis is used for examining data gathered from among Italian listed firms for 3 years (2021–2023), which provides us with 194 valid firm year-observations. We performed robustness checks also to ensure the accuracy of the results and to control for potential multicollinearity problems. Sustainability performance has been measured by resorting to third-party ESG ratings. The outcomes confirm that the quality of disclosure is positively linked to sustainability performance, which is also affected by the presence of ESG targets in sustainability reports and by the existence of compensation schemes linked to the company's sustainability performance. The last factor, however, exhibits a lower statistical significance in a dynamic perspective. The outcome will support managers and decision makers in improving company ESG disclosure to achieve better sustainability performance and higher ESG rating. To reach this target, it will be important also to assess the existence of well-defined ESG targets and adequate compensation schemes for the executive, to stimulate ESG compliance and encourage managers to engage in sustainability actions.

R. De Luca (✉) · O. Malandrino · V. Piluso · V. Cicalese
Department of Management and Innovation Systems, University of Salerno, Fisciano, Italy
e-mail: rodeluca@unisa.it

O. Malandrino
e-mail: ornellam@unisa.it

V. Piluso
e-mail: vpiluso@unisa.it

V. Cicalese
e-mail: vcicalese@unisa.it

A. Visvizi et al. (eds.), *Research and Innovation Forum 2025 Vol2*, Springer Proceedings in Complexity, https://doi.org/10.1007/978-3-032-09062-1_12

Keywords ESG · Sustainability · Governance · Agency problems · Information asymmetry · SDG 16

1 Introduction

Financial scandals that involved numerous companies at the beginning of the twenty-first century (i.e. WorldCom, Enron, etc.) and during the 2008/2009 financial crisis (i.e. Lehman Brothers) led to a significant loss of confidence and trust in businesses. Consequently, firms have been increasingly pressured to adopt sustainable and socially responsible strategies and behaviors, to protect their status and reputation [1].

Corporate sustainability is accomplished by integrating social and environmental performance into economic valuations, strengthening not just shareholder value, but also overall efficiency and sustainable growth. Therefore, the traditional perspective aimed at maximizing shareholder return [2] has drastically changed [3] and companies have become more sensitive and interested in the disclosure of non-financial information regarding their sustainability performance and ethical values [4]. Consequently, ESG (environmental, social, governance) adoption, awareness and reporting have recently become one of the most current and significant issues for businesses [5].

The diffusion of ESG compliance worldwide requires attention and contribution from all stakeholders, with the topmost being companies and investors [6].

Growing public interest and corporate awareness has led to a constant growth in the number of companies which are committed to sustainability actions and disclose their ESG information [7]. Various pieces of research point out the relevance of sustainability and ESG performance disclosure for companies and its importance in terms of reputation, corporate culture, danger of greenwashing and so on [8]. These elements, in turn, lead to a positive association between ESG score and financial performance, with specific regard to social and environmental "pillars" [9].

Therefore, firms are competing to provide sustainability reports to promote and communicate ESG commitments [10], also because investors are increasingly focusing their evaluations and their choices on non-financial disclosures [11].

The rising importance of ESG issues for investors implies that the most relevant factor in decision making is no longer mere financial information [12].

Some surveys among portfolio managers and investors confirm the tendency to integrate ESG data and information in their investment and financing choices [13], since they believe that good sustainability performance may help reduce firms' risks and cost of capital.

This trend is confirmed by Ng and Rezaee [14], whose investigation highlights that financial economic sustainability performance disclosures contribute to lowering the cost of equity and that ESG disclosure boosts this effect. As a consequence, investors tend to divest from businesses with poor results in terms of sustainability and ESG compliance [15]. Jain et al. [16] report that short sellers generally do not target companies with high ESG scores, whereas they aim at firms with low ESG scores.

The growing number of sustainability reports has attracted a significant increase in research related to sustainability assessment, but the connection between ESG disclosure and sustainability performance has yet to achieve broad and consistent understanding [17].

In Europe, sustainability disclosure has been further pushed by the issuance of Directive 2014/95/EU, according to which large companies are under obligation to disclose also non-financial information regarding corporate governance systems adopted, policies regarding employees and environmental practices and outcomes. Later on, the Corporate Sustainability Reporting Directive (CSRD) expanded the scope of non-financial information, requiring more companies to perform and communicate sustainability disclosures.

In our research we aim to find out how companies subject to these obligations are performing in terms of sustainability and which factors are affecting this performance the most.

The paper contributes to the literature in different ways. Firstly, most previous research has focused on the environmental aspects of sustainability, and in doing so offered a narrower perspective of the disclosure-performance relationship. Moreover, we provide hints also about the influence on sustainability performance of management commitment and compensation policies. Regarding the first subject, there are very few works that address the correlation between corporate commitment and sustainability performance, which is often treated simply as a moderating factor [18] and not as a main test variable. Moreover, managers' commitment to sustainability and ESG compliance is often treated as a dependent variable, whereas in our study it represents one of the most relevant independent variables to affect sustainability performance and not the other way around. Lastly, most previous studies have concentrated on single aspects of sustainability or ESG dedication, such as social commitment [19].

Regarding compensation policies, only a few studies have focused on proper ESG-related issues and the link between executive compensation and corporate social has been mainly focused on the US, with little evidence coming from European countries. Therefore, the purpose of the research is to fill this gap in the literature, resorting to an empirical analysis performed by analyzing companies listed on the Italian stock market. Moreover, the study is relevant since it is among the first to assess the simultaneous effect of sustainability disclosure, company commitment and management compensation schemes. Our results indicate that the level of sustainability disclosure is positively related to companies' actual sustainability performance. On the same wavelength, ESG scores are positively affected by managers' commitment to pursue not just economic objectives but also sustainability goals.

The remainder of the paper is structured as follows. In the next paragraph a literature review is carried out, focusing on the importance of sustainability and ESG factors for companies and investors and on the relevance of disclosure quality to company sustainability performance. We also analyze previous research regarding the influence of compensation schemes linked to ESG targets on sustainability performance.

Paragraph 3 explains the research design and the methodology used, whereas paragraph 4 presents the data analyses and the outcomes. Finally, paragraphs 5 and 6 are dedicated to the discussion of results, limitations and practical implications.

2 Literature Review

Corporate sustainability performance can be analyzed in the light of the "Triple Bottom Line" concept introduced by Elkington [20], according to whom firms integrate economic, environmental and social issues in their strategy, aiming at gaining good financial performance while protecting and supporting society and the environment for future generations.

Over the years, there has been a growing number of firms adopting sustainability strategies and sharing both qualitative and quantitative ESG data. This trend stems from the recognition of ESG information's significance by regulatory bodies, stock exchanges, and investors. Disclosing ESG data is considered essential not only for fostering sustainable growth within corporations but also for equipping the market with comprehensive metrics that extend beyond a company's financial risks, aiding in informed business and investment decisions [3].

It is worth noticing that various theoretical frameworks can provide possible reasons for corporations disclosing sustainability and ESG information beyond the obligations set by law, such as stakeholders, legitimacy, agency, and signaling (Table 1).

Table 1 Literature review: main theoretical frameworks analyzed

Theoretical frameworks
• Agency theory • Stakeholders theory • Legitimacy theory • Signaling theory • Resource-based view theory
Proponents
• Jensen and Meckling [21], Morris [22], El Ghoul et al. [23] • Freeman [25], Porter and Kramer [28] • Clarkson et al. [38], Guthrue and Parker [30], Deegan [31] • Dhaliwal et al. [49], Papoutsi and Sodhi [55] • Branco et al. [57], Lourenço et al. [58]

2.1 Agency Theory

In the light of the agency theory [21], moral hazard phenomena can emerge in case of severe information asymmetry, when managers are in a position of advantage and can decide to omit or alter relevant information shared with investors and shareholders. In order to lower information asymmetry and minimize principal-agent issues, companies can decide to increase the level of non-financial information disclosed.

In doing so, the principal attempts to reduce agency costs (i.e., information asymmetry) stemming from the separation between control and ownership. Therefore, a greater degree of non-financial information and ESG data disclosure will benefit the principal through higher transparency [9, 22]. All in all, good sustainability and ESG performance are likely to reduce information asymmetry [23], as proven also in very specific operations, such as Initial Public Offerings [24].

2.2 Stakeholders and Legitimacy Theory

The rising attention paid to sustainability and ESG compliance on the part of companies can be explained also in the light of stakeholder theory [25]. According to this framework, firms cannot aim solely at financial goals in favor of shareholders, but they need to take into account the needs and demands of all stakeholder groups [26].

Within this theoretical framework, in order to achieve corporate sustainability value in the long term, firms have to meet the needs of all stakeholders, including customers, communities, and other groups of interest that can affect or be affected by company activities and performance [25]. This reasoning is confirmed by Jensen [27] within his "enlightened value maximization" theory.

On the same wavelength, Porter and Kramer [28] define the concept of corporate shared value theory as a set of policies and operating practices by which companies create economic value through societal benefits. According to the authors, CSR increasingly becomes a driver of competitive advantage for businesses.

ESG disclosure helps stakeholders to assess and evaluate firms' commitment to sustainability practices and responsible business actions, especially on the part of ESG investors. Nevertheless, since many investors lack the time and specific knowledge to gauge the complex data conveyed by businesses, they often rely on ESG scores to appraise companies' sustainability performance [29]. Therefore, companies should introduce social and environmental factors into their strategy, improving their competitive position and their contribution to the communities in which they act.

Another driver for ESG disclosure lies in the legitimacy theory, which links the success of an organization with its social acceptance, leading companies to communicate some information (environmental commitments, community involvement and so on) to persuade society that their operations are legitimate and are creating value for society [30]).

This theory examines the connections between organizations and society as a whole [31], assuming the existence of a kind of social contract between them [32], which implies that all institutions, including businesses, operate within societal limits based on the need to provide desirable societal benefits and outcomes. A company's attention to sustainability and ESG factors depends on its resonance with societal values, whereas a misalignment between societal values and those of an entity can lead to a "legitimacy gap", which represents a threat and a danger to the organization's legitimacy [33].

Consistent with stakeholder theory and legitimacy theory, prior research suggests that firms with better corporate governance, such as intensive monitoring [34], higher corporate governance ratings [35], the presence of an environmental committee and a Chief Sustainability Officer, more independent and larger boards [36] tend to provide more sustainability disclosures to the public. For example, some studies provide evidence on the informativeness of standalone CSR reports by exhibiting an information asymmetry reduction after CSR reporting [37]. Moreover, Clarkson et al. [38] find that voluntary environmental disclosures are value-relevant and enhance firm value through signaling the firms' proactive environmental strategies to investors.

According to Devalle et al. [39], high ESG scores should indicate that a firm is contributing to the implementation of sustainable governance practices, as well as environmental and social development. Previous research has pointed out that increasing the quality and the magnitude of ESG disclosure can reduce information asymmetry about sustainability practices between investors and companies [40]. This disclosure can increase information transparency and boost stakeholders' confidence about firms' risk in the future [41, 42]. In some studies, environmental disclosure has been proven to be positively associated with financial and environmental performance [43].

2.3 Signaling Theory

In a competitive context characterized by increasing pressure from public opinion and social media, by engaging in ESG disclosure, firms can generate benefits and advantages such as better reputation, stronger brand image, higher employee motivation, better information transparency, deeper attention to the environment [44, 45]. Following these purposes, companies which pursue social legitimacy can engage in sustainability disclosure also on a voluntary basis to signal that their actions are compliant with laws and social expectations. The release of non-financial and sustainability information can help companies reduce adverse selection and information asymmetry issues [46, 47]. Also according to Dhaliwal et al. [48, 49], CSR reports do generate a reduction in information asymmetry and environmental information increases company value, acting as a signal for markets and investors [38]. Along with Al-Tuwaijri et al. [43], based on signaling theory, ESG disclosure can generate three positive effects: (1) improving perceptions of companies' accountability among the main stakeholders; (2) lowering information asymmetry suffered by external users of businesses' information and reports, among whom investors and other stakeholders; (3) signaling the firm's legitimacy and commitment to the benefit of society.

As a consequence of these benefits, companies are willing to incur financial costs to disclose their initiatives regarding sustainability and ESG compliance to stakeholders who otherwise could not be directly aware of them [50], since sustainability actions and results are often not readily available to shareholders and other stakeholders [51, 52].

Starting from signaling theory [53], according to many authors, companies which achieve high performance in terms of sustainability communicate and disclose more information regarding their actions compared to firms with poor sustainability performance [54, 55]. According to Surroca et al. [56], the reason that drives this behavior lies in the will to disclose sustainability information only when the benefits of such an activity surpass the related costs.

2.4 Resource-Based View Theory

Commitment to sustainability and ESG compliance can be analyzed in the light of resource-based view theory, which nowadays focuses especially on intangible factors to explain differences in companies' performance [56], among which human capital and reputation. Sustainability actions provide benefit from an internal point of view by supporting firms in developing corporate culture, know-how, new competences and resources [57]. Moreover, from an external perspective, corporate sustainability commitments improve and strengthen company reputation and build better relationships with all the relevant stakeholders [58].

2.5 Sustainability and Reporting Quality

Research on sustainability and ESG compliance has also analyzed their effect on financial reporting quality, although it has provided mixed results. According to Velte [59], companies with high ESG scores also provide better and more reliable financial disclosure, since it reduces and prevents opportunistic and unethical behavior on the part of the managers. Some Authors, on the other hand, point out that that ESG disclosure strategy could be an attempt to mislead and deceive stakeholders to strengthen a company's reputation even without providing any real sustainability results [60]. On the same wavelength, other authors highlight that in the same cases, managers who engage in earnings manipulation actions are more likely to show social and ethical commitment in order to deceive stakeholders and shareholders [61].

According to previous studies [48], managers pursue ESG compliance and high performance only when such actions generate a personal benefit for them, since companies which engage in ESG activities help gain the legitimacy of society and reduce the level of stakeholder control [62].

Conversely, different pieces of research assert that corporate social responsibility can become embedded in a firm's culture, leading to virtuous behaviors unrelated to potential agency conflicts [63]. Therefore, CSR and reporting quality can be positively related. This trend is confirmed also regarding, specifically, environmental actions, which are linked with higher disclosure quality, particularly with respect to discretionary provisions and earnings management [64]. Businesses engaged in socially responsible actions behave in a more transparent and reliable manner also in preparing their reports [65] and in their disclosure activities [66], pursuing higher reputation and strengthening relationships with their stakeholders [67]. Also Bozzolan et al. [68] confirm that companies which engage more in CSR policies are more likely to convey more trustworthy and transparent information than less CSR-oriented firms.

All in all, the connection between ESG disclosures and ESG scores has gathered significant attention among academics [56, 69, 70]. In the past, research has not been able to find a unanimous and consistent disclosure–performance relationship, since some studies found no significant association [71], and others indicate a negative association [72]. Only recently, researchers have started to find more significant positive association [43, 73, 74], pointing out that firms with higher levels of ESG disclosures tend to attain higher sustainability performance, gauged by ESG scores.

2.6 Sustainability Performance and Management Compensation

The importance of sustainability and ESG factors in affecting company financial performance and value have led to the introduction of executive compensation schemes based not just on earnings and share price, but also on the achievement of sustainability goals. Sustainability compensation packages have been gaining significance since they can represent a company's response to societal pressure and demand for sustainable and responsible behaviors, which require firms to go beyond mere financial objectives and to focus more on long-term goals [75].

Past research about the influence of compensation policies on firm performance have not gained unanimous consensus on how to reach an interest alignment between shareholders and management [76]. Providing the best incentive scheme for managers is one of the most relevant issues in financial economics, embedded as a principal-agent problem that companies face [77–79].

Research in the field has mainly focused on the effect of managers' compensation on companies' financial performance. Nowadays, the alignment between firm and management interests is even more difficult, since businesses have to meet the demands and needs expressed not only by shareholders but by a greater number of stakeholders with diverse requirements. Consequently, in recent years we have witnessed a new tendency, aimed at linking managers' compensation with sustainability matters, trying to encourage executives to carry out environmental and social activities [80, 81].

Mahoney and Thorne [82] confirm the relevance of the structure of managerial salaries in encouraging socially responsible actions. Also other studies [83] point out a positive correlation between executive wage schemes and corporate social performance.

Nevertheless, as highlighted in the introduction, previous research on the link between manager compensation and sustainability performance is still quite limited and almost none of the past studies analyze specific ESG-related compensation schemes.

3 Research Questions

The disclosure of ESG information has been proven to improve accountability and transparency within stakeholders [84, 85], since firms provide a source for comparison between company claims and their real sustainability performance [86, 87].

The quantitative facet of ESG/Sustainability disclosure refers to the amount of relevant information that businesses publicly report [84].

ESG disclosure, communicating non-financial information such as corporate governance, social responsibility and environmental actions, and other factors, has been recognized as a relevant means for enhancing company sustainability performance [88].

Recent studies [89] emphasized the link between ESG disclosure and sustainability performance, proxied by ESG scores, using an MLP algorithm.

Starting from the abovementioned reasoning, following previous research [90, 91], we state our first hypothesis as follows:

Hypothesis 1: *The degree of sustainability disclosure is positively correlated with corporate sustainability performance.*

We also aim at investigating the effect of manager commitment towards sustainability, which according to Muller and Kolk [92] has a positive impact on company sustainability performance. Therefore, we can formulate our second hypothesis as follows:

Hypothesis 2: *Management commitment to sustainability is positively linked to corporate sustainability performance*

Lastly our research focused on the influence of management compensation schemes on sustainability performance. As highlighted by previous research [54], we assume that ESG-related variable compensation policies can encourage managers to undertake socially and environmentally responsible actions. Hence, our third research hypothesis can be summarized as follows:

Hypothesis 3: *ESG-related management compensation schemes are positively linked to corporate sustainability performance*

4 Methodology

Considering a sample of Italian publicly listed companies, we analyzed their sustainability reports, which have been defined in many ways by academics and practitioners. According to Daub [93], this document should "contain qualitative and quantitative information on the extent to which the company has managed to improve its economic, environmental and social effectiveness and efficiency in the reporting period, and integrate these aspects in a sustainability management system". Reporting practices have developed because mere sustainable actions are often not effective enough to strengthen firm legitimacy and increase stakeholder awareness and sympathy: for companies it is necessary to communicate these practices and convey this information in a systematic manner through specific reports [94].

We gathered data from 94 Italian companies obliged by law to issue sustainability reports together with their financial statements, excluding banks and insurance companies, as they follow different financial reporting standards. In this country, the Legislative Decree 254/2016 implemented the Directive 2014/95/UE, compelling large companies to disclose comprehensive information about their ESG practices.

We collected data from 2021 to 2023 and, after correcting for missing or inaccurate information, our final sample is made up of 194 time-series observations.

In line with previous literature [89], our dependent variable is the ESG score, extracted by LSGE (former Refinitiv), one of the main ESG rating providers and it has been used in more than 1,200 academic articles over the past 15 years [95]. Refinitiv ESG data have also been mentioned by the World Economic Forum in 2019 [96] and they have been analyzed as one of the three key ratings providers in a recent OECD report [97]. This is also supported by other studies, according to which Refinitiv provides one of the most complete and broad ESG databases and has been recently used to conduct research on similar topics [98, 99].

One of the test variables is represented by the level and depth of sustainability disclosure. Most studies have used words, sentences or line counts to measure the quantity of disclosure [91, 100]. Several previous studies rely on these factors to assess the overall level of disclosure achieved by firms in their sustainability reports [101, 102]. Adopting the same approach, also following Darnall et al. [103], in our study the level of disclosure has been measured by the number of pages comprising each sustainability report. Another test variable is management commitment towards sustainability, which leads to the real implementation of an effective sustainability strategy to successfully integrate ESG factors into their choices, actions, policies, operations and decisions [104, 105].

Manager commitment towards ESG targets improves the likelihood of enhancing company performance from an environmental and social perspective, pushing the whole organizational culture in the direction of sustainability goals [106].

As in Rahman et al. [105], this factor is proxied by the presence of specific ESG targets within company sustainability reports, which has been treated as a dummy variable. Coherently with hypothesis 2, we expect this variable to attain a positive coefficient.

Finally, regarding manager compensation schemes, we intend to analyze their influence on company sustainability performance. Consistently with Baraibar-Diez et al. [75], we include in our model a dummy variable indicating whether the firm has implemented an ESG-related compensation policy. In accordance with Cohen et al. [107], we expect a positive relationship between this factor and firms' ESG scores.

In line with previous studies, among our control variables, we include profitability, Otman [108] points out that a firm's profitability has a positive impact on sustainability disclosures and in our model, we measured it using ROA [101] and the EBITDA/Sales ratio [109]. After testing for robustness and statistical significance, we decided to use only the latter indicator.

In the light of previous studies [110, 111], we hypothesize a positive correlation between profitability and sustainability performance, since high profitability and good financial performance could help companies fund investments aimed at improving social and environmental results [112]. Non profitable businesses are often forced to reduce their costs to improve their profitability, which becomes the most important priority, even at the expense of stakeholder expectations [113]. Moreover, financial performance may affect company willingness to convey proprietary information since only profitable businesses are able to exploit the benefits from additional disclosure compared to the costs of revealing potentially damaging sustainability information [114]. This hypothesis is also grounded in stakeholder and legitimacy theory [115].

We also included leverage among our control variables, to represent company risk, which has been proxied by debt/equity ratio, following Callan and Thomas [116]. According to different pieces of research [117, 118], firms with a lower level of debt invest more in CSR and sustainable activities than more indebted companies [119]. On the other hand, in moments of high leverage, banks and other financial stakeholders will be given priority over sustainability issues and targets [120]. Hence, as also suggested by Sharma et al. [121], we also assume a negative relationship between leverage and ESG. We included in our control variables also company size. Generally speaking, bigger companies are held accountable by a higher number of stakeholders and are subject to stronger pressure by public opinion, regulators and so on [122]. Larger companies should disclose more information [123] and achieve better performance regarding sustainability and ESG compliance. More diverse and numerous stakeholders, a larger market presence and a higher number of transactions lead to a greater probability of negative events [124]. Therefore, larger companies should be more willing than smaller firms to engage in sustainable actions to cover this increased risk. Chih et al. [125], investigating firms located in 34 different countries, highlight that bigger firms are more sustainability-oriented and achieve better results in this field. Therefore, we expect a positive relationship between size—which is related to stakeholder influence [126]—and sustainability performance. Following prior research [108], we calculated this variable as the natural logarithm of total assets.

The last control variable is growth, since previous research indicates that companies with stronger growth perspectives are more able to integrate sustainability activities into their strategies and operations [112, 127]. To account for this variable, we used price to book (P/B) ratio, which has been associated with growth companies, since investors are willing to pay a higher price in view of future opportunities and growth [91]. Therefore, consistent with previous research [42, 74] we hypothesize a positive relationship between Growth opportunities and sustainability performance.

The variables thus far mentioned have been examined using a time-series analysis, referring to data from 2021 to 2023 of Italian companies listed on the stock market. We can describe our regression model as follows:

$$\begin{aligned}\text{Sustainability performance}_{\text{it}} = & \alpha + \beta_1 \text{Disclosure}_{\text{it}} \\ & + \beta_2 \text{Commitment}_{\text{it}} + \beta_3 \text{Compensation}_{\text{it}} \\ & + \beta_4 \text{Size}_{\text{it}} + \beta_5 \text{Leverage}_{\text{it}} + \beta_6 \text{Profitability}_{\text{it}} \\ & + \beta_7 \text{Growth}_{\text{it}} + \varepsilon_{\text{it}}\end{aligned} \tag{1}$$

We also attempted to take into account lagged variables, but they did not exhibit high statistical significance, and the model did not gain overall goodness of fit as measured by R-squared. In the following table we provide a summary of the variables used, their construction and the kind of effect we expect on the dependent variable (Table 2).

Table 2 Variable description and expected effect

	Description	Expected value
Disclosure	Sustainability reports' number of pages	+
Commitment	Dummy variable to verify if managers have publicly set ESG targets in their reports (Yes = 1; No = 0)	+
Compensation	Dummy variable to verify if companies have defined a variable policy ESG-Related Compensation (Yes = 1; No = 0)	+
Size	Ln (total assets)	+
Leverage	Debt/equity ratio	−
Profitability	EBITDA/Sales ratio	+
Growth	Price/Book value ratio	+

5 Results

Descriptive statistics regarding ESG score display an average value around 61, ranging from 22 to nearly 95, with a standard deviation equal to 15. The length of ESG reports indicates substantial variance, ranging from 25 to 590 over the three-year period, generating a standard deviation of almost 90 (Table 3).

The outcomes emerging from our regression model are summarized in the following tables.

The level of disclosure indicates a positive coefficient, and it is statistically significant at a level of 1%, as we expected. This outcome allows us to confirm our first hypothesis, according to which companies providing more information about their sustainability behaviors actually exhibit better performance in terms of environmental, governance and social activities.

Firms with better sustainability performance, in sum, are also the ones that convey more information, in order to provide more transparency and lower information asymmetry issues, gaining more legitimacy and trust within society and among stakeholders. This outcome also proves that these companies do not engage in greenwashing or social washing practices, since sustainability reports are not used in a deceptive and opportunistic manner, but they reflect real engagement in sustainable behaviors and ESG compliance.

On the same wavelength, management commitment to sustainability goals is also positively related to the ESG score and it attains a statistical significance at a level of 1%, fulfilling our expectations and confirming our second hypothesis.

Executive commitment, in fact, promotes sustainable actions by leading to the allocation of the necessary financial and human resources towards an effective integration of ESG and sustainability issues within company strategy and operations [105]. Also in the light of stakeholder and legitimacy theory, communicating ESG targets without genuinely pursuing them could significantly harm companies, in terms of brand identity, reputation and trust on the part of investors and other stakeholders. Therefore, firms that set well defined goals tend to achieve a higher sustainability score.

Table 3 Descriptive statistics

	N	Minimum	Maximum	Mean	Std. Deviation
Sustainability performance	225	22.16	94.68	61.0144	15.31.409
Disclosure	221	25	590	163.63	89.422
Commitment	228	0	1	0.59	0.493
Compensation	228	0	1	0.60	0.492
Size	217	11.56	18.64	14.3142	1.62.722
Leverage	214	0.00	26.63	1.0441	2.39.793
Profitability	211	− 806.69	91.35	− 17.3259	117.03.669
Growth	207	− 1.056	20.974	2.52.237	2.885.565
Valid N (list wise)	194				

As for compensation schemes, even though the sign of the coefficient is positive as hypothesized, the absence of statistical significance prevents us from confirming our third research question. In our model, therefore, managerial bonus packages are less influential than their commitment and their reputation.

Regarding the control variables, all of them exhibit the sign we expected. Size is also statistically significant at 1%, confirming that bigger companies tend to achieve better results in terms of sustainability, probably because of the higher pressure they are exposed to by the media, public opinion, and other stakeholders in general.

Profitability indicates a positive sign as well, and statistical significance at 1% level, indicating that firms with stronger financial performance are better equipped to engage in sustainability actions, hence attaining better ESG scores.

The effect of leverage on sustainability performance is negative, as expected, although the result is not statistically significant enough to establish an actual relationship with the dependent variable. The same result emerges concerning the impact of growth, which presents a positive sign—as hypothesized—but it is not statistically significant.

As shown in Table 4, in order to validate our statistical analysis, we checked the model for potential multicollinearity issues using the variance inflation factor. Since all the VIF values are close to one, we can assume a very low risk of multicollinearity.

For the same reason we also performed a tolerance test, with a cutoff suggesting concern for values below 0.40. Also regarding this indicator, our data do not cast any doubts about collinearity problems [128].

In Table 5 we provide results regarding the goodness of fit of the whole model. The R-square exhibits good value, allowing us to state that the model has good predictive power about the phenomenon investigated. These statistics have been used also to

Table 4 Regression model results

Coefficients[a]

Model		Unstandardized coefficients		Standardized coefficients	t	Sig	Collinearity statistics	
		B	Std. Error	Beta			Tolerance	VIF
1	(Constant)	– 10.414	7.894		– 1.319	0.189		
	Disclosure	0.026	0.010	0.162	2.748	0.007	0.722	1.385
	Commitment	7.668	1.818	0.247	4.218	0.000	0.730	1.370
	Compensation	0.750	1.753	0.024	0.428	0.669	0.774	1.292
	Size	4.365	0.599	0.476	7.282	0.000	0.584	1.712
	Leverage	– 0.299	0.321	– 0.049	– 0.932	0.353	0.888	1.127
	Profitability	0.015	0.007	0.114	2.228	0.027	0.959	1.042
	Growth	0.396	0.319	0.065	1.240	0.217	0.906	1.104

[a]Dependent Variable: Sustainability performance

Table 5 Model summary

Model summary[b]				
Model	R	R Square	Adjusted R square	Std. Error of the estimate
1	0.732[a]	0.535	0.518	10.53.338

[a]Predictors: (Constant), disclosure, commitment, compensation, size, leverage, profitability, growth
[b]Dependent variable: Sustainability performance

check for possible reverse causality issues. In particular, we wanted to clarify if firms perform better because they disclose more, or if companies that already perform better tend to disclose more. Considering disclosure quality not as a test variable but as an independent variable, the model loses its predictive power (R square = 0.280), and other variables are not statistically significant anymore.

6 Implications, Discussion and Limitations

The aim of this paper was to examine whether a high level of sustainability disclosure actually signifies better sustainability performance and more sustainable activities. In sum, we tried to assess the disclosure-performance relationship in terms of sustainability matters and ESG compliance. Our investigation was carried out in light of different theoretical frameworks, among which especially stakeholder, legitimacy and agency theory. The study is relevant since it is among the first works to evaluate the simultaneous effect of sustainability disclosure, company commitment and management compensation schemes.

Our results demonstrate that there is a positive and significant correlation between the level of disclosure and firm sustainability performance. Since our sample is made up of companies obliged to issue reports about these matters, this can be a hint for law and policy-makers: if they want to push companies towards more responsible behaviors, they can try to broaden the scope of businesses forced to convey information regarding their sustainability performance and activities. Our findings also highlight that a publicly-stated commitment is an effective driver towards better results in terms of sustainability and ESG compliance, as stakeholders would hold companies accountable if such promises were not fulfilled. Consequently, reporting principles and guidelines could further encourage companies to disclose their ESG targets and their future actions, in order to improve their sustainability performance.

On the other hand, we have not been able to establish a significant relationship between management compensation schemes and company achievements in terms of sustainable actions. Therefore, these incentives do not appear to represent a strong driver towards sustainability performance.

Ong of the main limitations of the study regards the sample selected, composed of publicly-listed companies, which are subject to the obligation of issuing a sustainability report. Further research could be carried out using a cross-country analysis, expanding the sample to assess the effect of institutional factors. Moreover, future work could improve upon this study by analyzing unlisted firms, which disclose sustainability information on a voluntary basis. From a methodology point of view, we could perform a content analysis or similar techniques to assess the level and quality of sustainability disclosure. Moreover, we are aware that treating ESG targets and compensation as dummy variables may oversimplify managerial commitment and the complexity of incentive structures. Deeper and more robust analysis might require the use of a more precise measure, such as a continuous or a ratio variable.

Lastly, due to the subjective nature of ESG ratings across providers, further research could imply the use of more ESG rating scores, in order to strengthen and make our assessment more reliable.

References

1. Mutuc, E.B., Lee, J.-S., Tsai, F.S.: Doing good with creative accounting? Linking corporate social responsibility to earnings management in market economy country and business sector contexts. Sustainability **11**(4568), 1–20 (2019)
2. Friedman, M.: The social responsibility of business is to increase its profits. The New York Times Magazine September 13 (1970)
3. Alsayegh, M.F., Abdul Rahman, R., Homayoun, S.: Corporate economic, environmental, and social sustainability performance transformation through ESG disclosure. Sustainability **12**, 3910 (2020)
4. Rezaee, Z.: Business sustainability research: a theoretical and integrated perspective. J. Account. Lit. **36**, 48–64 (2016)
5. Githaiga, P.N.: Sustainability reporting, board gender diversity and earnings management: evidence from East Africa community. J. Bus. Socio-econ. Dev. **4**(2), 142–160 (2024)
6. Prior, D., Surroca, J., Tribó, J.A.: Are socially responsible managers really ethical? Exploring the relationship between earnings management and corporate social responsibility. Corp. Gov. Int. Rev. **16**(2), 160–177 (2008)
7. Weber, O.: Environmental, social and governance reporting in China. Bus. Strateg. Environ. **23**, 303–317 (2014)
8. Rezaee, Z.: Ling Tuo: Are the quantity and quality of sustainability disclosures associated with the innate and discretionary earnings quality? J. Bus. Ethics **155**(3), 763–786 (2019)
9. Carnini Pulino, S., Ciaburri, M., Magnanelli, B.S., Nasta, L.: Does ESG disclosure influence firm performance? Sustainability **14**(13), 7595 (2022)
10. Harymawan, I., Putra, F.K.G., Fianto, B.A., Wan Ismail, W.A.: Financially distressed firms environmental social and governance reporting in Indonesia. Sustainability **13**(10156), 1–25 (2021)
11. Cohen, J., Holder-Webb, L., Nath, L., Wood, D.: Retail investors' perceptions of the decision—usefulness of economic performance, governance and corporate social responsibility disclosures. Behav. Res. Account. **23**(1), 109–129 (2011)
12. Jadoon, I.A., Ali, A., Ayub, U., Tahir, M., Mumtaz, R.: The impact of sustainability reporting quality on the value relevance of corporate sustainability performance. Sustain. Dev. **29**(1), 155–175 (2021)
13. IIRC Institute: How Investors Integrate ESG: A Typology of Approaches (2017)
14. Ng, A.C., Rezaee, Z.: Business sustainability performance and cost of equity capital. J. Corp. Finan. **34**, 128–149 (2015)
15. Unruh, G., Kiron, D., Kruschwitz, N., Reeves, M., Rubel, H., Zum Felde, A.M.: Investing for a sustainable future. MIT Sloan Manage. Rev. **1**(57), 4 (2016)

16. Jain, P.K., Jain, A., Rezaee, Z.: Value-relevance of corporate social performance: evidence from short selling. J. Manage. Account. Res. **28**(2), 29–52 (2016)
17. Goyal, P., Rahman, Z.: Corporate sustainability performance and firm performance: literature review and future research agenda. Manage. Decis. **51**, 361–379 (2013)
18. Rahman, H.U., Zahid, M., Al-Faryan, M.A.S.: ESG and firm performance: the rarely explored moderation of sustainability strategy and top management commitment. J. Clean. Prod. **404**, 136859 (2023)
19. Arayakarnkul, P., Chatjuthamard, P., Treepongkaruna, S.: Board gender diversity, corporate social commitment and sustainability. Corp. Soc. Responsib. Environ. Manage. **29**(5), 1706–1721 (2022)
20. Elkington, J.: Cannibals with Forks: The Triple Bottom Line of the 21st Century. New Society Publishers, Stoney Creek, CT, USA (1998)
21. Jensen, M.C., Meckling, W.H.: Theory of the firm: managerial behavior, agency costs and ownership structure. J. Financ. Econ. **3**, 305–360 (1976)
22. Morris, R.D.: Signalling, agency theory and accounting policy choice. Account. Bus. Res. **18**(69), 47–56 (1987)
23. El Ghoul, S., Guedhami, O., Kwok, C., Mishra, D.: Does corporate social responsibility affect the cost of capital? J. Bank. Finan. **35**(9), 2388–2406 (2011)
24. Dyck, A., Lins, K., Roth, L., Wagner, H.: Do institutional investors drive corporate social responsibility? International evidence. J. Finan. Econ. **131**(3), 693–714 (2019)
25. Freeman, R.: Strategic management: a stakeholder theory. J. Manage. Stud. **39**, 1–24 (1984)
26. Donaldson, T., Preston, L.E.: Stakeholder theory: concepts, evidence, and implications. Acad. Manage. Rev. **20**(1), 65–91 (1995)
27. Jensen, M.C.: Value maximization, stakeholder theory, and the corporate objective function. J. Appl. Corp. Financ. **22**, 32–42 (2010)
28. Porter, M.E., Kramer, M.R.: Creating shared value. Harvard Bus. Rev. 62–77 (2011)
29. Kartal, M.T., Taşkın, D., Shahbaz, M., Depren, S.K., Pata, U.K.: Effects of environment, social, and governance (ESG) disclosures on ESG scores: investigating the role of corporate governance for publicly traded Turkish companies. J. Environ. Manage. **368**, 122205 (2024)
30. Guthrie, J., Parker, L.D.: Corporate social reporting: a rebuttal of legitimacy theory. Account. Bus. Res. **19**(76), 343–352 (1989)
31. Deegan, C.: Introduction: the legitimising effect of social and environmental disclosures–a theoretical foundation. Account. Auditing Account. J. **15**(3), 282–311 (2002)
32. Deegan, C., Rankin, M., Tobin, J.: An examination of the corporate social and environmental disclosures of BHP from 1983–1997. Account. Auditing Account. J. **15**(3), 312–343 (2002)
33. Gray, R., Owen, D., Adams, C.: Some theories for social accounting? A review essay and a tentative pedagogic categorisation of theorisations around social accounting. In: Freedman, M., Jaggi, B. (Eds.), Sustainability, Environmental Performance and Disclosures. Emerald, pp 1–54 (2009)
34. Mallin, C.A., Michelon, G., Raggi, D.: Monitoring intensity and stakeholders' orientation: how does governance affect social and environmental disclosure? J. Bus. Ethics **114**(1), 29–43 (2013)
35. Chan, M.C., Watson, J., Woodliff, D.: Corporate governance quality and CSR disclosures. J. Bus. Ethics **125**(1), 59–73 (2014)
36. Jizi, M.I., Salama, A., Dixon, R., Stratling, R.: Corporate governance and corporate social responsibility disclosure, evidence from the US banking sector. J. Bus. Ethics **125**(4), 601–615 (2014)
37. Peters, G.F., Romi, A.M.: Does the voluntary adoption of corporate governance mechanisms improve environmental risk disclosures? Evidence from greenhouse gas emission accounting. J. Bus. Ethics **125**(4), 637–666 (2014)
38. Clarkson, P.M., Fang, X., Li, Y., Richardson, G.D.: The relevance of environmental disclosures: are such disclosures incrementally informative? J. Account. Public Policy **32**(5), 410–431 (2013)

39. Devalle, A., Fiandrino, S., Cantino, V.: The linkage between ESG performance and credit ratings: a firm-level perspective analysis. Int. J. Bus. Manage. **12**(9), 53 (2017)
40. Sahlian, D.N., Popa, A.F., Banţa, V.C., Răpan, C.M., Chiriac, S.C.V.: The value relevance of financial performance and ESG scores during tumultuous economic periods. East. Eur. Econ. **62**(4), 528–546 (2023)
41. Schadewitz, H., Niskala, M.: Communication via responsibility reporting and its effect on firm value in Finland. Corp. Soc. Responsib. Environ. Manag. **17**(2), 96–106 (2010)
42. Lourenço, I.C., Branco, M.C., Curto, J.D., Eugénio, T.: How does the market value corporate sustainability performance? J. Bus. Ethics **108**(4), 417–428 (2012)
43. Al-Tuwaijri, S.A., Christensen, T.E., Hughes, K.: The relations among environmental disclosure, environmental performance, and economic performance: a simultaneous equations approach. Account. Organ. Soc. **29**(5–6), 447–471 (2004)
44. Brammer, S., Pavelin, S.: Factors influencing the quality of corporate environmental disclosure. Bus. Strateg. Environ. **17**(2), 120–136 (2008)
45. Hahn, R., Kuhnen, M.: Determinants of sustainability reporting: a review of results, trends, theory and opportunities in an expanding field of research. J. Clean. Prod. **59**, 5–21 (2013)
46. Healy, P.M., Palepu, K.G.: Information asymmetry, corporate disclosure, and the capital markets: a review of the empirical disclosure literature. J. Account. Econ. **31**(1–3), 405–440 (2001)
47. Lys, T., Naughton, J., Wang, C.: Signaling through corporate accountability reporting. J. Account. Econ. **60**(1), 56–72 (2015)
48. Dhaliwal, D.S., Radhakrishnan, S., Tsang, A., Yang, Y.G.: Non-financial disclosure and analyst forecast accuracy: international evidence on corporate social responsibility disclosure. Account. Rev. **87**(3), 723–759 (2012)
49. Dhaliwal, D.S., Li, O.Z., Tsang, A., Yang, Y.G.: Corporate social responsibility disclosure and the cost of equity capital: the roles of stakeholder orientation and financial transparency. J. Account. Publ. Policy **33**(4), 328–355 (2014)
50. Lyon, T.P., Maxwell, J.W.: Greenwash: corporate environmental disclosure under threat of audit. J. Econ. Manage. Strateg. **20**(1), 3–41 (2011)
51. Cuadrado-Ballesteros, B., Martínez-Ferrero, J., García-Sanchez, I.M.: Mitigating information asymmetry through sustainability assurance: the role of accountants and levels of assurance. Int. Bus. Rev. **26**(6), 1141–1156 (2017)
52. Clarkson, P., Li, Y., Richardson, G., Vasvari, F.: Revisiting the relation between environmental performance and environmental disclosure: an empirical analysis. Acc. Organ. Soc. **33**(4), 303–327 (2007)
53. Connelly, B.L., Certo, S.T., Ireland, R.D., Reutzel, C.R.: Signaling theory: a review and assessment. J. Manage. **37**(1), 39–67 (2011)
54. Mahoney, L.S., Thorne, L., Cecil, L., LaGore, W.: A research note on standalone corporate social responsibility reports: signaling or greenwashing? Crit. Perspect. Account. **24**(4), 350–359 (2013)
55. Papoutsi, A., Sodhi, M.S.: Does disclosure in sustainability reports indicate actual sustainability performance? J. Clean. Prod. **260**(4), 121049 (2020)
56. Surroca, J., Tribó, J.Á., Waddock, S.: Corporate responsibility and financial performance: the role of intangible resources. Strateg. Manage. J. **31**(5), 463–490 (2010)
57. Branco, M.C., Rodrigues, L.L.: Corporate social responsibility and resource-based perspectives. J. Bus. Ethics **69**(2), 1–15 (2006)
58. Lourenço, I.C., Branco, M.C.: Determinants of corporate sustainability performance in emerging markets: the Brazilian case. J. Clean. Prod. **57**(3), 134–141 (2013)
59. Velte, P.: The bidirectional relationship between ESG performance and earnings management-empirical evidence from Germany. J. Glob. Responsib. **10**(4), 322–338 (2019)
60. Yu, E.P., Van Luu, B., Chen, C.H.: Greenwashing in environmental, social and governance disclosures. Res. Int. Bus. Finan. **52**, 101192 (2020)
61. Gargouri, R.M., Shabou, R., Francoeur, C.: The relationship between corporate social performance and earnings management. Can. J. Adm. Sci. **27**(4), 320–334 (2010)

62. Garfatta, R.: Corporate social responsibility and earnings management: evidence from Saudi Arabia after mandatory IFRS adoption. J. Asian Finan. Econ. Bus. **8**(9), 189–199 (2021)
63. Calegari, M., Chotigeat, T., Harjoto, M.: Corporate social responsibility and earnings reporting. J. Curr. Res. Glob. Bus. **13**(20), 1–14 (2010)
64. Litt, B., Sharma, D., Sharma, V.: Environmental initiatives and earnings management. Manage. Audit. J. **29**(1), 76–106 (2014)
65. Chouaibi, Y., Zouari, G.: The effect of corporate social responsibility practices on real earnings management: evidence from a European ESG data. Int. J. Discl. Gov. **19**(4), 11–30 (2022)
66. Malik, M.: Value-enhancing capabilities of CSR: a brief review of contemporary literature. J. Bus. Ethics **127**(2), 419–438 (2015)
67. García-Sánchez, M., García-Meca, E.: CSR engagement and earnings quality in banks. The moderating role of institutional factors. Corporate Social Responsibility and Environmental Management 24(2), 145–158 (2017).
68. Bozzolan, S., Fabrizi, M., Mallin, C.A., Michelon, G.: Corporate social responsibility and earnings quality: international evidence. Int. J. Account. **50**(4), 361–396 (2015)
69. Rajesh, R., Rajendran, C.: Relating environmental, social, and governance scores and sustainability performances of firms: an empirical analysis. Bus. Strateg. Environ. **29**(3), 1247–1267 (2019)
70. Eng, L.L., Fikru, M., Vichitsarawong, T.: Comparing the informativeness of sustainability disclosures versus ESG disclosure ratings. Sustain. Account. Manage. Policy J. **13**(2), 494–518 (2022)
71. Freedman, M., Wasley, C.: The association between environmental performance and environmental disclosure in annual reports and 10Ks. Adv. Publ. Interest Acc. **3**, 183–193 (1990)
72. Brown, N., Deegan, C.: The public disclosure of environmental performance information—a dual test of media agenda setting theory and legitimacy theory. Account. Bus. Res. **29**(1), 21–40 (1998)
73. Clarkson, P., Li, Y., Richardson, G.D., Vasvari, F.P.: Revisiting the relation between environmental performance and environmental disclosure: an empirical analysis. Acc. Organ. Soc. **33**(4–5), 303–327 (2008)
74. Clarkson, P., Li, Y., Richardson, G.D., Vasvari, F.P.: Does it really pay to be green? Determinants and consequences of proactive environmental strategies. J. Account. Publ. Policy **30**(2), 122–144 (2011)
75. Baraibar-Diez, E., Odriozola, M.D., Fernández Sánchez, J.L.: Sustainable compensation policies and its effect on environmental, social, and governance scores. Corp. Soc. Responsib. Environ. Manage. **26**(6), 1457–1472 (2019)
76. Devers, C.E., Cannella, A.A., Jr., Reilly, G.P., Yoder, M.E.: Executive compensation: a multidisciplinary review of recent developments. J. Manage. **33**(6), 1016–1072 (2007)
77. Coughlan, A., Schmidt, R.: Executive compensation, management turnover, and firm performance. J. Account. Econ. **7**(1–3), 43–66 (1985)
78. Zajac, E., Westphal, J.: The costs and benefits of managerial incentives and monitoring in large U.S. corporations: when is more not better? Strateg. Manage. J. **15**, 121–142 (1994)
79. Craighead, J., Magnan, M., Thorne, L.: The impact of mandated disclosure on performance-based CEO compensation. Contemp. Account. Rev. **21**(2), 369–397 (2004)
80. Oh, W.Y., Hyun, S., Park, S.: Top management team pay dispersion and corporate social action: empirical evidence from U.S. banks. In: Manos, R., Drori, I. (Eds.) Corporate Responsibility, pp. 140–161. Palgrave Macmillan, London (2016)
81. Haque, F.: The effects of board characteristics and sustainable compensation policy on carbon performance of UK firms. Br. Account. Rev. **49**, 347–364 (2017)
82. Mahoney, L., Thorne, L.: An examination of the structure of executive compensation and corporate social responsibility: a Canadian investigation. J. Bus. Ethics **69**(2), 149–162 (2006)
83. Ji, Y.: Top management team pay structure and corporate social performance. J. Gen. Manage. **40**(3), 3–20 (2015)

84. Dando, N., Swift, T.: Transparency and assurance minding the credibility gap. J. Bus. Ethics **44**(2), 195–200 (2003)
85. Williams, L.S.: The mission statement: a corporate reporting tool with a past, present, and future. J. Bus. Commun. **45**(2), 94–119 (2008)
86. Bewley, K., Li, Y.: Disclosure of environmental information by Canadian manufacturing companies: a voluntary disclosure perspective. Adv. Environ. Account. Manage. **1**(1), 201–226 (2000)
87. Kassinis, G., Panayiotou, A.: Visuality as greenwashing: the case of BP and deep water horizon. Organ. Environ. **31**(1), 25–47 (2018)
88. van Duuren, E., Plantinga, A., Scholtens, B.: ESG integration and the investment management process: fundamental investing reinvented. J. Bus. Ethics **138**(3), 525–533 (2016)
89. Kartal, M.T., Kılıç Depren, S., Pata, U.K., et al.: Modeling the link between environmental, social, and governance disclosures and scores: the case of publicly traded companies in the Borsa Istanbul sustainability index. Finan. Innov. **10**, 80 (2024)
90. Duan, K., Qin, C., Ma, S., Lei, X., Hu, Q., Ying, J.: Impact of ESG disclosure on corporate sustainability. Financ. Res. Lett. **78**, 107134 (2025)
91. Herbohn, K., Walker, J., Loo, H.Y.M.: Sustainability disclosure and performance. Abacus **50**(4), 422–459 (2014)
92. Muller, A., Kolk, A.: Extrinsic and intrinsic drivers of corporate social performance. Evidence from foreign and domestic firms in Mexico. J. Manage. Stud. **47**(1), 1–26 (2010)
93. Daub, C.L.: Assessing the quality of sustainability reporting: an alternative methodological approach. J. Clean. Prod. **15**(1), 75–85 (2007)
94. Datta, P.R., Gopalakrishna-Remani, V.E., Bozan, K.A.: The impact of sustainable governance and practices on business performance: an empirical investigation of global firms. Int. J. Sustain. Soc. **7**(2), 97–120 (2015)
95. Berg, F., Fabisik, K., Sautner, Z.: Is history repeating itself? The (Un)Predictable past of ESG ratings. European Corporate Governance Institute—Finance Working Paper 708/2020 (2021)
96. World Economic Forum: Leaders need hard data to make the hard decisions about sustainability, World Economic Forum Annual Meeting (2019)
97. Boffo, R., Patalano. R.: ESG Investing: practices, progress and challenges. OECD, Paris (2020)
98. Demers, E., Hendrikse, J., Joos, P., Lev, B.: ESG did not immunize stocks during the COVID-19 crisis, but investments in intangible assets did. J. Bus. Financ. Acc. **48**, 433–462 (2021)
99. Mooneeapen, O., Abhayawansa, S., Khan, N.M.: The influence of the country governance environment on corporate environmental, social and governance (ESG) performance. Sustain. Account. Manage. Policy J. **13**(1), 953–985 (2022)
100. de Sola Pool, I.: Tracking the flow of information. Science **221**(4611), 609–613 (1983)
101. Roca, L.C., Searcy, C.: An analysis of indicators disclosed in corporate sustainability reports. J. Clean. Prod. **20**(1), 103–118 (2012)
102. Patten, D.M., Zhao, N.: Standalone CSR reporting by US retail companies. Account. Forum. **38**(2), 132–144 (2014)
103. Darnall, N., Ji, H., Iwata, K., Arimura, T.H.: Do ESG reporting guidelines and verifications enhance firms' information disclosure? Corp. Soc. Responsib. Environ. Manage. **29**(5), 1214–1230 (2022)
104. Asif, M., Searcy, C., Zutshi, A., Ahmad, N.: An integrated management systems approach to corporate sustainability. Eur. Bus. Rev. **23**(4), 353–367 (2011)
105. Rahman, H.U., Zahid, M., Khan, M.: Corporate sustainability practices: a new perspective of linking board with firm performance. Total Qual. Manage. Bus. Excell. **33**(7–8), 929–946 (2021)
106. Jahanshahi, A.A., Brem, A.: Sustainability in SMEs: top management teams behavioral integration as a source of innovativeness. Sustainability **9**(10), 1–16 (2017)
107. Cohen, S., Kadach, I., Ormazabal, G., Reichelstein, S.: Executive compensation tied to ESG performance: international evidence. J. Account. Res. **61**, 805–853 (2023)

108. Otman, K.A.M.: Corporate governance and firm performance in listed companies in the United Arab Emirates. Doctoral Dissertation, Victoria University (2014)
109. Reber, B., Gold, A., Gold, S.: ESG disclosure and idiosyncratic risk in initial public offerings. J. Bus. Ethics **179**(3), 867–886 (2022)
110. Lassala, C., Apetrei, A., Sapena, J.: Sustainability matters and financial performance of companies. Sustainability **9**(9), 1498 (2017)
111. Friede, G., Busch, T., Bassen, A.: ESG and financial performance: aggregated evidence from more than 2000 empirical studies. J. Sustain. Finan. Invest. **5**(4), 210–233 (2015)
112. Roberts, R.W.: Determinants of corporate social responsibility disclosure: an application of stakeholder theory. Acc. Organ. Soc. **17**(6), 595–612 (1992)
113. Artiach, T., Lee, D., Nelson, D., Walker, J.: The determinants of corporate sustainability performance. Account. Finan. **50**(1), 31–51 (2010)
114. Cormier, D., Magnan, M., Van Velthoven, B.: Environmental disclosure quality in large German companies: economic incentives, public pressures or institutional conditions? Eur. Account. Rev. **14**(1), 3–39 (2005)
115. Berman, S.L., Wicks, A.C., Kotha, S., Jones, T.M.: Does stakeholder orientation matter? The relationship between stakeholder management models and firm financial performance. Acad. Manage. J. **42**(5), 488–506 (1999)
116. Callan, S.J., Thomas, J.M.: Corporate financial performance and corporate social performance: an update and reinvestigation. Corp. Soc. Responsib. Environ. Manage. **16**, 61–78 (2009)
117. Brammer, S., Millington, A.: Does it pay to be different? An analysis of the relationship between corporate social and financial performance. Strateg. Manage. J. **29**(12), 1325–1343 (2008)
118. Wahba, H.: Exploring the moderating effect of financial performance on the relationship between corporate environmental responsibility and institutional investors: some Egyptian evidence. Corp. Soc. Responsib. Environ. Manage. **15**, 361–371 (2008)
119. Kim, Y., Park, M.S., Wier, B.: Is earnings quality associated with corporate social responsibility? Account. Rev. **87**(3), 761–796 (2012)
120. Bansal, P., Roth, K.: Why companies go green: a model of ecological responsiveness. Acad. Manage. J. **43**(4), 717–736 (2000)
121. Sharma, P., Panday, P., Dangwal, R.C.: Determinants of environmental, social and corporate governance (ESG) disclosure: a study of Indian companies. Int. J. Discl. Gov. **17**(4), 208–217 (2020)
122. Branco, M.C., Rodrigues, L.L.: Corporate social responsibility and resource-based perspectives. J. Bus. Ethics **69**(2), 111–132 (2006)
123. Gao, S.S., Heravi, S., Xiao, J.Z.: Determinants of corporate social and environmental reporting in Hong Kong: a research note. Account. Forum. **29**(2), 233–242 (2005)
124. Godfrey, P.C., Merrill, C.G., Hansen, J.M.: The relationship between corporate social responsibility and shareholder value: an empirical test of the risk management hypothesis. Strateg. Manage. J. **30**(4), 425–445 (2009)
125. Chih, H.L., Chih, H.H., Chen, T.Y.: On the determinants of corporate social responsibility. International evidence on the financial industry. J. Bus. Ethics **93**, 115–135 (2010)
126. Ullmann, A.A.: Data in search of a theory: a critical examination of the relationships among social performance, social disclosure, and economic performance of U.S. firms. Acad. Manage. Rev. **10**(3), 540–557 (1985)
127. McWilliams, A., Siegel, D.: Corporate social responsibility and financial performance: correlation or misspecification? Strateg. Manage. J. **21**(5), 603–609 (2000)
128. Allison, P.: Multiple Regression: A Primer. Pine Forge Press, Thousand Oaks (1999)

Enhancing Circular Economy in Intermediate Labor Markets—A Case Study in Finland

Sanna-Maaria Siintoharju, Jari Jussila, and Anne-Mari Järvenpää

Abstract Intermediate labor markets (ILMs) have emerged as a significant mechanism in the transition towards circular economy (CE), facilitating the integration of marginalized groups into the workforce while promoting sustainable practices. ILMs serve as a bridge between traditional labor markets and those seeking to engage in circular economy activities. These markets often focus on providing employment opportunities for disadvantaged groups, including the long-term unemployed and those with intellectual or physical disabilities. A case study in Finland examined how municipalities, CE organizations, Work Integration Social Enterprises (WISEs), and educational institutions can collaborate to promote ILMs and social entrepreneurship. The study used a workshop format to explore practical strategies, guided by the question: *"What can be done in practice to promote the intermediate labor market and social entrepreneurship in the circular economy?"* Several innovative practices emerged. Instead of relying on national salary subsidies, companies could directly hire through ILMs. Training programs and apprenticeships focused on end-of-life textile sorting and reuse were highlighted, along with ideas for repurposing marketing materials and using surplus materials more efficiently. In conclusion, ILMs can support the transition to a circular economy by providing pathways for marginalized groups into sustainable employment. They facilitate the development of skills necessary for circular economic practices while also addressing social inequalities. As such, a comprehensive approach that integrates ILMs into the broader framework of the circular economy is needed for achieving sustainable development goals.

Keywords Intermediate labor markets · Social entrepreneurship · Circular economy · SDG8

S.-M. Siintoharju (✉) · J. Jussila · A.-M. Järvenpää
Häme University of Applied Sciences (HAMK), Hämeenlinna, Finland
e-mail: sanna-maaria.siintoharju@hamk.fi

A. Visvizi et al. (eds.), *Research and Innovation Forum 2025 Vol2*, Springer Proceedings in Complexity, https://doi.org/10.1007/978-3-032-09062-1_13

1 Introduction

A circular economy is one major objective of today's societies, [1] and it offers numerous business opportunities for the private sector. The private sectors (companies and start-ups) key role in supporting transition towards circular economy has been acknowledged [2]. However, government and education system collaboration with intermediate labor markets and social enterprises in circular economy transformation has received less attention from the literature [3].

Recycling and other utilization of waste have an essential role in achieving a circular economy and the waste hierarchy is the guiding principle of waste regulation [4]. For instance, the general waste legislation in Finland (Waste Act 646/2011) has mandated municipalities to collect consumer textile waste and forward it for further use. The aim is to reduce the environmental impact of textiles by collecting not only textiles that can still be reused but also those that are no longer usable as such but can still be utilized as material. Textile waste presents a significant challenge, particularly in its management, as it requires extensive manual labor. One solution that municipalities are now implementing is intermediate labor markets.

Intermediate labor markets (ILMs) exist between the fully employed and unemployed employment categories, and in some countries are used to distinguish between the regular and non-regular employment [5]. In Japan, manufacturing companies during the global financial crisis reduced many non-regular employees, such as part-time workers and temporary workers to adjust for the economic recession, which caused significant social problems [5], but also sparked public discussion on the phenomena. Classifying employees into regular and non-regular was found insufficient [6], and the category of "limited regular employee" [7] was established to describe employees who possess constraints and more limited comprehensiveness compared to regular employees [5]. In Finland, ILMs are considered as services designed to enhance job opportunities for individuals facing challenges in securing stable employment [8], which could be in analogous manner labeled "intermediate employees" and "limited intermediate employees".

One potential employment and business opportunity is social entrepreneurship. Social entrepreneurship emerged as a special type of venture that emphasizes the creation of social value over the maximization of profits [9]. The Social Enterprises Act, introduced in Finland January 2004, defines a social enterprise as one that employs individuals with reduced work capacity or a mix of such individuals and long-term unemployed persons, making up at least 30% of its workforce [10]. According to the act, individuals with reduced work capacity are jobseekers whose ability to secure, maintain, or advance in employment is significantly diminished due to a recognized disability, illness, or impairment.

2 Methods and Materials

A case study was selected as the research methodology. To obtain a broader understanding of the current state of ILMs, social entrepreneurship, and best practices in the end-use of textile waste in Finland, phone interviews were carried out. Notes were taken during the calls, which included one interview with a representative from Northern Finland, two from Central Finland, one from Western Finland, and five from the Kanta-Häme region. Simultaneously, a literature review was conducted to define key concepts and to summarize existing research on the subject.

SCOPUS database was investigated with the search strings: TITLE-ABS-KEY ("circular economy") and TITLE-ABS-KEY ("intermediate labor markets"), and TITLE-ABS-KEY ("circular economy") and ("social entrepreneurship"). The first search string yielded 0 search results and the second 25 search results. In total out of the 25 discovered articles, five articles addressed the textile industry that was in the scope of this study. Since no articles were found that covered intermediate labor markets and circular economy, reports were used as additional information sources.

Intermediate labor markets (ILMs) aim to improve employment prospects for those struggling to secure stable jobs [8]. Introduced in Finland's 2006 Ministry of Labor report as "intermediate employees" [11], ILMs provide transitional pathways to competitive employment through labor and social policy measures [12]. They offer structured support—such as internships, coaching, and rehabilitative programs—targeting long-term unemployed individuals or those with limited access to traditional jobs [12, 13]. The *Paths to Employment* report highlights ILMs' role in aiding individuals with diminished social capabilities due to prolonged unemployment, similar to Japan's "limited regular employees" [7, 11]. Secondary barriers may include minor mental health issues or substance misuse [14].

Entrepreneurship drives economic development, and social entrepreneurship (SE) is increasingly valued for its social and economic benefits, though no universal definition exists [15, 16]. In Finland, SE focuses on employing individuals with reduced work capacity and the long-term unemployed; the 2004 Social Enterprises Act requires such groups to make up at least 30% of staff [17]. In Latvia, social enterprises are limited companies that create social impact in areas like education, environmental protection, and social services; four of 176 enterprises operate in the textile sector [18]. In Romania, Staicu and Pop found limited collaboration, lack of shared spaces, and low awareness of circular economy practices in the textile industry [19].

2.1 Case Study

An explorative case study was carried out to generate innovative ideas for enhancing circular economy, with a particular focus on textile waste and social entrepreneurship opportunities. The key question guiding our research was: "*What practical steps can be taken to support the intermediate labor market and social entrepreneurship in the circular economy?"* To address this research question, this paper is structured as follows: in methods and materials, we define the key concepts based on literature review and present the case study and data collection. In results we outline major themes and ideas for supporting ILM and SE in the circular economy. In discussion and conclusions, we draw together the study's findings and propose future research direction.

A case study was designed and carried out to examine how municipalities, circular economy organizations, WISEs, and educational institutions can collaborate to promote innovative circular economy practices (Table 1). A workshop was used as a data collection method [20], which involved seven participants from relevant organizations. Ideas were gathered through sticky notes and group discussions.

The goal of the workshop was to ideate concrete ways in which ILMs, and SE could be utilized in the future to support municipalities' green transition. The ideation included an individual phase of writing the ideas on sticky notes and then a discussion and further ideation based on the joint discussion. The sticky notes were organized into themes by the first author, and the entire workshop was recorded and transcribed. The second author examined the categorizations, and the ideas derived from the transcripts and formed consensus on the findings with the first author. All three authors were observing the discussions during the workshop.

Table 1 Case organizations

Participating organization	Core business area	Role of participant
Work center	Non-profit, intermediate labor market provider	Responsible work coach
Recycled textiles workshop	Non-profit, social entrepreneurship	Workshop leader
Municipality	Municipality employment and integration services	Employment director
Employment care association	Non-profit that helps matching workers and employers	Development manager
Educational federation of municipalities	Vocational education and private sector complementary training	Manager of education
Waste management association	Circular economy, waste management and materials recycling	Communications manager
University of applied sciences	Design department	Lecturer

Source Authors

3 Results

The first identified central theme from the workshop included the need for new circular economy business models. New business models from abroad could also be adopted, for example, from Austria, where there is electronics recycling store in Vienna. We can also look at responsibly operating companies, like Patagonia, as examples. From different business models, it should be noted that the secondary labor market never competes with local businesses and should not be feared by businesses [9]. Additionally, it would be good to identify the needs for recycling in society, but how does the waste hierarchy relate to this? News campaigns could help inform the broader public on this point. We already have good examples of new business models in Finland. One social enterprise is offering jobs for reduced work capacity persons by making new products out of used leaflets or vinyl fabrics.

In the context of emerging circular economy business models in Finland, the phone interviews revealed a notably distinct operational approach. Rather than relying on government-provided wage subsidies, this particular social enterprise directly employed individuals ILMs. Furthermore, the enterprise had advanced its business model by entering into a formal agreement with a local vocational institution. Through this collaboration, employees received on-the-job training as part of their employment. The training corresponded to a specific module within a broader vocational qualification framework.

Second central theme was reusing and recycling end-of-life textiles. In Finland, sorting discarded textiles is expensive for the municipalities. When talking about clothing, people should be informed and guided to buy products that can be recycled. Clothing repair is also a good practice, whereas ordering new items from China is a growing trend in consumerism that challenges the repair, reuse and recycling sectors [21]. There is a clear need to better inform households about proper recycling practices. A major challenge is consumers placing non-recyclable, moldy, or contaminated items in recycling bins. Even organic waste recycling proves difficult. In some cases, recycling may do more harm than good if emissions from the process outweigh its benefits. For example, in Lapland, Finland, recycling has been reduced for this reason.

Third, several ideas were centered around what should be the main message to the public. Importantly, public communication aimed at households should inform people that materials should be used as they are (for example, paper should be sent to paper recycling, not made into compostable bags) and, if that's not possible, then the material should be recycled and burned for energy. Recycling is essentially simple and easy, but people still need to be educated. Environmental education for children should be prioritized, and their training should be enhanced. Often, it's children who ask their parents, "Why don't you recycle?" Awareness of recycling among students should also be increased, which can be done by involving students more in local circular economy projects and targeting thesis topics on the subject. In addition, education should already include the idea that when any product is being designed, its second life should be considered from the beginning. This could include instructions along with the product. Additionally, student projects should involve visits to local recycling companies, as on-site visits make things more tangible.

Fourthly, in addition to the previously discussed topics, the root causes of textile waste and the role of education were addressed. As part of this fourth theme, the discussion also raised questions about how product life cycle considerations—regardless of the material—could be integrated into design education across different disciplines and educational levels. The central question was how this perspective could be incorporated into curricula at various stages of education. The clothing and textile industry was naturally used as a practical example. It was considered how design and fashion education could include training on what a product might become after its initial use. One proposed idea was that garments could be labeled in advance with information about how they can be recycled or what the designer has envisioned for the product's "next life."

4 Discussion and Conclusions

Making use of ILMs in circular economy is novel research area, and only a few empirical studies exist on SE in circular economy [18, 19, 22]. Like the other few empirical studies on SE in circular economy this study has limitations. Like [19] the findings are subjective in nature and reflect the views of few individuals. More qualitative, quantitative, and mixed-method studies are needed to gain better understanding of ILMs and SE in circular economy. To the best of the author's knowledge, this is the first study investigate ILMs and circular economy.

Regarding SE in circular economy this study complements previous study on social enterprises in clothing and textile production [18]. In addition, to custom made swimsuits out of old fishing nets, upcycled clothing, water colored, and graphics printed clothing, and posture-correcting t-shirt [18], this study introduced vinyl fabrics as material that SE has made successful business in upcycling to shopping bags and other accessories. Furthermore, it was uncovered that in addition to providing circular economy education in higher education [22, 23], it is important to educate wider public and especially children in circular economy due to positive spillover effects.

One novel idea discovered during research was integrating education to solve the employment problem in the ILMs. Instead of rehabilitating the ILMs in Finland, we could hire a jobseeker [cf. 10] to circular economy businesses and train them to become circular economy experts at the same time. In Finland curriculums are regularly inspected and this allows flexibility. SE in the circular economy could also serve as a short-term option, such as summer jobs. Furthermore, in higher education, especially in design programs, it would be beneficial to enhance students' ability to consider the "next life" of products already in the design phase [22].

Acknowledgements This research was supported by the European Regional Development Project Participatory Design of Source Separation in Recycling—KOTILO (A80574).

References

1. Aceleanu, M.I., Şerban, A.C., Suciu, M.C., Biţoiu, T.I.: The management of municipal waste through circular economy in the context of smart cities development. IEEE Access. **7**, 133602–133614 (2019)
2. Baraniewicz-Kotasińska, S., Visvizi, A., Guazzo, G.M.: Conceptualizing business sector's involvement in the smart city: Towards a typology of engagement modes. In: Visvizi, A., Troisi, O., Grimaldi, M. (eds.) ICT, The Business Sector and Smart Cities, pp. 33–50. Routledge, London (2025)
3. Troisi, O., Visvizi, A., Grimaldi, M.: Rethinking innovation through industry and society 5.0 paradigms: a multileveled approach for management and policy-making. Eur. J. Innov. Manage. **27**, 22–51 (2024). https://doi.org/10.1108/EJIM-08-2023-0659
4. Ympäristöministeriö: Kierrätyksestä kiertotalouteen : Valtakunnallinen jätesuunnitelma vuoteen 2027. Ympäristöministeriö, Helsinki (2022)
5. Hirano, M.: Limited regular employees and boundary of employment: an analysis by the three-layered labor market model. In: Japanese Management in Change: The Impact of Globalization and Market Principles, pp. 123–139. Springer, Tokyo (2015)
6. Sato, H.: Diversifying employment categories and balancing wages across employment categories. Organ. Sci. **41**, 22–32 (2008)
7. Sato, H.: Limited regular employees and unlimited non-regular employees—A new issue of personnel management. In: Collected Research Papers of the No. 42 National Convention of Japan Society of Human Resource Management, pp. 201–208 (2012)
8. Mäkipää, K.: Työvoiman aktivointipolitiikka ja välityömarkkinat Euroopassa (2010)
9. López-Cózar-Navarro, C., Priede-Bergamini, T.: Social entrepreneurship and related concepts: the path of opportunity to foster new ventures. In: Handbook of Research on Nascent Entrepreneurship and Creating New Ventures, pp. 273–294. IGI Global Scientific Publishing (2021)
10. Kauko-Valli, S., Niittykangas, H.: Sosiaalinen yrittäjyys: näkökulmia ja tuloksia. University of Jyväskylä (2009)
11. Tupo 2-työryhmä: Väyliä työhön. Työministeriö, Helsinki (2006)
12. Alatalo, J., Nio, I., Tuomaala, M.: Työ-ja elinkeinoministeriön lyhyen aikavälin työmarkkinaennuste. Ministry of Employment and the Economy. Helsinki (2011)
13. Vintola, M.: Lisäpalveluilla palkkatukityöstä ja työvoimakoulutuksesta vaikuttavampaa Tampereen seudun välityömarkkinoilla (2011)
14. Jolkkonen, A., Roivas, S.: Välityömarkkinoiden laajuus ja kehittämistarve Pohjois-Karjalassa. University of Joensuu (2008)
15. Abu-Saifan, S.: Social entrepreneurship: definition and boundaries. Technol. Innov. Manage. Rev. **2**, 22–27 (2012). https://doi.org/10.22215/timreview/523
16. Rahim, H.L., Mohtar, S.: Social entrepreneurship: a different perspective. Int. Acad. Res. J. Bus. Technol. **1**, 9–15 (2015)
17. Hanikka, K., Korpela, P., Mähönen, A., Nyman, C.: Yrityksen yhteiskuntavastuu ja sosiaalinen yritystoiminta. Kestävää työllisyyttä sosiaalisissa yrityksissä–projekti. Savion Kirjapaino, Kerava (2007)
18. Atstja, D., Cudečka-Puriņa, N., Vesere, R., Abele, L., Spivakovskyy, S.: Challenges of textile industry in the framework of Circular economy: case from Latvia. In: International Conference on Sustainable, Circular Management and Environmental Engineering (ISCMEE 2021. EDP Sciences (2021)
19. Staicu, D., Pop, O.: Mapping the interactions between the stakeholders of the circular economy ecosystem applied to the textile and apparel sector in Romania. Manage. Mark. **13**, 1190–1209 (2018)
20. Thoring, K., Mueller, R., Badke-Schaub, P.: Workshops as a research method: guidelines for designing and evaluating artifacts through workshops. In: Proceedings of the 53rd Hawaii International Conference on System Sciences, pp. 5036–5045. Maui (2020)

21. Llorente-González, L.J., Vence, X.: How labour-intensive is the circular economy? A policy-orientated structural analysis of the repair, reuse and recycling activities in the European Union. Resour. Conserv. Recycl. **162**, 105033 (2020)
22. Baeza, C., Quinn, E.: Transforming the fashion industry by: the evolution of design & merchandising education. In: ICERI2021 Proceedings, pp. 7643–7649. IATED (2021)
23. Mishra, O.: Twirl Store: pioneering sustainable textile practices by closing the loop. Emerald Emerg. Mark. Case Stud. **14**, 1–29 (2024)

Governance Structures in Digital Entrepreneurial Ecosystems: The Impact on Strategic Choices

Alberto Bongiorno, Katarzyna Szopik-Depczyńska, Carlotta D'Alessandro, and Maria Cristina Cinici

Abstract The digital paradigm leverages the power of collaboration and collective intelligence to drive the development of innovative entrepreneurial ventures. Within this context, the concept of Digital Entrepreneurial Ecosystems (DEEs) has emerged as a robust framework, integrating digital entrepreneurship into traditional entrepreneurial ecosystems to create dynamic and supportive environments. Governance structures play a crucial role in these ecosystems by fostering interaction, collaboration, and innovation, ultimately enhancing ecosystem performance. By establishing rules and guiding decision-making processes, they aim to ensure effective coordination among actors while promoting trust and transparency to leverage the advantages of digital transformation. Despite the growing significance of DEEs, there remains a critical gap in understanding how different governance structures influence DEE outcomes. This study addresses this gap through a Qualitative Comparative Analysis (QCA). QCA merges qualitative and quantitative data to enable the analytical evaluation of distinct cases, yielding comparative outcomes. The research focuses on Italian DEEs, characterized by diverse governance structures. In particular, it investigates the interaction between governance attributes, regulatory mechanisms, and ecosystem outcomes, uncovering how governance structures impact strategic decision-making and strategy development. The findings offer an overview of the critical elements that affect ecosystem efficiency and resilience.

A. Bongiorno (✉) · C. D'Alessandro · M. C. Cinici
Department of Economics, University of Messina, Messina, Italy
e-mail: alberto.bongiorno@unime.it

C. D'Alessandro
e-mail: carlotta.dalessandro@unime.it

M. C. Cinici
e-mail: mariacristina.cinici@unime.it

K. Szopik-Depczyńska
Department of Business Management, Institute of Management, University of Szczecin, Szczecin, Poland
e-mail: katarzyna.szopik-depczynska@usz.edu.pl

A. Visvizi et al. (eds.), *Research and Innovation Forum 2025 Vol2*, Springer Proceedings in Complexity, https://doi.org/10.1007/978-3-032-09062-1_14

Keywords Digital entrepreneurial ecosystem · Entrepreneurship · Qualitative comparative analysis · Governance · SDG 9

1 Introduction

The concept of Entrepreneurial Ecosystems (EEs) has attracted growing attention in research, policy, and professional circles over the past decade [1, 2]. This interest is driven by their capacity to foster entrepreneurial success and drive economic growth by creating jobs, increasing the gross domestic product, and improving long-term productivity [3]. Furthermore, technological advancements have revolutionized collaboration, resource management, product development, and demand–supply coordination while facilitating the establishment of new standards and solutions [4]. Consequently, established enterprises and startups leverage digital innovations and global connectivity through standardized protocols, such as the internet and mobile web, to upgrade their infrastructure and adapt to the evolving digital landscape [5]. To understand contemporary EEs, it is first necessary to examine how digitalization shapes value creation, delivery, and capture across economic and societal dimensions [6]. Building upon this, the concept of DEEs has emerged, integrating EEs with digital infrastructure, resources, and user engagement [7]. The interpretation and analysis of ecosystems as artificial systems leads to several governance problems. First, delineating the boundaries of an EE becomes a key concern. This, in turn, raises questions about the ecosystem's governance structure: How can resources be allocated efficiently within these boundaries? Which actors should be responsible for managing this allocation and distribution? Given that corporate governance research centers on addressing such issues, its insights are highly relevant to understanding the governance of ecosystems [8]. Indeed, while research is recognizing the significance of DEEs [9, 10], an unresolved question remains: *How do governance structures influence the strategic decision-making and overall effectiveness of DEEs?* Existing studies have explored the impact of digitalization on entrepreneurship and ecosystem interconnectivity, but they have not produced a cohesive understanding of governance models within DEEs [11, 12]. Furthermore, applying the metaphor of an "ecosystem" remains incomplete without delineating its boundaries and the mechanisms that govern it. To fill this gap, this study employs a QCA on Italian DEEs to compare and examine how different governance structures influence outcomes and strategic choices. The findings highlight how different organizational structures shape the dynamics of DEEs, influence decision-making processes, and affect ecosystem effectiveness. This paper is structured as follows. The next section highlights the theoretical background on DEEs. Following, the methodological section explains the steps undertaken in applying QCA to ensure an analytical approach. The subsequent sections present and interpret our findings, examining their implications. Finally, the conclusion synthesizes the insights and suggests future research directions to advance understanding DEEs.

2 Theoretical Background

The ecosystem metaphor for entrepreneurship was introduced by James Moore [13], which suggests the existence of a socioeconomic system where equilibrium or near-equilibrium emerges as individuals pursue their interests or fulfill their needs, with minimal external control over specific processes [14]. This beneficial interdependence encompasses collaborative and competitive interactions among autonomous, dispersed, and integrated participants, each advancing their objectives within the ecosystem, collectively adding to its complexity [15]. Their work contributes to creating the basis for understanding ecosystem dynamics, but does not offer a standardized methodology to evaluate such complexity across different contexts. A recent contribution [16] discusses the transformative influence of digital technologies on entrepreneurship, arguing that these technologies redefine global market structures. While such work recognizes broad patterns, it often lacks granularity in addressing sectoral or regional differences and tends to treat digitalization as a homogeneous driver of change.

Moreover, digital entrepreneurship yields individual and societal benefits, such as increased convenience and improved work models [17]. Such improvements have been connected with the prominent idea of Digital Ecosystems (DE), a decentralized, modular systems that move away from traditional hierarchical value chains [18]. These systems are described as composed of digitally linked modular organizations [19]. While this description is compelling, it assumes a level of interoperability and trust among actors that may not reflect real-world frictions. At the heart of these transformations lies Digital Entrepreneurship (DES), defined as the creation and reconfiguration of businesses through digital innovation [20]. While foundational, this work remains relatively descriptive and could benefit from more attention to structural constraints or power asymmetries in digital platforms. To bridge existing gaps, the DEE framework has been proposed, integrating DEs and EEs into a unified model [7]. This framework represents a significant theoretical advancement by linking micro-level entrepreneurial activity with macro-level digital infrastructure and policy environments. Nonetheless, its practical applicability remains underexplored, especially regarding how different institutional settings mediate the framework's components. Furthermore, recent studies have extended this discussion to include the role of users and platforms within the DEE, yet the analysis remains limited in addressing governance dynamics or the evolving role of public institutions [21] (Fig. 1).

Fig. 1 The DEE's puzzle. *Source* Authors

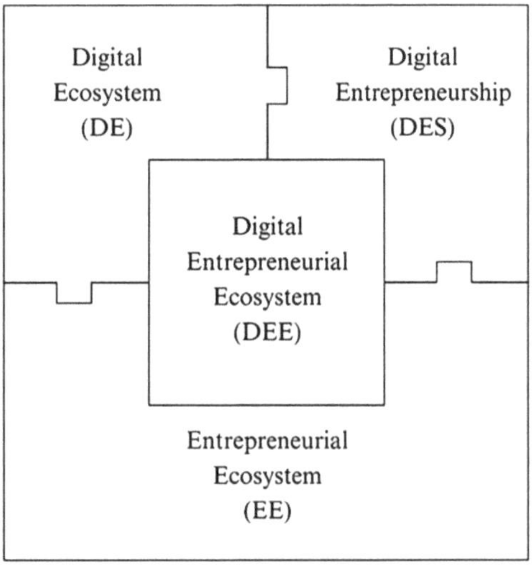

3 Methodology

3.1 Research Design

This study applies a Qualitative Comparative Analysis (QCA) to explore causal relationships between variables referred to as "conditions" within this methodology. Developed by Charles Ragin [22] for empirical analyses with few cases, this method takes advantage of the strengths of both qualitative and quantitative research methods while reducing some of their natural weaknesses. Unlike qualitative methods, which focus on detailed analysis of a few cases, QCA allows researchers to compare multiple cases in studies with a medium-sized sample [23]. In the entrepreneurial context, it has seen several applications. For instance, it has been employed to investigate how institutional factors influence entrepreneurship development [24], to reveal how various combinations of explanatory conditions explain both the presence and the absence of entrepreneurial intention [25] and mixed with the Necessary Condition Analysis to identify and examine the specific combination of elements within an EEs that foster high levels of entrepreneurial activity in both urban and non-urban settings [26].

Drawing on this branch of literature that leverages QCA to understand complex causal dynamics in entrepreneurship, this study adopts a similar methodological approach. In particular, a mvQCA is a variation of the traditional one, developed to manage intricate datasets beyond simple binary classification. Instead of limiting conditions to just two options (presence or absence), it accommodates multiple values for each condition [27].

3.2 Data Collection and Dataset Creation

The data for this study were collected from the European Digital Innovation Hubs (EDIHs) project, the Italian Ministry of Enterprises, and the Made in Italy website [28]. This enabled the identification of characteristics of the DEEs, such as their composition, governance structures, and strategic objectives. Then, the research focused on quantitative and qualitative data, allowing for a systematic evaluation of patterns, trends, and influential factors in the digital transition of enterprises within DEEs. The collected data were then organized into Excel worksheets. Once all the data were collected, the study was conducted in three key steps:

Step 1: Identifying measures of the conditions and the outcome

Outcome measures and causal conditions correspond to our study's dependent and independent variables. In this context, Ragin [29] advises researchers to operationalize theoretical concepts related to conditions and outcomes by complying with standard scientific criteria while drawing on their understanding of the empirical context under investigation. Different combinations and configurations that define the governance structure of DEEs and analyze their impact on the decision-making process have been assessed to fulfill the research aim. Moreover, given that the DEE's composition is strongly related to strategic decisions and overall performance, we set them as causal conditions. Utilizing the framework introduced by Osterwalder and Pigneur [30], we explain each building block below:

SME Support: refers to the mechanisms and initiatives facilitating SMEs' growth, sustainability, and scalability. This includes financial aid, mentorship programs, regulatory guidance, access to digital infrastructure, and networking opportunities.

Knowledge Transfer: encompasses how expertise, skills, and innovations are disseminated within and across DEEs. This involves collaboration between academic institutions, research centers, and businesses to facilitate technological adoption and best practices.

Innovation Management: Concern to the structured processes and governance mechanisms that support the development, implementation, and scaling of new ideas and technologies. It includes R&D activities, intellectual property management, funding opportunities, and regulatory frameworks that ensure a sustainable and competitive innovation ecosystem.

Environment: encompasses the external factors and ecosystem conditions that influence entrepreneurial success and digital transformation. This includes policy frameworks, market dynamics, technological infrastructure, and socio-economic conditions.

Step 2: Calibration

The mvQCA analysis necessitates calibrating measures and converting them into set membership scores, which range from 0.0 (indicating exclusion from the set) to 1.0 (signifying full inclusion) for fuzzy values [31] and either 0 or 1 for crisp values [23]. In particular, this step involved several phases. First, the DEE evaluations have been normalized to ensure that the data were prepared for analysis using the fsQCA (2.5) software. Then, data were calibrated for our cases using five-value fuzzy (0, 0.2, 0, 6, 0.8, and 1) and crisp set (0,1). Therefore, we incorporated three intermediate values to reflect partial membership, ensuring a comprehensive representation of the cases. In this regard, a membership score of 1 was assigned to instances characterized by strong governance frameworks, structured collaboration mechanisms, and defined strategic roles within the DEE. Differently, cases with minimal governance coordination, loosely structured partnerships, and fragmented strategic alignment were given a score of 0. Intermediate values of 0.2, 0.6, and 0.8 were employed to capture varying governance structuring and coordination across different conditions. Following this classification, qualitative data were coded according to these calibration criteria, ensuring a structured approach to analyzing governance structures and their impact on decision-making (see Fig. 2).

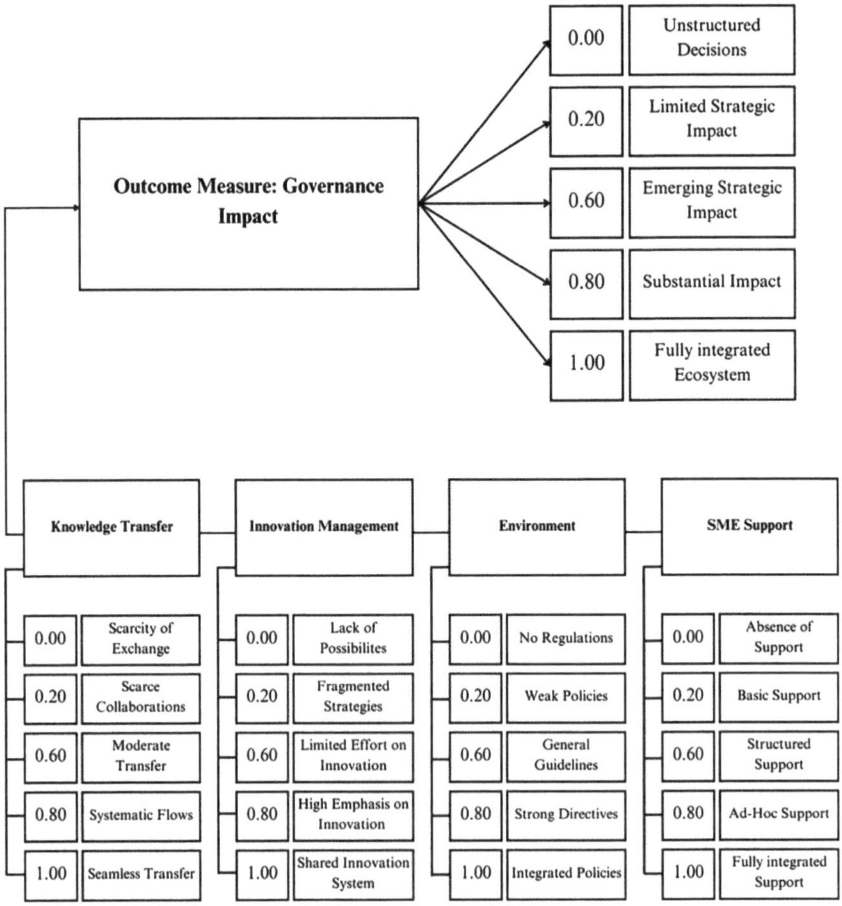

Fig. 2 Criteria for membership scores and coding. *Source* Authors

Step 3: Truth Table

The fsQCA software automatically generates a truth table from the calibrated data. This allows the assessment of case distributions across all possible combinations of causal conditions measured through coverage metrics. Using four variables, we derived 16 possible theoretical configurations (24). A consistency threshold of 0.8 was established, with cases meeting or exceeding this threshold assigned a value of 1 in the value proposition column, while those below 0.8 were assigned a value of 0.

4 Results

The findings show that the pathway reflecting effective governance relies on SME Support combined with Knowledge Transfer. This configuration exhibits a high raw coverage of 0.798 and a consistency of 0.796 (Table 1), indicating that strong support mechanisms for SMEs and effective knowledge transfer strategies are dominant factors in achieving governance success. Several cases, including B, L, M, N, and P (Fig. 3), demonstrate high membership in this pathway. Further, the second pathway involves Innovation Management, SME Support, and the absence of a strong environmental factor (~Environment). This configuration, with a raw coverage of 0.489 and a consistency score of 0.870 (Table 2), indicates that a favorable environmental context is not always a prerequisite for governance effectiveness. Instead, when innovation management and SME support are well integrated, the absence of an environmental factor does not hinder successful governance. M, U, AQ, and cases exemplify this configuration (Fig. 3). The third pathway comprises Innovation Management, Environmental Policies, and Knowledge Transfer. Although this solution presents a lower raw coverage of 0.305 (Table 1), it exhibits a high consistency score of 0.962, indicating that when these three factors align, they reliably lead to governance success. The cases, O, V, AP, and Z exhibit strong membership in this pathway (Fig. 3). Knowledge Transfer emerges as the most critical determinant of governance success among the identified conditions. Its presence correlates with stronger governance structures, suggesting that ecosystems prioritizing efficient knowledge-sharing mechanisms are more likely to succeed.

Overall Solution Coverage and Necessity of Conditions

Table 1 Configuration for governance impact: consistency and coverage

	Consistency	Coverage
Innovation management	0.706422	0.791096
SME support	0.871560	0.734536
Environment	0.519878	0.885417
Knowledge transfer	0.926606	0.694954

Source Authors

Digital Ecosystems	Innovation Management	SME support Development	Environment	Knowledge Transfer	Governance Impact
A	1,00	1,00	0,20	0,20	0,60
B	1,00	1,00	0,20	1,00	0,80
C	1,00	0,80	0,20	0,80	0,70
D	0,20	1,00	0,60	0,20	0,50
E	0,20	0,60	0,20	1,00	0,50
G	0,00	1,00	0,20	0,80	0,50
H	0,20	0,00	0,20	0,20	0,15
I	0,20	0,00	0,00	1,00	0,30
L	0,20	1,00	0,00	1,00	0,55
M	1,00	1,00	0,00	1,00	0,75
N	0,00	1,00	0,80	1,00	0,70
O	1,00	0,20	0,80	1,00	0,75
P	0,00	1,00	0,60	1,00	0,65
S	0,20	1,00	0,80	0,00	0,50
T	1,00	1,00	0,20	1,00	0,80
U	1,00	1,00	0,00	1,00	0,75
V	1,00	0,00	0,80	1,00	0,70
Z	0,60	1,00	0,80	0,80	0,80
AA	0,60	1,00	1,00	1,00	0,90
AE	0,20	0,00	0,00	1,00	0,30
AG	1,00	0,80	0,00	0,80	0,65
AI	0,00	0,80	1,00	1,00	0,70
AL	1,00	1,00	0,20	1,00	0,80
AN	0,00	0,20	0,00	1,00	0,30
AP	1,00	1,00	0,80	1,00	0,95
AQ	1,00	1,00	0,00	1,00	0,75

Fig. 3 Calibration matrix. *Source* Authors

Table 2 Configuration for governance impact: solutions

	Solution 1	Solution 2	Solution 3
Knowledge transfer	●		●
SME support	●	●	●
Innovation management		●	
Environment		⊗	●
Consistency			

Source Authors

The overall solution coverage of 0.920 suggests that these three configurations explain 92% of the variance in governance outcomes, demonstrating the robustness of the identified pathways. Furthermore, the necessary conditions analysis confirms that Knowledge Transfer is the most critical factor, with a consistency of 0.927 and coverage of 0.695. This finding implies that efficient knowledge-sharing mechanisms must be in place for an ecosystem to achieve strong governance outcomes. In addition, with a consistency of 0.872 and coverage of 0.735, SME Support highlights that governance structures lacking SME support are less likely to perform effectively. However, while Innovation Management is a significant condition (consistency: 0.706, coverage: 0.791), it does not independently guarantee governance success, emphasizing the need for complementary factors. Differently, the Environmental configuration shows lower consistency (0.520), suggesting that they are not always necessary for effective governance. Further, Table 2 summarizes the configurations or pathways identified as "sufficient" through the conducted analyses, following the procedures outlined earlier.

This table format, recommended by Ragin and Fiss [32], effectively illustrates the results. Black circles (●) signify the presence of a condition, while white circles (⊗) indicate its absence. In addition, blank cells represent the "do not care" condition, meaning that a particular factor is not considered in the solution.

5 Discussion

Governing an ecosystem involves managing relationships to gain a competitive advantage, foster collaboration, sustain motivation, and create an entrepreneurial network [8]. A critical aspect of governance structures is how authority and responsibilities are structured. One of the common distinctions is the identification of centralized and decentralized models. This refers to allocating power within certain boundaries, especially in decision-making, policy development, and execution [33]. Both of the typologies have advantages and disadvantages. Indeed, while centralized models might be quicker in short-term decision-making, they often lack diverse viewpoints, which can restrict the organization's ability to navigate complex problems [34]. Differently, a decentralized governance system gives participants full collective governance power, facilitating the inclusion of their perspectives in the decision process. However, a "too wide" distribution of governance can also decrease the probability of collective action and slow down decision-making [35]. Our findings highlight that governance structures that exhibit high value for SME development and facilitate knowledge-sharing mechanisms are central in enhancing ecosystem performance. As innovative SME utilize digital technologies and transformations to achieve decentralized governance to promote agility, collaboration, and better allocation of resources, driving a growing trend away from hierarchical structures and toward decentralized ones. Moreover, leveraging digital technology as a platform can provide services that enhance the coordination and integration of independent entities, allowing them to maximize their resources for innovation and transactions, thereby creating value [36]. Such coordination also serves as a "fair balance of interests" among diverse stakeholders, particularly those of different scales, such as small enterprises and large corporations. This, in turn, fosters inclusivity in governance and decision-making. It is crucial to establish a culture of openness, transparency, and engagement in meetings and communication [37]. These principles drive ecosystems, where collaboration among various entities depends on the flow of knowledge and ideas. For this reason, fostering a knowledge-sharing community, ecosystems can enable stakeholders to exchange information efficiently, driving innovation and collective problem-solving [38]. This, in turn, has strengthened collaboration within organizations and promoted cooperation across different entities [39]. On this basis, the digital transformation can also lead to the Schumpeterian [40] concept of creative destruction, where entire industries and traditional business models are disrupted, paving the way for the emergence of entirely new models driven by innovations [41]. Innovation represents a tool for entrepreneurship, making the two concepts inherently interconnected [42, 43]. However, the dynamic landscape of digital innovation also presents obstacles and complexities for entrepreneurs, such as regulatory barriers, securing funding, outdated and obscure technologies, and sociopolitical factors [44]. Nevertheless, while influential factors shape the choice of governance models, there is no "unique" optimal structure that can be applied across all ecosystems. Indeed, each ecosystem is embedded within a context defined by its specific stakeholder composition, technological infrastructure, regulatory environment, and strategic objectives. These contextual differences demand governance structures that are tailored to fit the ecosystem's distinct characteristics and needs. As a result, governance is not about fitting into a centralized or decentralized model but about finding a balance that supports collaboration and innovation and fosters overall performance improvements.

6 Conclusion

This study has explored the role of governance structures in DEEs, examining how different models influence the decision-making process and overall ecosystem performance. Using a QCA approach applied to Italian DEEs, our findings reveal several configurations in which governance features produce different but potentially efficient pathways to ecosystem effectiveness. In particular, DEEs governed by strong frameworks prioritizing SME support, digital infrastructure development, and collaborative knowledge-sharing tend to adopt a decentralized structure, performing better across strategic and operational dimensions. However, the results also underscore that no single governance archetype guarantees success. Instead, governance styles, from more hierarchical, state-driven models to highly decentralized ones, are central to understanding how ecosystems adapt and evolve within different technological, economic, and regulatory contexts. This variability has significant implications. First, it suggests that DEE governance should be tailored to local conditions and stakeholder capabilities, acknowledging that different governance styles can be equally efficient if well-aligned with contextual factors. Second, it emphasizes the importance of flexible and adaptive governance mechanisms that can evolve alongside digital transformation and market dynamics. Nevertheless, limitations do exist. First, its exclusive focus on Italian DEEs may limit the replicability of the findings, as governance structures can differ significantly across geographic and industrial contexts. Second, the reliance on specific projects and government sources could introduce bias into the data, potentially affecting the objectivity of the findings. Future research should expand the empirical scope by including diverse regional and sectoral cases and incorporating a broader range of data sources to validate and refine the configurations identified in this study.

Acknowledgements Co-financed by the Minister of Science under the "Regional Excellence Initiative" Program.

References

1. Song, A.K.: The digital entrepreneurial ecosystem—a critique and reconfiguration. Small Bus. Econ. **53**(3), 569–590 (2019)
2. Wurth, B., Stam, E., Spigel, B.: Toward an entrepreneurial ecosystem research program. Entrep. Theory Pract. **46**(3), 729–778 (2022)
3. Hajli, N., Baydarova, I., Nisar, T.: Digital entrepreneurial ecosystem: the role of the sharing economy in driving innovation. Entrepreneurship Reg. Dev. **37**(5–6), 785–815 (2025)
4. Elia, G., Margherita, A., Passiante, G.: Digital entrepreneurship ecosystem: how digital technologies and collective intelligence are reshaping the entrepreneurial process. Technol. Forecast. Soc. Chang. **150**, 119791 (2020)
5. Bharadwaj, A., El Sawy, O.A., Pavlou, P.A., Venkatraman, N.V.: Digital business strategy: toward a next generation of insights. MIS Q. **1**, 471–482 (2013)
6. Autio, E., Nambisan, S., Thomas, L.D., Wright, M.: Digital affordances, spatial affordances, and the genesis of entrepreneurial ecosystems. Strateg. Entrep. J. **12**(1), 72–95 (2018)
7. Sussan, F., Acs, Z.J.: The digital entrepreneurial ecosystem. Small Bus. Econ. **49**, 55–73 (2017)
8. Colombo, M.G., Dagnino, G.B., Lehmann, E.E., Salmador, M.: The governance of entrepreneurial ecosystems. Small Bus. Econ. **52**, 419–428 (2019)
9. Li, W., Du, W., Yin, J.: Digital entrepreneurship ecosystem as a new form of organizing: the case of Zhongguancun. Front. Bus. Res. China **11**, 1–21 (2017)
10. Bejjani, M., Göcke, L., Menter, M.: Digital entrepreneurial ecosystems: a systematic literature review. Technol. Forecast. Soc. Chang. **189**, 122372 (2023)
11. Troisi, O., Visvizi, A., Grimaldi, M.: The different shades of innovation emergence in smart service systems: the case of Italian cluster for aerospace technology. J. Bus. Ind. Mark. **39**(6), 1105–1129 (2024)
12. Moran, P., Simoni, M., Megaro, A., Sarno, D.: Innovative solution bundling–unbundling: a paradox in the digital era. In The International Research & Innovation Forum, pp. 465–474. Springer International Publishing, Cham (2023)
13. Moore, J.F.: Predators and prey: a new ecology of competition. Harvard Bus. Rev. **71**(3), 75–86 (1993)
14. Isenberg, D.J.: Applying the ecosystem metaphor to entrepreneurship: uses and abuses. Antitrust Bull. **61**(4), 564–573 (2016)
15. Stam, E., Spigel, B.: Entrepreneurial Ecosystems, vol. 16, No. 13, pp. 1–15). USE Discussion paper series (2016)
16. Kraus, S., Palmer, C., Kailer, N., Kallinger, F.L., Spitzer, J.: Digital entrepreneurship: a research agenda on new business models for the twenty-first century. Int. J. Entrep. Behav. Res. **25**(2), 353–375 (2019)
17. Fernandes, C., Ferreira, J.J., Veiga, P.M., Kraus, S., Dabić, M.: Digital entrepreneurship platforms: Mapping the field and looking towards a holistic approach. Technol. Soc. **70**, 101979 (2022)

18. Valdez-De-Leon, O.: How to develop a digital ecosystem: a practical framework. Technol. Innov. Manage. Rev. **1**(9), 8 (2019)
19. Barykin, S.Y., Kapustina, I.V., Kirillova, T.V., Yadykin, V.K., Konnikov, Y.A.: Economics of digital ecosystems. J. Open Innov. Technol. Mark. Complex. **6**(4), 124 (2020)
20. Zhao, F., Collier, A.: Digital entrepreneurship: research and practice (2016)
21. Purbasari, R., Muttaqin, Z., Sari, D.S.: Digital entrepreneurship in pandemic Covid 19 era: the digital entrepreneurial ecosystem framework. Rev. Integr. Bus. Econ. Res. **10**, 114–135 (2021)
22. Ragin, C.C.: The comparative method: moving beyond qualitative and quantitative strategies. University of California Press. (2014)
23. Saridakis, C., Zaefarian, G., Ganotakis, P., Angelidou, S.: A step-by-step guide of (fuzzy set) qualitative comparative analysis: from theory to practice via an implementation in a B2B context. Ind. Mark. Manage. **107**, 92–107 (2022)
24. Sendra-Pons, P., Comeig, I., Mas-Tur, A.: Institutional factors affecting entrepreneurship: a QCA analysis. Eur. Res. Manage. Bus. Econ. **28**(3), 100187 (2022)
25. Pérez-Macías, N., Medina-Molina, C., Gismera-Tierno, L.: Gamification in the development of entrepreneurial intentions: a QCA analysis. Entrepreneurship Educ. **5**(3), 343–365 (2022)
26. Calispa-Aguilar, E.: The determinants of entrepreneurship in urban and non-urban regions: a fuzzy-set QCA approach. Entrepreneurial Bus. Econ. Rev. **12**(2), 143–156 (2024)
27. Di Paola, N., Chari, S., Iannacci, F., Kraus, S.: Configurational theory in business and management research: status quo and guidelines for the application of qualitative comparative analysis (QCA). Technol. Forecast. Soc. Chang. **211**, 123907 (2025)
28. MIMIT: Poli Europei di Innovazione Digitale—European Digital Innovation Hubs (EDIH) (2024). Retrieved from https://www.mimit.gov.it/it/incentivi/poli-europei-di-innovazione-digitale-european-digital-innovation-hubs-edih
29. Ragin, C.C.: Fuzzy-Set Social Science. University of Chicago Press (2000)
30. Osterwalder, A., Pigneur, Y.: Business Model Generation: a Handbook for Visionaries, Game Changers, and Challengers. John Wiley & Sons (2010)
31. Abbate, T., Cesaroni, F., Cinici, M.C., Villari, M.: Business models for developing smart cities. A fuzzy set qualitative comparative analysis of an IoT platform. Technol. Forecast. Soc. Chang. **142**, 183–193 (2019)
32. Ragin, C.C., Fiss, P.C.: Net effects analysis versus configurational analysis: an empirical demonstration. Redesigning Soc. Inquiry: Fuzzy Sets Beyond **240**, 190–212 (2008)
33. Gaskell, J., Stoker, G.: Centralized or decentralized: which governance systems are having a "Good" pandemic? Democratic Theory **7**(2), 33–40 (2020)
34. Awoyemi, O., Attah, R.U., Basiru, J.O., Leghemo, I.M., Onwuzulike, O.C.: Revolutionizing corporate governance: a framework for solving leadership inefficiencies in entrepreneurial and small business organizations. Int. J. Multi. Res. Updates **6**(1), 45–52 (2023)
35. Chen, Y., Richter, J.I., Patel, P.C.: Decentralized governance of digital platforms. J. Manage. **47**(5), 1305–1337 (2021)
36. Ofe, H.A., Sandberg, J.: The emergence of digital ecosystem governance: an investigation of responses to disrupted resource control in the Swedish public transport sector. Inf. Syst. J. **33**(2), 350–384 (2023)
37. Darking, M., Nachira, F., Nicolai, A., Dini, P., Le Louran, M., Leon, L.: Understanding the role of governance in the context of digital ecosystems. Digit. Bus. Ecosyst. **1**, 78–82 (2007)
38. Jones, P., Ratten, V.: Knowledge spillovers and entrepreneurial ecosystems. Knowl. Manage. Res. Pract. **19**(1), 1–7 (2021)
39. Chen, A., Lin, Y., Mariani, M., Shou, Y., Zhang, Y.: Entrepreneurial growth in digital business ecosystems: An integrated framework blending the knowledge-based view of the firm and business ecosystems. J. Technol. Transf. **48**(5), 1628–1653 (2023)
40. Schumpeter, J.A.: The theory of economic development: an inquiry into profits, capital, credit, interest, and the business cycle. University of Illinois at Urbana-Champaign's Academy for Entrepreneurial Leadership Historical Research Reference in Entrepreneurship. (1934)
41. Kraus, S., Roig-Tierno, N., Bouncken, R.B.: Digital innovation and venturing: an introduction into the digitalization of entrepreneurship. RMS **13**(3), 519–528 (2019)

42. Soriano, D.R., Huarng, K.H.: Innovation and entrepreneurship in knowledge industries. J. Bus. Res. **66**(10), 1964–1969 (2013)
43. Drucker, P., Maciariello, J.: Innovation and Entrepreneurship. Routledge (2014)
44. Kreiterling, C.: Digital innovation and entrepreneurship: a review of challenges in competitive markets. J. Innov. Entrepreneurship **12**(1), 49 (2023)

The Concept of Productivity and Developing Countries: The Case of Morocco

Anass Bendarkawi

Abstract Productivity holds a universally recognized importance for a country's prosperity, whether in developed or developing countries. Moreover, the level of techno-economic development of a company, an industry, or a nation is often assessed through its productivity, as productivity plays a crucial role in economic growth and projection. This is why developing countries are increasingly interested in, and even actively engaging in, research related to productivity. It is within this framework that we are undertaking this research, which aims to evaluate how the evolution of factors influencing business productivity and determining the pace of growth and economic development can contribute to strengthening the competitiveness of Moroccan companies. This research is based on a qualitative study conducted with a group of Moroccan economic and political decision-makers, with the goal of better understanding this phenomenon and highlighting their perceptions and analyses of productivity factors within a specific context or situation (the case of Morocco).

Keywords Productivity · Economic and political decision-makers · Moroccan companies

1 Introduction

Productivity is one of the key indicators of a given economy's growth and development, revealing the push and pull factors of the economy and its capacity to catch up with other economies. It is also an important indicator that businesses can draw from to boost their competitiveness [1, 2].

However, in this broader context, productivity continues to pose a major challenge for developing countries, particularly in Morocco. Labor productivity in Morocco is approximately 30% lower than the average of other emerging markets [3], constraining both national economic growth and business competitiveness. Total

A. Bendarkawi (✉)
Research Laboratory in Strategy and Organizational Management, National School of Business and Management, Hassan First University, Settat, Morocco
e-mail: bendarkawi.anass@uhp.ac.ma

A. Visvizi et al. (eds.), *Research and Innovation Forum 2025 Vol2*, Springer Proceedings in Complexity, https://doi.org/10.1007/978-3-032-09062-1_15

factor productivity remains low in strategic sectors, such as manufacturing [4], due to limited investment in innovation, vocational training deficits, and persistent institutional constraints.

These challenges are exacerbated by a global environment marked by globalization, trade liberalization, and intensified competition, where a country's ability to enhance productivity largely determines the sustainability of its economic development. This issue is especially pressing for countries like Morocco [5, 6].

Productivity in transition economies is often linked to complex political and economic processes. In this sense, the analysis of productivity in a rapidly transforming context must take into account not only economic reforms but also the political and social contexts that influence long-term competitiveness and stability [7]. These dynamics are particularly relevant in the case of Morocco, where institutional and economic reforms are ongoing in the context of transition and adaptation to new global realities.

In light of these dynamics, international literature identifies numerous productivity determinants within developing countries [8, 9]. Recent empirical studies conducted in Asia [10] and Africa [11] emphasize the complexity of the factors influencing productivity and the need for approaches that take regional specificities into account. Yet, despite this growing body of research, relatively few studies have focused on the productivity determinants specific to Moroccan firms, which operate in a distinctive institutional and structural context [12, 13].

This gap leads to the formulation of a research problem: identifying the main factors influencing the productivity of Moroccan firms and determining how these factors can be leveraged to strengthen their competitiveness.

To address this question, this study aims to provide a deeper understanding of how various institutional, managerial, and organizational dynamics shape productivity within Moroccan firms, and how these dynamics can serve as strategic levers for enhancing competitiveness in the country's current economic environment.

Methodologically, the study adopts a qualitative methodology based on interviews with key economic and political decision-makers, providing in-depth insights into the interactions between internal organizational practices and broader institutional frameworks. This approach seeks to offer a novel and complementary perspective to predominantly quantitative studies, and to contribute meaningfully to the debate on strategies for boosting the productivity of Moroccan firms.

To that end, this section frames the study through a literature review that highlights the multidimensional complexity of productivity, an analysis of the Moroccan context, and a clarification of the research problem along with the guiding questions.

1.1 Literature Review: Productivity as a Complex and Multidimensional Concept

The term productivity was first used in the late eighteenth century [14], and has remained a key element of public discourse since, although its definition remained vague for a long time. In the collective imagination, productivity is often associated with notions such as progress, economic growth, and future prosperity [15]. Over time, it has become a central concept in economic analysis and a major driver of economic progress [16].

Productivity is a key indicator of the techno-economic development of a company, sector, or country, serving as a benchmark for policymakers, business leaders, and economic decision-makers [17]. It is one of the few concepts commonly recognized across various schools of economic thought, including Marxist and liberal currents [18].

Productivity began to be defined more precisely in the early twentieth century as the ratio between outputs and production factors [19]. By the mid-century, the concept of "human labor productivity" emerged, referring to output per unit of time and per worker in a given sector [20]. This definition was later expanded into a broader concept: productivity as the ratio of total output to all resources mobilized for production [21, 22].

Despite its apparent simplicity, productivity is a multifaceted concept shaped by economic, social, technological, and organizational dynamics [23]. It is understood as a social process that encompasses research, technology, education, management, infrastructure, labor, and social relations [24].

From this broader perspective, recent studies have deepened the concept of multi-factor productivity, which incorporates not only capital and labor but also innovation, work organization, and institutional quality [25, 26]. This approach highlights the wide range of determinants that influence the productive performance of a country or a firm.

Productivity plays a key role in economic growth theories, particularly within neoclassical approaches that emphasize the fundamental contribution of technological progress to long-term development. Total Factor Productivity (TFP) is widely used to capture the residual effects of innovation, knowledge accumulation, and organizational efficiency beyond the direct inputs of capital and labor [27, 28].

Building on this theoretical framework, a substantial body of empirical research has identified a consistent set of levers that enhance firm-level productivity. Innovation, human capital development through education and training, effective managerial practices, and investments in both tangible and intangible assets are regularly associated with improved productivity outcomes [29, 30]. The effective interplay between human resources and physical capital is thus recognized as a key determinant of productivity, especially in developing and transitioning economies [31, 32].

Furthermore, recent technological advancements particularly digitalization, artificial intelligence, and Industry 4.0—are now regarded as major levers for improving productivity. In this context, the adoption of digital technologies and automation in public administrations can significantly reduce time and costs while transforming work processes, even in data-limited environments. While this digital shift improves productivity by reducing costs and increasing efficiency, it also presents challenges related to change management, training, and worker reintegration [33].

In parallel with these technological changes, recent studies emphasize the relevance of the Smart City Competitiveness Index (SMCI) as a robust tool for assessing the impact of technology on urban competitiveness and the productivity of public services. By quantifying the effects of innovation policies and investments, the SMCI highlights the role of advanced technologies in optimizing urban management. It demonstrates that the determinants of productivity now extend beyond technological factors to include public policy, data governance, and the adaptability of economic actors to digital transformation [34].

However, they also present significant challenges, especially in developing countries: skills adaptation, digital transition, unequal access to technology, and weak integration of innovation into productive systems [35, 36].

These developments underscore the need for a renewed, context-sensitive understanding of productivity—one that takes into account local specificities and the institutional conditions within which firms operate.

Nevertheless, despite the abundance of literature on productivity, several areas remain underexplored, particularly in the context of developing countries such as Morocco. Most existing studies favor quantitative approaches based mainly on macroeconomic or sectoral data [37]. While these analyses offer valuable insights into broader trends, they often fall short in capturing the underlying mechanisms and the perspectives of key stakeholders who shape productivity dynamics.

Moreover, in Morocco, qualitative studies that explore the views of economic and political decision-makers are still scarce, despite the critical role these actors play in designing and implementing productivity-enhancing policies and strategies. This gap strongly justifies the need for a qualitative approach to better understand the evolving determinants of productivity within Moroccan enterprises.

1.2 Empirical Context: The Case of Morocco

The choice of Morocco is also justified by the availability of institutional mechanisms aimed at improving firm competitiveness, such as the Industrial Acceleration Plan, the Investment Charter, and the support provided by public development agencies. Nevertheless, recent national reports and international assessments (World Bank, Organization for Economic Co-operation and Development) continue to highlight organizational, managerial, and innovation-related constraints that hinder productivity growth.

The productivity of Moroccan firms remains significantly lower than that observed in comparable economies, with a gap exceeding 60% relative to the average of OECD member countries [38]. This situation reflects persistent structural weaknesses that

Table 1 Capital productivity ratio—international comparison (2018–2024)

Country	Capital productivity ratio	Period
Morocco	0.55–0.57	2018–2024
Turkey	0.62–0.66	2018–2024
Malaysia	0.68–0.71	2018–2024
Poland	0.65–0.67	2018–2024

Source World Bank. HCP

impede the development of the national productive fabric and highlights the urgency of structural transformation.

Despite sustained investment efforts, this productivity gap persists. Table 1 provides a comparative overview of capital productivity levels between Morocco and selected emerging countries for the period 2018–2024 [39, 40].

This data highlights Morocco's lag in productive efficiency, particularly in terms of capital utilization.

In parallel, Morocco's economic growth has largely relied on the accumulation of traditional production factors labor and capital rather than on gains in efficiency. As shown in Table 2, the average annual growth of Total Factor Productivity (TFP) has remained modest, particularly during the 2000s and 2010s [41, 42].

On the other hand, Table 3 indicates that labour productivity, measured as Gross Domestic Product (GDP) per hour worked, remains substantially lower in Morocco, estimated at approximately USD 44.7 (Purchasing Power Parity, PPP, 2017 constant), which is nearly 55% below the OECD average [43, 44].

Despite some convergence over the past decade, the persistent gap emphasizes the need to address structural barriers affecting labour efficiency.

In this regard, the persistently low productivity of Moroccan firms is largely explained by structural constraints, including limited investment in continuous training and a mismatch between workforce skills and labor market needs. Over

Table 2 Evolution of total factor productivity (TFP) in Morocco

Period	Average annual growth of TFP (%)
1960–1980	+ 1.01
1980–1990	+ 0.89
1990–2000	+ 2.10
2000–2019	≈ + 0.70

Source UNIDO

Table 3 Moroccan labour productivity—GDP per hour worked

Year	Indicator	Morocco (USD, PPP 2017 constant)	Gap versus OECD average (%)
2022	GDP per hour worked	≈ $ 44.7	≈ − 55

Source World Bank, OECD

40% of firms report difficulties in recruiting qualified workers, while fewer than 10% of SMEs provide regular employee training [45]. This deficit is exacerbated by the dominance of the informal sector, which represents more than 80% of economic units [46]. In addition, weak innovation capacity—illustrated by R&D spending of only 0.7% of GDP and Industry 4.0 adoption below 5%—further hinders productivity gains [47, 48].

Consequently, despite incremental progress, Morocco still struggles to close the productivity gap with emerging and developed countries, raising concerns about the effectiveness of current drivers in enhancing firm competitiveness and national economic growth.

1.3 Research Problem and Research Questions

Increased market liberalization has intensified domestic competition, making the economic environment more uncertain, especially amid rapid global changes. In this context, Moroccan companies find themselves less protected against international competitive pressures. Therefore, improving productivity becomes not only desirable but essential. Productivity thus emerges as a key issue. Understanding the core determinants of Moroccan firms' productivity appears essential to designing effective support policies, enhancing the national economic fabric, and boosting companies' competitiveness.

Accordingly, the main research question is:

To what extent does the evolution of factors influencing firms' productivity, and thereby determining the pace of economic growth and development, contribute to strengthening the competitiveness of Moroccan companies?

The study is guided by the following research questions:

- **RQ1:** What are the main determinants influencing the productivity of Moroccan firms?
- **RQ2:** How can these factors be mobilized as levers to enhance competitiveness in Morocco's current economic context?

Despite sustained efforts and ongoing reforms, the productivity of Moroccan firms remains significantly below that of comparable economies. While macroeconomic and quantitative approaches highlight persistent structural inefficiencies, they offer limited insight into the underlying causes of low productivity. To address this gap, the present study adopts a qualitative perspective grounded in the perceptions of economic and political decision-makers. It aims to provide a comprehensive, contextualized analysis that complements existing approaches and deepens the understanding of firm-level productivity determinants in Morocco.

2 Methods and Materials

Selecting an appropriate data collection technique tailored to the research objective is a crucial preliminary step before adopting a suitable data analysis method [49]. This study employs a qualitative approach based on semi-structured interviews conducted with Moroccan economic and political decision-makers. This methodological choice aims to explore their perceptions and analyses in depth, emphasizing the richness and relevance of the data collected rather than statistical representativeness. The objective is to provide a nuanced contextual understanding of the determinants of productivity in Moroccan enterprises.

The methodological approach unfolds in two successive phases:

- **Phase 1:** Conducting interviews and collecting data.
- **Phase 2:** Processing and analyzing qualitative data.

2.1 Conducting Interviews and Data Collection

The semi-structured interview guide was developed based on the study's objectives and specificities. Participants were selected through purposive sampling, complemented by a preliminary negotiation phase with each participant. All interviews were fully recorded and transcribed verbatim. The sample includes 10 diverse profiles, representing multiple levels of responsibility. The total duration of the interviews amounts to 290 min, ensuring the richness and coherence of the collected data. The diversity of profiles is essential to guarantee a nuanced and progressive analysis, minimizing biases associated with excessive homogeneity [50]. This methodological choice aims to ensure the relevance and quality of the information gathered, relying on actors capable of providing a comprehensive and nuanced understanding of the determinants of productivity within the Moroccan context.

Although the number of interviewees is limited, this is consistent with the principles of purposive sampling commonly used in qualitative research. The goal is not statistical representativeness but rather to gain in-depth and context-rich insights from individuals who are strategically positioned to inform the research question [51]. This targeted approach ensures that the data collected is both relevant and analytically meaningful, despite the small sample size [52].

Selection and Justification of Interviewees

The selection of interviewees is based on rigorous methodological criteria aimed at ensuring the validity and relevance of the data collected. Three main criteria were established:

1. **Level of strategic responsibility.**
2. **In-depth knowledge of the Moroccan productive fabric.**
3. **Direct experience with productivity-related issues.**

The interviewees were selected for their strategic roles in shaping Morocco's economic and institutional landscape. They included ministers overseeing key portfolios—economy, finance, industry, employment, and technology—actively involved in structural reforms and sectoral strategies related to competitiveness, innovation, and human capital. A representative of the Economic, Social and Environmental Council (CESE) contributed expertise on public policy adaptation, employment, and vocational training.

From the private sector, leaders from the General Confederation of Moroccan Enterprises (CGEM), the Center of Young Business Leaders (CJD), and trade union officials provided insights into organizational practices, performance management, labor relations, and competitiveness challenges. Finally, experts from McKinsey & Company and sectoral consultants offered strategic perspectives on industrial planning, technological upgrading, and the interaction between institutional frameworks and firm-level practices.

2.2 *Processing and Analyzing Qualitative Data*

Interview data from economic and political decision-makers were analyzed using the thematic analysis framework [50]. The interview transcripts were subjected to systematic analysis. First, a thorough reading enabled the identification of key elements. Second, a thematic grouping process was undertaken to identify convergences and recurring patterns in participants' discourse [53]. This approach involves identifying, coding, and organizing themes through iterative reading of transcripts. It follows a progressive, constructive logic aimed at structuring the corpus into coherent thematic clusters aligned with the research problem. The themes were defined based on similarities in participant narratives, facilitating a rigorous and contextualized interpretation. Coding was conducted using RQDA software, which ensured structured and systematic data management, reinforcing the study's scientific rigor. Cross-validation sessions were conducted to compare interpretations and ensure consistency.

To address limitations related to the small sample size, a triangulation strategy was applied, combining multiple data sources to enhance the credibility and robustness of the findings [54, 55]. This involved cross-referencing stakeholders' perceptions with secondary data and theoretical insights from three complementary sources: (1) semi-structured interviews with key stakeholders; (2) analysis of official reports (High Commission for Planning, World Bank, UNIDO); and (3) a review of recent academic literature on productivity in developing economies.

3 Results

This section summarizes the key findings from the thematic analysis of interviews with economic and political decision-makers. Five major themes emerged, offering a contextualized understanding of the main factors shaping the productivity of Moroccan firms, supported by illustrative excerpts reflecting diverse stakeholder perspectives.

Table 4 Productivity determinants: themes and interview excerpts

Theme	Analytical summary and illustrative excerpts
1. Productivity planning, measurement, and evaluation: persistent challenges	Respondents noted that productivity measurement and planning remain fragmented across Moroccan firms. The lack of standardized tools and reliable quantitative data undermines performance assessment and hinders the formulation of long-term productivity strategies. Most initiatives remain focused on short-term outputs, with insufficient attention to structural improvements *[Only the business owner is responsible for making forecasts based on competitor behavior and market trends] [Few firms have structured productivity programs. In some cases, unions initiated such efforts, but these remain isolated practices]*
2. Innovation and technology adoption	Despite their recognized potential, innovation and digital technology adoption remain limited across most Moroccan firms, except in a few dynamic sectors—such as automotive and aeronautics—that have effectively integrated Industry 4.0 principles. This lag is mainly attributed to the absence of an innovation-oriented culture, low R&D investment, and the limited diffusion of digital tools throughout production and management processes. Firms that have implemented digital transformation strategies report significant productivity gains, highlighting the strategic value of innovation. However, structural and financial barriers continue to constrain most SMEs *[Factory 3.0 aims to reduce costs and improve product quality by introducing so-called smart factories, enabling greater flexibility across the plant and the entire company—this is part of what we call the Fourth Industrial Revolution] [In the absence of innovation, productivity gains cannot be achieved, as productivity is inevitably limited by product obsolescence]*
3. Human capital and the skills gap: a structural barrier to productivity	There is a persistent mismatch between workforce skills and labor market needs, undermining firm performance. While large firms benefit from training programs and public support, small and microenterprises remain largely excluded. Social dialogue, especially in employer–union relations, is also underused despite its potential to enhance human capital mobilization *[The rate of continuing training in Moroccan companies remains very low, especially among micro and small enterprises that do not hire human resources managers] [In Morocco, there is a human resources issue concerning the qualification level of workers, which affects their productivity, along with social factors such as the work environment and employer–employee relations that hinder productivity]*

(continued)

Table 4 (continued)

Theme	Analytical summary and illustrative excerpts
4. Governance and managerial practices: strategic levers for sustainable productivity	Governance and managerial practices are pivotal to firm productivity in Morocco. Respondents pointed to a lack of structured management systems, with informal, top-down decision-making often replacing strategic planning and performance monitoring. Promoting transparent governance, inclusive leadership, and accountability mechanisms is essential to improving resource allocation, resilience, and long-term productivity *[Leadership positions within the company must be granted based on competence and merit] [At the first sign of crisis, companies resort to cutting salaries and dismissing workers, without questioning the manager's effectiveness, competence, or whether the company's structure is well organized]*
5. Structural barriers and institutional inertia: external constraints on productivity	Respondents underscored the negative effects of excessive bureaucracy, outdated regulations, and institutional inertia, which hinder organizational agility and long-term investment. The informal sector was also highlighted as a key distortion factor, creating unfair competition and limiting the growth potential of formal enterprises *[When a company seeks to grow, it faces administrations that are either non-functional or not up to date with current laws.] [The informal sector undermines the competitiveness of formal enterprises and poses a serious issue of tax injustice and market distortion.]*

4 Discussion

This discussion offers a deeper understanding of the mechanisms influencing productivity in Moroccan firms.

4.1 Structural and Institutional Constraints

The Moroccan state was often viewed as a necessary facilitator of productivity reform. International experiences—such as national productivity boards in Europe or productivity centers in Africa—suggest that dedicated institutions can help coordinate efforts, disseminate best practices, and build technical capacity [56].

Institutional and structural constraints significantly undermine productivity at the macro level, including a complex administrative environment, inconsistent policy coordination, and the widespread presence of the informal sector. There is a clear need for a more supportive institutional framework to reduce bureaucratic obstacles, enhance infrastructure, and provide accessible programs and public–private coordination to support formal enterprises.

4.2 Toward an Integrated and Sector-Sensitive Strategy

The empirical results point to the multidimensional nature of productivity determinants in Morocco. The diversity of the Moroccan productive structure requires a differentiated approach. For example, productivity in offshoring depends primarily on human capital and working conditions, while productivity in manufacturing hinges more on equipment and process optimization.

This implies that a one-size-fits-all strategy is unlikely to be effective. Instead, sector-specific interventions—grounded in reliable data and co-designed with industry actors—are necessary to unlock productivity gains across the economy.

4.3 Innovation and Technology: A Selective Modernization

While flagship sectors such as automotive and aeronautics have adopted digital technologies and Industry 4.0 practices, most SMEs face structural and financial barriers that hinder sustained innovation efforts. Several participants linked productivity gains to capital deepening—namely the acquisition of modern machinery—but pointed to the obsolescence of production equipment in traditional industries such as textiles as a major constraint. These findings echo existing research highlighting the role of technological adoption in driving productivity in developing economies [57]. However, the Moroccan case shows that capital investment alone is insufficient without complementary managerial and organizational innovations [58].

4.4 Human Capital and Skills Mismatch

The results highlight a structural deficit in continuous vocational training, particularly among SMEs and microenterprises, which often lack dedicated human resources departments and access to training funds. In addition, the limited adaptability of the education system exacerbates the mismatch between available skills and the evolving needs of productive sectors [59]. However, this study underscores the specificity of the Moroccan context, where training efforts are concentrated only in large firms.

4.5 Labor Conditions and Social Dialogue

Poor working conditions, insufficient social protection, and the lack of structured social dialogue constitute significant barriers to productivity [60]. Productivity gains cannot be achieved without decent work, which includes fair wages, job security, and effective communication channels between employees and management.

However, the empirical evidence presented here also shows that union representation remains weak in many Moroccan firms and that collective bargaining mechanisms are often underutilized. Strengthening institutional frameworks for social dialogue and ensuring effective enforcement of labor standards thus appear essential levers for unlocking the productivity potential of the workforce.

4.6 *Managerial Practices and Organizational Processes*

The absence of measurement tools, process optimization mechanisms, and strategic planning frameworks hinders productivity. Many firms rely on informal decision-making processes, with limited use of indicators or dashboards to monitor performance.

A renewal of managerial culture, based on engineering approaches to process design, work step rationalization, and systematic productivity evaluation, is necessary.

These findings align with the literature on high-performance work systems, which link structured managerial practices to increased efficiency and output [61], and highlight the need for targeted support to professionalize management in the Moroccan context.

4.7 *Limitations of the Study*

The purposive sampling method facilitated access to hard-to-reach profiles, notably high-level stakeholders. However, it may have introduced a selection bias by favoring participants from similar networks, thereby limiting the diversity of perspectives and the overall representativeness of the results. In addition, administrative constraints and limited availability of decision-makers posed challenges to data collection and affected the richness of the empirical material.

The interpretive nature of qualitative analysis also introduces a degree of subjectivity. To address this, rigorous thematic coding was conducted using RQDA software. This was further reinforced by methodological triangulation, incorporating secondary sources such as official reports, previous studies, and national statistics.

Finally, due to the qualitative design of this research, it was not possible to quantitatively assess the impact of the identified productivity factors.

5 Conclusion

In contrast to dominant macroeconomic or quantitative approaches, our study offers an original contribution by uncovering the often-overlooked underlying dynamics that influence firm-level productivity. Drawing also on institutional reports and recent academic literature, the study provides a contextualized understanding of the challenges within a national framework marked by multiple reforms and unfinished sectoral modernization, yet still facing persistent structural constraints.

The recommendations derived from this study can be articulated along three complementary levels:

At the policy level, it is crucial to institutionalize productivity governance by establishing a National Productivity Council that brings together representatives from government, employers' organizations, and trade unions. This body should coordinate cross-sectoral initiatives, disseminate best practices, and ensure consistency between national strategies and firm-level realities. Public authorities must simplify administrative procedures, strengthen inter-ministerial coordination, and implement targeted fiscal incentives and accessible financing mechanisms, particularly for SMEs and high-potential sectors. Furthermore, reforming vocational training to better align with labor market needs is crucial to addressing persistent structural skill gaps.

At the enterprise level, firms should aim to professionalize their managerial practices by investing in leadership development, modern management tools, and long-term strategic planning. Embracing digital transformation is imperative; this entails not only adopting new technologies but also strengthening organizational readiness and ensuring continuous workforce training. Companies are also encouraged to adopt participatory and flexible governance models.

At the institutional level, more concerted efforts are needed to combat informality, reinforce the rule of law, and modernize the industrial base through infrastructure development and regulatory reforms. Strengthening mechanisms for social dialogue and enforcing labor standards are also vital to improving workforce motivation, stability, and retention.

Improving productivity requires a coordinated national strategy that links firm-level transformation with institutional modernization. While this research provides an initial foundation, further efforts are needed. In particular, future studies should adopt mixed-methods approaches that combine qualitative insights with quantitative analysis in order to assess the relative weight of each determinant and validate findings across sectors.

References

1. Solow, R.M.: Technical change and the aggregate production function. Rev. Econ. Stat. **39**(3), 312–320 (1957). https://doi.org/10.2307/1926047
2. Jorgenson, D.W., Griliches, Z.: The explanation of productivity change. Rev. Econ. Stud. **34**(3), 249–283 (1967). https://doi.org/10.2307/2296675

3. Organisation for Economic Co-operation and Development (OECD): Labor productivity in emerging economies. OECD, Paris (2023). https://doi.org/10.1787/74623e5b-en
4. UNID: Industrial Development Report 2021–The future of industrialization in a post-pandemic world. UNIDO, Vienna (2021)
5. OECD: Economic Outlook for Emerging Countries: Morocco. OECD, Paris (2021). https://doi.org/10.1787/66c5ac2c-en
6. HCP: Situation économique nationale en 2023. HCP Publications, Rabat (2023)
7. Visvizi, A., Żukrowska, K.: Transition and transformation: the case of Poland'. In: Visvizi, A., Matysek-Jedrych, A., Mroczek-Dąbrowska, K. (eds.) Poland in the Single Market: Politics, Economics, the Euro, pp. 13–32. Routledge, London (2021). https://doi.org/10.4324/9780429285301-2
8. World Bank: World Development Report: Finance for an Equitable Recovery. World Bank, Washington(2022). https://doi.org/10.1596/978-1-4648-1730-4
9. Aghion, P., Bergeaud, A., Van Reenen, J.: A Toolkit of Policies to Promote Innovation. NBER Working Paper, 2838 (2021). https://doi.org/10.3386/w28381
10. OECD/APO.: Identifying the Main Drivers of Productivity Growth: A Literature Review. OECD Publishing, Paris, 163–184 (2022). https://doi.org/10.1787/00435b80-en
11. Lee, J., Kim, S.: Determinants of productivity in African emerging markets. Afr. Econ. Rev. **34**(1), 102–119 (2022)
12. El Mansour, H., Tazi, H., Chafik, A.: Structures et dynamiques productives au Maroc: Un nouveau regard par la théorie des réseaux. Research papers & Policy papers on Economic Trends and Policies, Policy Center for the New South (2024)
13. Chemingui, M-A.: Productivity performance in developing countries: Country case studies: Morocco, UNIDO, Vienna (2020)
14. Quesnay, F.: Analysis of the Arithmetical Formula of the Economic Table of the Distribution of Annual Expenses of an Agricultural Nation. J. Agri. Comm. Fin. **5**(3), 11–41 (1766)
15. Schaller, F., Vincent, L-A.: Critical Essay on the Concept of Productivity. Eco. Rev. **18**(5), 865–868 (1967)
16. Dominique, A., Roux, B.: La productivité autrement. 1st ed, L'Harmattan, Paris (1994)
17. Prokopenko, J.: Managing Productivity: A Practical Manual, 1st edn. ILO, Geneva (1990)
18. Fourastié, J.: Richesse et progrès: Essai sur la répartition et sur les différences de productivité. Gauthier-Villars, Paris (1952)
19. Aftalion, A.: Les trois notions de la productivité et les revenus: les variations comparées des trois notions de la productivité. Revue d'économie Politique **25**(3), 345–369 (1911)
20. Jany-Catrice, F.: La performance totale: nouvel esprit du capitalisme?. Presses Universitaires du Septentrion, 81–82 (2012)
21. Fourastié, J.: Éléments d'économie et de prospective - Des Trente Glorieuses à l'aube du troisième millénaire. Ellipses, Paris (1997)
22. Economic and Social Council: Analysis of Productivity, its Determinants and Outcomes in an International Context. Luxembourg (2018)
23. Vincent, L.A.: Définition et mesure de la productivité. J de la Soc de statis **100**, 35–46 (1959)
24. International Labour Organization: La mesure de la productivité du travail. Studies and Documents, Geneva (1969)
25. Autor, D.: The Task Approach to Labor Markets: an overview. J Labour Market Res **46**, 185–199 (2013). https://doi.org/10.1007/s12651-013-0128-z
26. Bloom, N., Brynjolfsson, E., Foster, L., Jarmin, R., Megha, P., Saporta-Eksten, I., Van Reenen, J.: American Economic Review **109**(5), 1648–83 (2017). https://doi.org/10.1257/aer.20170491
27. Solow, R.: A contribution to the theory of economic growth. Q. J. Econ. **70**(1), 65–94 (1956). https://doi.org/10.2307/1884513
28. Solow, R.: Technical change and the aggregate production function. Rev. Econ. Stat. **3**, 312–320 (1957). https://doi.org/10.2307/1926047
29. OECD: Compendium of Productivity Indicators 2019. OECD Publishing, Paris (2019). https://doi.org/10.1787/b2774f97-en

30. Syverson, C.: What determines productivity? J. Econ. Lit. **49**(2), 326–365 (2011). https://doi.org/10.1257/jel.49.2.326
31. Hall, R.E., Jones, C.I.: Why do some countries produce so much more output per worker than others? Q. J. Econ. **114**(11), 83–116 (1999). https://doi.org/10.1162/003355399555954
32. Bloom, N., Van Reenen, J.: Measuring and explaining management practices across firms and countries. Q. J. Econ. **122**(4), 1351–1408 (2007). https://doi.org/10.1162/qjec.2007.122.4.1351
33. Bonomi Savignon, A., Zecchinelli, R., Costumato, L., Scalabrini, F.: Automation in public sector jobs and services: a framework to analyze public digital transformation's impact in a data-constrained environment. Transf. Gov. People Process. Policy **18**(1), 49–70 (2024). https://doi.org/10.1108/TG-04-2023-0044
34. Visvizi, A., Wosiek, R., Malik, R.: The smart city competitiveness index (SMCI): conceptualization, modelling, application—An evidence-based insight. Land Use Policy **148**(C) (2025). https://doi.org/10.1016/j.landusepol.2024.107408
35. Brynjolfsson, E., McAfee, A.: The Second Machine Age: Work, Progress, and Prosperity in a Time of Brilliant Technologies. Norton & Company, **14**(11) (2014)
36. Aghion, P., Bergeaud, A., Boppart, T., Klenow, P.J., Li, H.: Missing Growth from Creative Destruction. Am Eco Rev. **109**(8), 2795–2822 (2019). https://doi.org/10.1257/aer.20171745
37. World Bank: Global Economic Prospects. World Bank, Washington (2022). https://doi.org/10.1596/978-1-4648-1843-1
38. OECD: OECD Compendium of Productivity Indicators 2024. OECD Publishing, Paris (2024). https://doi.org/10.1787/b96cd88a-en
39. HCP.: Annuaire statistique 2023. HCP Publications, Rabat (2023)
40. World Bank: World Development Indicators. World Bank, Washington (2024)
41. UNIDO.: Industrial Development Report 2005–Capacity Building for Catching-up: Historical, Empirical and Policy Dimensions. UNIDO, Vienna (2005).
42. UNIDO: Industrial Development Report 2009–Breaking in and moving up: New industrial challenges for the bottom billion and the middle-income countries. UNIDO, Vienna (2009)
43. World Bank: Morocco Economic Update: The Recovery is Running Dry. World Group, Washington (2022)
44. OECD.: OECD Economic Surveys: Morocco 2024: Economic Assessment. OECD Publishing, Paris (2024). https://doi.org/10.1787/80777ea7-en
45. African Development Bank: Report on Skills Development and Training in African (SEPA) Action Plan. ADB Publishing, Abidjan (2022)
46. World Bank: Morocco Economic Update Unlocking the potential of the private sector to spur growth and job creation. World Bank, Washington, DC (2024)
47. UNESCO: Maroc: rapport d'évaluation de l'état de préparation à l'intelligence artificielle. UNESCO Publishing, Rabat (2024)
48. CGEM: Economic Outlook Décembre 2023. CGEM Publications, Casablanca (2023)
49. Amari Cherif, A.: Enjeux de la méthode qualitative. Communication at the International Congress organized by ISEOR, Jean Moulin University Lyon 3 and the Academy of Management, Lyon (2007)
50. Savall, H., Zardet, V.: Recherche en Sciences de Gestion: Approche Qualimétrique, observer l'objet complexe. Economica (2004)
51. Braun, V., Clarke, V.: Using thematic analysis in psychology. Qual. Res. Psychol. **3**(2), 77–101 (2006). https://doi.org/10.1191/1478088706qp063oa
52. Glaser, B.G., Strauss, A.L.: The Discovery of Grounded Theory: Strategies for Qualitative Research. Aldine Publishing Company, Chicago (1967)
53. Thiétart, R.A.: Méthodes de Recherche En Management. 3e éd., Dunod, Paris (2007)
54. Denzin, N.K.: The Research Act: A Theoretical Introduction to Sociological Methods. Routledge, New York (2009). https://doi.org/10.4324/9781315134543
55. Patton, M.Q.: Enhancing the quality and credibility of qualitative analysis. Health Serv. Res. **34**(5 Pt 2), 1189–1208 (1999)

56. ILO: Productivity Ecosystems for Decent Work: Glossary of frequently used terminology. ILO, Geneva (2023)
57. Glossary of frequently used terminology. ILO, Geneva (2019)
58. Cirera, X., Maloney, W.F.: The Innovation Paradox: Developing-Country Capabilities and the Unrealized Promise of Technological Catch-Up. World Bank, Washington (2017). https://doi.org/10.1596/978-1-4648-1160-9
59. OECD: Science, technologie et innovation: Perspectives de l'OCDE 2021 (version abrégée): Affronter la crise et saisir les opportunités. OECD Publishing, Paris (2021). https://doi.org/10.1787/1a6d0f4c-fr
60. Hanushek, E.A., Woessmann, L.: The role of cognitive skills in economic development. J. Econ. Lit. **46**(3), 629–668 (2012). https://doi.org/10.1257/jel.46.3.607
61. ILO: Decent Work and Productivity: Department of Enterprises. ILO, Geneva (2020)
62. Bloom, N., Van Reenen, J.: Why do management practices differ across firms and countries? J. Econ. Perspect. **1**, 203–224 (2010). https://doi.org/10.1257/jep.24.1.203

Developing Eco-innovations in Manufacturing Companies in Finland

Anne-Mari Järvenpää and Jari Jussila

Abstract Eco-innovations aim to decrease the negative environmental impact and increase resource efficiency. It is seen as a critical means to achieve the carbon neutral circular economy society. Developing eco-innovations in the manufacturing industry could be an impactful activity in terms of green transition. In fact, manufacturing industry produces a great deal of CO_2 emissions. On the other hand, the manufacturing industry is not able to take advantage of sustainable manufacturing technologies nor digital technologies. In this paper, we conduct a qualitative and multiple case study in Finland. We seek answers for what actions manufacturing SMEs are taking in terms of green and digital transition. The interview data were collected in January 2025 and was constructed as six case descriptions for content analysis. The potential for developing eco-innovations in these case companies relates to the data-based tools to help improve operational efficiency of case companies. University is seen as potential development partner in this kind of development, but the need to adjust the level of communication to the level of small companies is highlighted.

Keywords Circular economy · Eco-innovation · Manufacturing industry · SME · Digitalization · SDG9 · SDG12

1 Introduction

The transition toward a regenerative circular economy needs eco-innovations to make the planet healthier, improve the economy and society, and improve companies' social responsibility [1] and green reputation in the eyes of their customers. This is crucial for manufacturing companies, which are often perceived as a significant source of pollution.

Highlighting the importance of the manufacturing industry in the green transition is the fact that it produces 30% of CO_2 emissions, and is unable to take advantage of sustainable manufacturing technologies or basic digital technologies to collect and

A.-M. Järvenpää (✉) · J. Jussila
Häme University of Applied Sciences, Hämeenlinna, Finland
e-mail: anne-mari.jarvenpaa@hamk.fi

A. Visvizi et al. (eds.), *Research and Innovation Forum 2025 Vol2*, Springer Proceedings in Complexity, https://doi.org/10.1007/978-3-032-09062-1_16

analyze data [2] even in Europe. The main barriers to developing eco-innovation are identified as a lack of experts, RDI funding, and ecosystem activities [3]. The research gap is a lack of company-level understanding on how to enable a circular economy by utilizing digital technologies for value creation and the efficient use of resources [4].

The eco-innovation index measures countries' performance and shows where to focus development activities. Finland was chosen as the context of this study as the country is considered an eco-innovation leader in the EU [5], but recording below the EU-average progress [6]. In fact, the material footprint is high and resource productivity is low among EU countries [6].

This research aims to increase understanding of the development of eco-innovations in manufacturing companies, promoting a circular economy and sustainable development. We conducted an explorative case study [7] of six manufacturing small and medium-sized enterprises (SMEs) in Finland. The case study explores the needs and capabilities of manufacturing companies to co-create environmentally friendly innovations, leveraging data and digital technologies in collaboration with other companies, stakeholders, and universities. The research question is: *What actions are manufacturing SMEs taking in terms of the green and digital transition to create sustainable solutions?*

This paper provides background on the concept and the role of eco-innovation in the green transition. The research part was conducted as an explorative multiple case study. Research data were collected through thematic interviews, which were then elaborated into interview-specific case descriptions for content analysis. These descriptions are presented in the results section. In the conclusion section, we draw together the study's findings for further action.

2 Eco-innovations and the Manufacturing Industry

Innovation has been comprehensively defined as "production or adoption, assimilation, and exploitation of a value-added novelty in economic and social spheres; renewal and enlargement of products, services, and markets; development of new production methods; and the establishment of new management systems. It is both a process and an outcome" [8, p. 1157]. The Finnish standard that defines terms and fundamental concepts of innovation management defines innovation similarly as a new or changed entity, either realizing or redistributing value [9], value-added being an essential component in both definitions. Eco-innovation is a specific type of innovation that emphasizes its ecological dimension, resulting "throughout its life cycle in a reduction of environmental risk, pollution, and other negative impacts of resources use (including energy use) compared to relevant alternatives" [10, p. 7].

Regarding the process aspect, it has been identified that eco-innovations demand creating new process designs [11], and enabling the use of alternative materials by creating dematerialization processes and returning obsolete products via bridge systems and reverse logistics [12]. A characteristic of eco-innovation processes is the recovery of waste, materials, and energy, which remain in the used products at the end of their life cycles [13]. This necessitates closed-loop processes, where waste should be sustainably managed as a technical or biological resource that can be returned to industrial processes or the biosphere [14].

Concepts of "eco-innovation," "green innovation," "environmental innovation," and "sustainable innovation" are sometimes used interchangeably by researchers to describe innovations that have reduced negative impact on the environment [1]. However, Schiedering et al. [15] emphasize that, among these concepts, only eco-innovation encompasses a life-cycle-wide perspective (for material flow reduction).

While it is commonly stated that digital technologies can enable a circular economy, there is a lack of understanding at the company level of how to improve resource flows and value creation [4]. In general, digitalization can be seen as a tool to increase efficiency and the opportunity to create innovation, while sustainability stands for practices to minimize environmental impact [16].

In manufacturing companies, the focus of eco-innovations is on eco-design and engineering, decision-making behaviors and options, industrial trends and indicators, and energy efficiency [17]. Manufacturing companies seek to connect environmental benefits with improvements in overall eco-efficiency [18] that simultaneously achieve economic, environmental, and social sustainability [17]. Nevertheless, the literature states that managers suffer from a lack of understanding of strategies and mechanisms for creating sustainable solutions [19].

Finding opportunities for developing eco-innovations can be increased by accessing stakeholders' information to identify their needs, increasing research and development programs, and identifying better technologies and ways to close the resource and material loops [14]. Manufacturing companies are not competitive by themselves, but as part of collaboration networks, as eco-innovations require collaboration with other companies, universities, and research institutions to improve skills and resources, and facilitate ideas [17]. Previous research has identified three main barriers to developing eco-innovations in Finland: lack of experts, RDI funding, and ecosystem activities [3].

3 Methods

This research is conducted as a qualitative and explorative multiple case study [7] to determine the needs of manufacturing companies for a green and digital transition, particularly in reducing the environmental impact caused by their operations. The aim is to identify challenges that can be leveraged as opportunities for collaboration and ecosystem activities.

We conducted thematic interviews with six Finnish industry representatives in January 2025. The interviewed companies operate in the metal industry, the building product industry, the brewery sector, local business development, and the entrepreneur association. Interviews were recorded, transcribed, and summarized as case descriptions presented in the Results section. Interview themes were (1) Current development actions toward a green transition, (2) The role of data and digitalization in development activities, and (3) What kind of collaboration would enable successful development?

The content of the case descriptions was analyzed to compare the cases and identify common themes. Six cases were selected to conduct a first-round pilot study to generate insight from the state of case companies. This study will continue with a second round of interviews, utilizing more specific interview questions, and involving additional case companies.

4 Results

In this section, we summarize the six expert interviews as six cases to describe the views of the local manufacturing industry for a green and digital transition and the need for collaboration with the local university. In the last section, we show the results of the content analysis.

4.1 Case A

Corporate responsibility requirements for large companies are beginning to affect small manufacturing companies. They are often situated in the middle of the value chain and must adapt to regulations, even if they do not directly affect companies of their size. There is a need to promote transparency and take concrete measures regarding sustainable products and production processes. The challenge is that SMEs often lack sufficient resources and competence. Companies must understand what can be done and practical ideas to exploit existing opportunities. The local university already assists companies with reporting and data visualization of the current state, but further development efforts are also needed. For example, in the carbon footprint calculation, it is difficult to obtain the necessary scope 3 information from the value chain actors.

Collaboration is vital for the emergence of significant innovations. Green transition and digitalization are a cross-cutting theme for products, services, and processes—this is where collaboration pays off and is rarely a competitive activity among companies. Collaboration between universities and companies is essential and should be increased by activating companies. Nevertheless, it must be based on the needs of companies; otherwise, it will not last long. The local university plays a vital role, especially in developing competence and communicating current research results to companies at a level they can understand.

4.2 Case B

For companies, green demands come from customers, staff, and legislation. Companies that exceed the legislative minimum requirements will surely gain efficiency, increase sales margins, or improve employee satisfaction. Entrepreneurs' daily lives are focused on things necessary for business, and everything else is done at a minimum level. The challenge for companies is competence: in our region, there are many small companies where the entrepreneur's substance competence is emphasized. It is unlikely that SMEs will invest in more responsible or greener processes without a clear business advantage. Entrepreneurs are not necessarily interested in reducing carbon emissions.

Companies should learn how to utilize data effectively. The university can assist companies, but they must be able to adopt the business from a data perspective and minimize the burden on the entrepreneur as much as possible. University actors should observe the company's everyday life to understand how to utilize data more effectively. For example, the company's data could be reviewed and built into the customer life cycle, which would be an immediately valuable tool for sales and customer relationship management.

Local universities play a crucial role in developing expertise and vitality in the region by attracting young talent and fostering the emergence of new companies. The university is ahead of companies, but it is crucial to understand the concept of "one step backward" and the entrepreneur's context; otherwise, the entrepreneur will not understand.

4.3 Case C

This company takes responsibility seriously but needs to make its commitment more visible. The responsibility manual and workbook are necessary to guide the company's responsibility work. Responsibility work requires a place to collect and publish data, monitor operations, and guide development work. Lack of expertise is a challenge that can be overcome by learning by doing and through collaboration. Courses, training, and guidance are needed for responsibility work. This could be a topic for regional collaboration.

The university is expected to offer training, development projects, and student collaboration to combine digitalization and responsibility. It can facilitate collaboration between companies to bring synergies and apply development funding to the region. However, even though there are already collaboration opportunities between the university and companies, there is a need for more approachable opportunities.

4.4 Case D

This company's customers demand that the products' carbon footprint be as accurately known as possible. The challenge is to collect data for calculation and, above all, to obtain data along the supply chain. Some companies are willing to share the required data, while others are not. This company conducts ongoing development activities for data collection, addressing the principles of double materiality.

The local university is expected to possess the best research knowledge and be at the forefront of the green and digital transition. Learning environments in the context of education are considered essential. Regarding local collaboration, the university can act as a catalyst by coordinating development activities and providing a platform for addressing important topics. The university should facilitate the sharing of experiences and know-how between companies at a general level. This is viewed as a valuable and achievable development opportunity to enhance understanding and facilitate easier development for companies.

4.5 Case E

This company is working to enhance its overall operational efficiency to improve productivity. Sustainable development and a circular economy are viewed as efficient uses of resources throughout the business, as "all waste is waste, and it is wasteful!" The challenge is the limited resources of a small company, as there is not enough time to do everything. Particularly, there is a need for resources to implement development ideas. University students with the latest knowledge are seen as a potential resource for developing quality guidelines incorporating 3D design and production efficiency using Lean methodology. On the other hand, staff well-being as a social side of responsibility can be improved by Lean processes: reducing rush helps reduce sick leave. This company is interested in what can be measured in terms of staff well-being. For all this, they need data collection, analysis, and easy-to-understand, visualized information for staff in real-time.

Facilitated collaboration between local companies is viewed as an interesting opportunity to develop processes through study visits and workshops with people from different production environments. With the help of the university, the development ideas could be demonstrated, e.g., by simulating a workstation or components moving through the factory. Collaboration with university students is expected to be facilitated in a manner that is not burdensome to the company.

4.6 Case F

The industry has many demands; it is energy-intensive, has poor material recyclability, and high water usage. Given its long payback period, the green transition requires significant external pressure for a poorly profitable sector to invest in. Another limiting factor is an old-fashioned sector with aging equipment and old owners. This company is competing with cheaper countries that have new factories. The new thing is closing the loop with the material supplier by converting material waste back into raw material. They can provide their customers with a greener product as a competitive advantage. Still, customers in Finland do not recognize it as a competitive advantage, although it is common in Europe. The benefits of greener products need to be demonstrated in numbers for the entire chain, and there is a need to help companies find easy-to-implement opportunities.

This company is struggling with old technology, saying, "We don't need artificial intelligence; common sense is enough for now." They are developing production time management to smooth out peaks in electricity consumption by not running energy-intensive machines at full load. Waste is money in this industry; therefore, waste streams need to be measured, and how much money flows out of the factory as a side-stream must be shown. The challenge is to implement the measurement. Data may be collected continuously, but not combined or utilized in any way. Efficiency in material use begins with production planning and optimization. The challenge in developing processes is that there is a need to dive deep into processes before being able to develop them, and this is not an easy task for university collaboration. Instead, data processing and guidance could be a collaboration opportunity, as historical information provides a clear picture of the future, showing how waste streams are scaled up when production increases.

4.7 Content Analysis

Activities toward a green transition relate to responsibility reporting and developing operational efficiency. Limited resources and competences of SMEs are a challenge in many ways [20]. Business owners' attitudes toward green values can be a barrier, as investments required by the green transition will not be made without clear business benefits. At the same time, minimizing waste is seen as a motivation for the operational efficiency and profitability of business management.

In responsibility reporting, SMEs are challenged by data availability along the supply chain. From the perspective of improving operational efficiency, the data-related challenge is competence. Data and digitalization are thought to be tools in the green transition, but even today, SMEs are still learning how to use data. Nevertheless, it is not necessarily clear to SMEs what benefits could be achieved through data analysis and visualization—they need an outside perspective on how data could be utilized in their day-to-day operations. The university is seen as playing a role in

Table 1 Case content analysis

	Activities toward the green transition	Role of data and digitalization	Need for learning and collaboration
Case A	Responsibility reporting	Carbon footprint calculation	Knowledge transfer
Case B	Operational efficiency	Data-based tool to guide operations	Knowledge transfer
Case C	Responsibility reporting	Carbon footprint calculation	Joint learning
Case D	Responsibility reporting	Carbon footprint calculation	Joint learning
Case E	Operational efficiency	Data-based tool to guide operations	Knowledge transfer
Case F	Operational efficiency	Data-based tool to guide operations	Joint learning

producing practical information to support companies in the green and digital transition. It is highlighted that the university must communicate in a manner that is accessible to SMEs, allowing them to understand and absorb new knowledge. Conversely, the university is expected to play a role in providing joint learning opportunities for companies to share experiences and brainstorm solutions.

Case content analysis (Table 1) shows that at the moment, the green activities relate to either responsibility reporting or improving operational efficiency. Data and digitalization's role lies in calculating the carbon footprint or providing operations guidance with data-based tools. The need for learning relates to either knowledge transfer between the company and university or joint learning opportunities.

Content analysis reveals that companies undertaking activities for responsibility reporting, quite obviously, consider the role of data and digitalization in calculating their carbon footprint. Companies seeking operational efficiency consider that data and digitalization provide opportunities to create data-based tools to guide operational efficiency development.

5 Discussion and Conclusions

As the results show, SMEs tend to be more practical than innovation-driven, and they are more interested in resource efficiency and cost savings when seeking opportunities to develop a circular economy and resource efficiency—this aligns with previous research in Finland [2].

According to the industry representatives we interviewed, manufacturing SMEs' actions in terms of green transition relate to developing operational efficiency (e.g., reducing waste) or responsibility reporting. We noted that entrepreneurs' attitudes toward the green transition can pose a challenge and emerge as a lack of investment or development activities.

The role of digitalization lies quite obviously in responsibility reporting, as well as data analysis and visualizations for operational efficiency. The great challenge for small companies is their limited resources and capabilities to drive the green transition through the digital transition.

The potential to start developing eco-innovations in these case companies relates to a data-based tool to increase the company's efficiency and, in this manner, boost the development of greener production. The known challenges or barriers are a lack of strategies and mechanisms to create sustainable solutions [19] and a lack of company-level understanding of how digital tools can enable sustainable solutions and value creation [4]. Here, the concepts of Innovation 5.0 and Society 5.0 are approaches to develop human-centered digital and sustainable innovations [21, 22].

Considering these challenges with the identified Finnish barriers [3] of a lack of experts, RDI funding, and ecosystem activities, the recommendations for further activities can be formed as follows. Companies' capabilities can be increased by providing facilitated organizational learning opportunities and joint development opportunities [23] as ecosystem activities. University students are considered essential resources for companies, but the collaboration opportunities must not burden a company. An important research and development opportunity is identifying companies' stakeholders' needs to find development opportunities [14]. An important reminder for the university is the need to adjust communication to the level that small companies with limited resources and capabilities can absorb. This kind of ecosystem collaboration could be developed and conducted by action research. Further research could explore the needs of knowledge transfer and organizational learning regarding eco-innovation development.

Acknowledgements This research was supported by the European Regional Development project *DAVE—Data-driven circular economy R&D eco-innovation system and demonstration environment (A80798).*

References

1. Díaz-García, C., González-Moreno, Á., Sáez-Martínez, F.J.: Eco-innovation: insights from a literature review. Innovation **17**(1), 6–23 (2015). https://doi.org/10.1080/14479338.2015.1011060
2. Salminen, K., Kostia, S., Ijas, M., Siivonen, J., Kiviniemi, E., Aho, M.: Boosting resource efficiency and circular economy of the manufacturing companies in Finland. In Sustainable Manufacturing as a Driver for Growth, Kohl, H., Seliger, G., Dietrich, F., Mur, S. (Eds.). Springer Nature Switzerland, Cham, pp. 363–371. https://doi.org/10.1007/978-3-031-77429-4_40
3. Autio, V., Raudaskoski, A., Johansen, M.R.: Eco-Innovation Country Profile 2022: Finland. Policy Brief, Sep. 2022. [Online]. Available: https://circabc.europa.eu/ui/group/96ccdecd-11b4-4a35-a046-30e01459ea9e/library/0b9f1a2a-2b3e-4afc-abdb-3db8258107b9/details

4. Ranta, V., Aarikka-Stenroos, L., Väisänen, J.-M.: Digital technologies catalyzing business model innovation for circular economy—multiple case study. Resour. Conserv. Recycl. **164**, 105155 (2021). https://doi.org/10.1016/j.resconrec.2020.105155
5. European Commission: EU eco-innovation index 2024. European Commission, 2024. Accessed: Feb. 13, 2025. [Online]. Available: https://doi.org/10.2777/4878812
6. European Environment Agency: Circular economy country profile 2024—Finland. Europea Environment Agence, ETC-CE Report 2024/Finland, Mar. 2024. Accessed: Feb. 19, 2025. [Online]. Available: https://www.eea.europa.eu/en/topics/in-depth/circular-economy/country-profiles-on-circular-economy/circular-economy-country-profiles-2024/finland_2024-ce-country-profile_final.pdf
7. Yin, R.K.: Designing case studies. Qual. Res. Methods **5**(14), 359–386 (2003)
8. Crossan, M.M., Apaydin, M.: A multi-dimensional framework of organizational innovation: a systematic review of the literature. J. Manage. Stud. **47**(6), 1154–1191 (2010)
9. SFS Finnish Standards: Innovation Management. Fundamentals and vocabulary (ISO 56000:2025: en), 2025. [Online]. Available: https://sfs.fi/en/
10. Kemp, R., Pearson, P.: Final report of the MEI project measuring eco innovation. UM-Merit, Maastricht, 2007. [Online]. Available: http://www.merit.unu.edu/MEI/deliverables/MEI%20D15%20Final%20report%20about%20measuring%20eco-innovation.pdf
11. Sehnem, S., Campos, L.M., Cazella, C.F.: Circular business models: level of maturity. Manage. Decis. **57**(4), 1043–1066 (2019)
12. Ponte, B., Framinan, J.M., Cannella, S., Dominguez, R.: Quantifying the bullwhip effect in closed-loop supply chains: the interplay of information transparencies, return rates, and lead times. J. Prod. Econ. **230**, 107798 (2020)
13. Park, J., Sarkis, J., Wu, Z.: Creating integrated business and environmental value within the context of China's circular economy and ecological modernization. J. Clean. Prod. **18**(15), 1494–1501 (2010). https://doi.org/10.1016/j.jclepro.2010.06.001
14. Prieto-Sandoval, V., Jaca, C., Santos, J., Baumgartner, R.J., Ormazabal, M.: Key strategies, resources, and capabilities for implementing circular economy in industrial small and medium enterprises. Corp. Soc. Responsib. Environ. Manage. **26**(6), 1473–1484 (2019). https://doi.org/10.1002/csr.1761
15. Schiederig, T., Tietze, F., Herstatt, C.: Green innovation in technology and innovation management–an exploratory literature review. Rd Manage. **42**(2), 180–192 (2012). https://doi.org/10.1111/j.1467-9310.2011.00672.x
16. Tiron-Tudor, A., Deliu, D.: The sustainability performance of SMEs in industry 5.0: a proposal for a synergetic digital sustainable development maturity model (SDSDMM). East. Eur. Econ. **1**, 45 (2024). https://doi.org/10.1080/00128775.2024.2442306
17. Janahi, N.A., Durugbo, C.M., Al-Jayyousi, O.R.: Eco-innovation strategy in manufacturing: a systematic review. Clean. Eng. Technol. **5**, 100343 (2021). https://doi.org/10.1016/j.clet.2021.100343
18. Afshari, H., Searcy, C., Jaber, M.Y.: The role of eco-innovation drivers in promoting additive manufacturing in supply chains. Int. J. Prod. Econ. **223**, 107538 (2020). https://doi.org/10.1016/j.ijpe.2019.107538
19. Salim, N., Ab Rahman, M.N., Abd Wahab, D.: A systematic literature review of internal capabilities for enhancing eco-innovation performance of manufacturing firms. J. Clean. Prod. **209**, 1445–1460 (2019). https://doi.org/10.1016/j.jclepro.2018.11.105
20. Järvenpää, A.-M., Kunttu, I., Mäntyneva, M.: Using foresight to shape future expectations in circular economy SMEs. Technol. Innov. Manage. Rev. **10**(7), 41–50 (2020). https://doi.org/10.22215/timreview/1374
21. Celenta, R., Cucino, V., Feola, R., Parente, R.: Towards innovation 5.0: the role of corporate entrepreneurship. In Research and Innovation Forum 2023, Visvizi, A., Troisi, O., Corvello, V. (Eds.). Springer International Publishing, Cham, pp. 451–463. https://doi.org/10.1007/978-3-031-44721-1_34
22. Troisi, O., Visvizi, A., Grimaldi, M.: Rethinking innovation through industry and society 5.0 paradigms: a multileveled approach for management and policy-making. Eur. J. Innov. Manage. **27**(9), 22–51 (2023). https://doi.org/10.1108/EJIM-08-2023-0659

23. Järvenpää, A.M., Jussila, J., Kunttu, L., Kunttu, I.: Organizational learning practices to develop digitalization capabilities in circular economy SMEs. J. Circ. Econ. (2025). https://doi.org/10.55845/ULPA5543

Catalyzing Digital Innovation in AgTech: The Case of XFARM as a Digital Innovation Ecosystem

Domenica Barile, Giustina Secundo, and Pasquale Del Vecchio

Abstract This study examines how digital innovation ecosystems (DIEs), supported by advanced technologies, foster collaboration and co-innovation among stakeholders. The research focuses on XFARM, an Ag-Tech start-up, as a case study to explore its contributions to digital transformation in the agricultural sector. A qualitative methodology is employed, adopting a single case study approach and utilizing both primary and secondary data sources. The theoretical framework draws on innovation ecosystem theory, particularly the works of Adner (Harvard Bus Rev, 2006) [2] and Granstrand and Holgersson (Technovation, 2020) [22], to analyze interdependencies and dynamics within the DIE orchestrated by XFARM. XFARM integrates advanced technologies—including IoT sensors, AI-driven analytics, and satellite-based remote sensing—to deliver real-time data on crop conditions, optimize resource usage, and enhance decision-making efficiency for farmers. Through strategic partnerships with international organizations, XFARM has driven innovations such as smart irrigation systems and sustainable agricultural practices. These advancements align with sustainability goals while improving agricultural competitiveness. The study offers a unique perspective on how a digital platform like XFARM can act as a catalyst for innovation, fostering a collaborative network involving farmers, technology providers, research institutions, and governmental bodies. XFARM's success underscores the critical role of DIE orchestration in enabling co-innovation. By embedding digital tools into traditional agricultural practices, XFARM illustrates the transformative potential of technology. Nonetheless, challenges remain, including cultural barriers to technology adoption and the high

D. Barile (✉) · G. Secundo · P. Del Vecchio
Department of Engineering, LUM University Giuseppe Degennaro, Casamassima (Bari), Italy
e-mail: barile@lum.it

G. Secundo
e-mail: secundo@lum.it

P. Del Vecchio
e-mail: delvecchio@lum.it

A. Visvizi et al. (eds.), *Research and Innovation Forum 2025 Vol2*, Springer Proceedings in Complexity, https://doi.org/10.1007/978-3-032-09062-1_17

initial investment required by small-scale farmers. Theoretical and practical implications are discussed, highlighting the significance of interdependencies and collaborative dynamics in achieving sustainable innovation and demonstrating the efficacy of this analytical framework within digital ecosystems.

Keywords Innovation ecosystem · AgTech · Digital technologies · Start-up · SDGs · Goal 9

1 Introduction

Agriculture has long been a cornerstone of human civilization, driving food production and supporting various industries. In recent years, technological advancements have reshaped the agricultural landscape, enhancing productivity, sustainability, and competitiveness. The so-called "third revolution in agriculture" [28, 30] has been propelled by digital innovation ecosystems (DIEs), which foster collaboration among diverse stakeholders, including farmers, technology providers, and research institutions. These ecosystems facilitate the co-creation of innovative solutions by leveraging cutting-edge technologies such as artificial intelligence (AI), the Internet of Things (IoT), and satellite-based remote sensing.

The digital revolution has introduced new forms of agricultural innovation that rely more heavily on information and communication technologies (ICT) than on traditional agricultural knowledge [9, 24]). These digital solutions have enabled firms to increase efficiency and competitiveness while promoting more sustainable practices [18]. Although prior research has investigated the motivations behind agribusinesses adopting such technologies, a gap remains in understanding how DIEs facilitate the large-scale commercialization of radical innovations.

Historically, agricultural innovations have emerged within traditional supply chains, often dominated by large firms [13, 30]. However, the rise of ICT-based solutions has opened new opportunities for smaller firms—particularly start-ups—to introduce disruptive innovations. These AgTech start-ups [28] use advanced technologies to enhance agricultural production and distribution. Given their limited financial, physical, and knowledge resources [25], start-ups derive substantial benefits from participating in innovation ecosystems (IEs), which provide access to essential resources, improve risk management, and facilitate market connections [2, 5].

A prominent example is California's AgTech hub, home to leading start-ups, business accelerators, and venture capital firms committed to revolutionizing agricultural production [28]. While the growing body of literature has emphasized the benefits of digital technologies (DTs) in agriculture—ranging from improved productivity to enhanced decision-making [18, 24]—less attention has been paid to the systemic role of DIEs in enabling the orchestration and scalability of these innovations within the AgTech domain.

Most existing studies concentrate on firm-level drivers of technology adoption or focus on innovation within isolated organizational contexts, rather than exploring how multiple actors interact within dynamic, interdependent ecosystems to co-create and diffuse innovations (32). Furthermore, empirical research is limited on how agricultural start-ups—particularly platform-based AgTech firms—serve as ecosystem orchestrators, coordinating stakeholders, managing interdependencies, and driving sustainable innovation [2, 33]. This research gap is particularly significant, as the ecosystem perspective provides a holistic understanding of innovation through collaboration, co-creation of value, and shared capabilities [34].

This study contributes to addressing this gap by investigating the case of XFARM, an AgTech digital platform, and illustrating how it orchestrates an agricultural IE through technological integration and stakeholder collaboration. XFARM equips farmers with digital tools to optimize operations by integrating data from IoT sensors, AI-based analytics, and remote sensing technologies. The platform fosters collaboration among a diverse range of ecosystem actors, including farmers, technology developers, and research institutions.

Despite the significant opportunities presented by digitalization, several challenges persist. Technology adoption is influenced by factors such as farm size, economic capacity, and cultural attitudes towards innovation [26]. Moreover, the financial sustainability of these solutions remains uncertain, particularly for small and medium-sized farms that face high implementation costs and require training [6, 23]. While DTs are increasingly being adopted in agriculture, few empirical studies explore how AgTech companies function as ecosystem orchestrators in scaling sustainable digital transformation.

This study seeks to fill this gap by addressing the following research question:

> How do collaboration dynamics, digital technology adoption, and co-innovation processes within AgTech ecosystems impact agricultural firms' innovation, competitiveness, and sustainability?

Through an in-depth examination of XFARM, a platform operating in Italy, the study offers both theoretical and practical insights into the orchestration of AgTech IEs. Drawing from IE theory [2, 22], the analysis focuses on how stakeholder interdependencies shape digital technology (DT) adoption and influence innovation outcomes. Adner's framework also provides a lens for understanding ecosystem coordination and value creation, with an emphasis on co-innovation [1].

The paper is structured as follows: first, a theoretical overview of IEs and AgTech developments is presented, with attention to models explaining collaboration and technology adoption in agriculture. Next, the XFARM case study is analyzed to explore the internal and external ecosystem dynamics that have contributed to its success. Insights from interviews with XFARM managers further enrich the understanding of strategic approaches to digital innovation challenges. The paper concludes by distilling key lessons from XFARM's experience and discussing future directions for AgTech ecosystems.

By examining these dynamics, this study aims to advance the understanding of how IEs drive agricultural transformation. Collaboration among stakeholders, adoption of DTs, and co-innovation emerge as critical levers for ensuring the sustainability and competitiveness of agricultural enterprises in an evolving market landscape.

2 Theoretical Background

2.1 Digital Platform and Innovation Ecosystem

Innovation is increasingly conceptualized as a phenomenon rooted in ecosystems composed of autonomous actors who collaborate to create and capture value through complementarities and structured interdependencies [7]. The concept of an IE has emerged to describe the interactions among firms, research institutions, and other stakeholders that co-evolve to generate innovation and competitive advantage [22]. Unlike traditional innovation systems, ecosystems place greater emphasis on dynamic relationships among actors and the governance of complementarities and interdependencies [17].

In the context of agricultural innovation, several categories of actors are involved [35], including:

(a) agricultural producers;
(b) the education and training system;
(c) the agricultural research system;
(d) research and innovation agencies;
(e) credit agencies;
(f) the rural extension and technical assistance system;
(g) companies supplying inputs, equipment, and services;
(h) producer and business organizations;
(i) agroprocessors;
(j) exporters;
(k) government institutions; and
(l) end consumers.

Additionally, enterprise generation mechanisms such as business incubators, accelerators, coworking spaces, and open laboratories play a significant role [36]. Central to this landscape are agricultural technology-based startups—often referred to as agtechs, agritechs, or agrifoodtechs—which are recognized for their agility, knowledge, and boldness in bringing disruptive innovations to the field.

A key theme in the literature on IEs is the role of digital platforms as facilitators of value creation and interaction among diverse actors [10, 14]. These platforms, often supported by open innovation practices, enable knowledge sharing and foster new idea generation through collaborative mechanisms [12]. The rapid advancement of DTs has led to the emergence of platform-based ecosystems, where value is co-created through the integration of advanced tools and interactions among heterogeneous participants [14].

Different types of actors play crucial roles in the emergence and development of IEs. Dedehayir et al. [17] identify four main categories:

- Ecosystem leaders—orchestrators who guide innovation and set strategic directions;
- Direct value creation actors—such as suppliers, assemblers, and users, who contribute directly to value generation;
- Value creation support actors—including experts and sponsors, who provide essential resources and expertise;
- Entrepreneurial ecosystem actors—such as regulators and investors, who shape the ecosystem's structure and policies.

This framework provides insight into how various actors contribute to the formation and evolution of IEs. Moreover, an IE's adaptability hinges on its ability to manage interdependencies and leverage the modularity of digital platforms for coordinated value creation [7].

Within business ecosystems, Start-up Collaboration Units (SCUs) have gained prominence as intermediaries between large corporations and innovative startups [16]. These units help mitigate information asymmetries, facilitate knowledge transfer, and manage cultural tensions between incumbents and new entrants. The success of an IE depends on the effective orchestration of key players who must balance openness with control to enable co-evolution [22].

An emerging perspective in the literature focuses on the governance of digital ecosystems, particularly on how interactions among actors are managed to optimize collective innovation [14]. While digital platforms have the potential to democratize access to innovation, they also introduce challenges related to power asymmetries and the equitable distribution of value [14]. The long-term resilience and dynamism of an IE will depend on its capacity to navigate these tensions and adapt governance structures to the evolving digital environment.

2.2 AgTech Startups, Digital Technologies and Innovation Ecosystems

AgTech startups are emerging as pivotal actors in agricultural innovation, offering technological solutions aimed at enhancing productivity and sustainability. The digitization of agriculture—propelled by advances in AI, the IoT, and robotics—is transforming the sector [11]. These innovations not only increase agricultural competitiveness but also foster the development of IEs that integrate startups, research institutions, and technology firms.

By leveraging emerging technologies and open innovation practices, AgTech startups actively respond to climate and economic challenges, thereby contributing to global food security [26]. Operating within dynamic and interconnected ecosystems, these startups collaborate with incubators, accelerators, and investors to bring innovative solutions to market [27]. A notable example is the AgTech ecosystem in California, where startups, universities, and accelerator programs work together to deploy advanced technologies, such as smart irrigation systems [28].

DTs serve as the primary growth enablers for AgTech startups. The integration of IoT sensors, blockchain, and AI is catalyzing a shift toward more automated and intelligent agricultural practices [4]. These technologies support real-time crop monitoring, optimize resource use, and enhance environmental sustainability. A particularly significant advancement is the deployment of autonomous agricultural robots for crop management. These robotic solutions reduce labor reliance and enhance operational precision, although their adoption is often constrained by concerns about high upfront costs and the need for specialized training.

AgTech IEs thrive through collaborations among startups, universities, farms, and investors. In Brazil, for instance, the AgTech ecosystem benefits from robust support provided by entrepreneurial networks and innovation communities, which facilitate access to financial and technological resources [8]. Regional clusters and startup accelerators play a critical role in fostering localized solutions tailored to specific market needs. In Europe, initiatives such as innovation hubs in France promote the adoption of DTs by facilitating cooperation between companies and research institutions, thus accelerating the sector's digital transition [11].

Despite the promising opportunities enabled by DTs, AgTech startups continue to face significant challenges. Limited access to funding, regulatory constraints, and cultural resistance to technological change remain persistent barriers [37, 38]. Moreover, scalability poses a major hurdle, as many startups encounter difficulties in expanding their solutions due to insufficient infrastructure and strategic support. Another key issue is the integration of DTs within traditional agricultural business models. Long-term viability of AgTech startups depends on their capacity to develop sustainable business models, such as digital platforms and subscription-based services.

AgTech startups are reshaping the agricultural landscape by harnessing DTs within structured IEs. Their success increasingly relies on collaboration among diverse stakeholders, supported by digital platforms and accelerator initiatives. Although scalability and technology adoption present ongoing challenges, the outlook for AgTech startups remains optimistic, with a growing influence on the future of global agriculture.

3 Research Method

This study employs a single exploratory case study methodology. The single case study represents a robust and context-sensitive research strategy for examining complex phenomena or underexplored processes [31], particularly when assessing the relevance of theoretical and conceptual frameworks [19]. From this perspective, a single case study is particularly well-suited for addressing 'how' and 'why' questions related to contemporary developments.

The case selection follows [20] guidance on aligning theoretical interests with empirical inquiry to deepen the understanding of DIEs. The choice aims to enhance theoretical insights by anchoring them in the real-world dynamics of a digital startup operating within a platform-based environment.

This study focuses on xFarm as a representative case to explore its role as a driver of digital transformation in the AgTech sector. The case study approach allows for a detailed examination of the complex interactions that characterize IEs, particularly within the domain of digital agriculture. The research seeks to uncover the mechanisms through which xFarm fosters co-innovation and supports the adoption of advanced DTs in farming.

3.1 Data Collection and Analysis

The data collection process took place from January 2023 to December 2024 and combine primary and secondary sources to ensure a comprehensive understanding of the xFarm ecosystem. Primary data were collected through semi-structured interviews conducted with xFarm managers and key stakeholders, focusing on their motivations, strategies, and challenges related to the development and implementation of the xFarm platform. Specifically, the interviews explored:

- The history and evolution of xFarm.
- Its organizational structure and strategic perspectives.
- The technological innovations developed and their impact on agricultural practices.
- Collaborations between xFarm and various ecosystem actors (e.g., farmers, research institutions, and technology providers).
- Barriers and facilitators of technology adoption in agriculture.

Using an inductive and iterative methodology, this research provides a systematic analysis of the xFarm DIE while ensuring reliability and theoretical soundness [15, 20, 21].

Semi-structured interviews were conducted, each lasting between 40 and 70 min, with key figures in the company, including the Education & Academy Manager and the AgTech Education Specialist [3, 29]. These interviews offered rich qualitative insights into xFarm's role in driving digital transformation in agriculture within an IE.

In addition to interviews, secondary data were collected from various sources, including archived documents (e.g., reports, industry publications, and company white papers), Social media communications and online discussions related to xFarm, Corporate websites and other publicly available materials.

These secondary sources complemented the primary data, offering a broader perspective on xFarm's activities and impact within the AgTech sector.

The data analysis followed a qualitative content analysis approach to identify key themes and patterns within the collected data. The steps of the analysis included:

- Transcription and Data Cleaning of all the recorded interviews that were fully transcribed and anonymized to ensure confidentiality.
- Coding and Thematic Analysis of transcripts and secondary data that were analyzed using thematic coding techniques to identify recurring themes related to innovation dynamics, ecosystem collaboration, and technological adoption.
- Triangulation fundamental in order to enhance data reliability, findings from interviews were cross-referenced with secondary sources to verify key insights.
- Interpretation and Framework Application of the IE theory from the author Adner's [2] and Granstrand and Holgersson [22] that was applied to examine the interdependencies between ecosystem actors and assess how xFarm orchestrates innovation through digital tools.

Through this methodological approach, the study aims to uncover the critical success factors that have enabled xFarm to emerge as a driver of digital transformation in agriculture provides a robust foundation for analyzing DIEs in AgTech. Additionally, it highlights key challenges that persist in the sector, particularly in relation to technology adoption barriers, economic sustainability, and the need for coordinated stakeholder engagement.

4 Findings

The analysis of xFarm reveals several key factors that have contributed to the platform's success and its widespread adoption within the agricultural sector. These factors include technological innovation, strategic market integration, developmental challenges, and the platform's overall impact on agricultural efficiency and sustainability.

A central theme emerging from the analysis is the presence of strong technological interdependencies within the AgTech ecosystem. These interdependencies transcend physical processes and traditional innovation, encompassing digital elements such as data management and the integration of external technologies. In the case of xFarm, these interconnections are critical to the platform's functionality. Its digital infrastructure—including IoT sensors, AI systems, and data management tools—must effectively interface with the technologies and capabilities of other ecosystem stakeholders to ensure seamless operation and value co-creation (see Table 1).

4.1 Technological Innovation and Digital Integration

A defining characteristic of XFARM is its ability to integrate multiple emerging technologies into a single, unified platform. The system employs IoT sensors to

Table 1 XFARM technological solutions and their impact

Technology/service	Description	Reported impact
Smart irrigation (Ocmis)	Real-time water management based on soil data	Up to 20% reduction in water usage
Satellite imaging (SpaceSense)	Crop health and weather predictive analytics	Early risk detection, improved resilience
Decision support system (Univ. Bari)	AI-based support for agronomic decisions	Optimized fertilizer use, yield increase (est. 12%)
User-friendly mobile interface	Easy-to-use dashboard for farmers	Higher engagement among low-digital-literacy users

Source Authors

collect real-time data from crops, which is then analyzed using AI and machine learning to generate actionable insights for farmers.

For example, XFARM has developed smart irrigation systems in partnership with Ocmis, optimizing water use based on real-time soil moisture data. Additionally, satellite imagery from SpaceSense enhances crop monitoring, providing predictive analytics on weather conditions and crop health. These innovations have significantly advanced precision farming, reducing resource waste and increasing productivity.

Moreover, XFARM's modular platform allows users to customize digital tools according to their specific agricultural needs. This adaptability has played a crucial role in fostering the adoption of DTs, particularly in regions where traditional farming methods still dominate.

4.2 XFARM's Innovation Ecosystem

To analyze XFARM's role within the DIE, this study adopt IE theory from the author Adner [2] and Granstrand and Holgersson [22], which emphasizes the importance of interdependencies and ecosystem orchestration. Successful IEs rely on the coordinated participation of multiple stakeholders, each contributing complementary resources and capabilities [2].

Adner's framework highlights three critical aspects:

- Interdependence Management: The value of an innovation depends not only on the focal firm's actions but also on the alignment and contributions of other ecosystem actors. XFARM exemplifies this by integrating its platform with third-party technologies, including IoT devices, AI-driven analytics, and satellite monitoring, ensuring seamless functionality.
- Orchestration and Coordination: The ability to coordinate multiple stakeholders—farmers, research institutions, and technology providers—determines an IE's success. XFARM plays a central role in facilitating collaboration and knowledge exchange through its platform.

- Value Creation and Capture: The ecosystem must ensure value distribution among all participants. XFARM enables farmers to optimize resources, while technology providers and agribusinesses benefit from enhanced data analytics and market access.

Findings suggest that XFARM's ability to manage interdependencies, coordinate actors, and distribute value across the ecosystem has been instrumental in its success as an AgTech innovator. An interpretative framework about XFARM's IE is presented in Fig. 1.

The XFARM IE Framework, illustrated in Fig. 1, presents a multi-layered model that encapsulates the company's strategic approach to orchestrating digital transformation in agriculture. At its core is the DT and Platform Layer, which constitutes the operational backbone of the ecosystem. This foundational layer integrates a suite of enabling technologies—including IoT devices (e.g., xTrap for pest monitoring), AI, machine learning, and satellite-based remote sensing—to support precision agriculture and data-driven decision-making.

Surrounding the core is the Innovation and Sustainability Layer, which emphasizes continuous research and development. This layer comprises internal innovation initiatives, access to public and European R&D funding mechanisms (e.g., the EIC Accelerator), and collaborative partnerships with academic institutions such as the University of Bari. It reflects XFARM's strategic commitment to sustainability and co-creation of knowledge within the AgTech space.

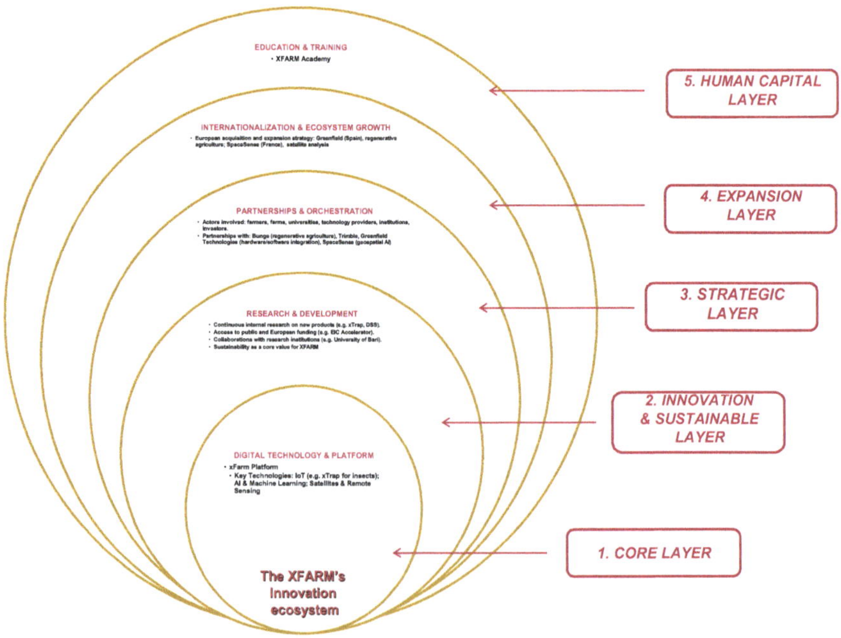

Fig. 1 XFARM's innovation ecosystem. *Source* Authors

The Strategic Layer linked to partnerships and orchestration activity encompasses XFARM's coordination with a wide array of ecosystem actors, including farmers, universities, technology providers, institutional stakeholders, and investors. Strategic alliances with firms such as Bunge, Trimble, and SpaceSense have expanded the platform's capacities in fields like regenerative agriculture and geospatial AI, enhancing the overall resilience and adaptability of the ecosystem.

The Expansion Layer reflects XFARM's internationalization strategy, including strategic acquisitions and collaborations in Spain (e.g., Greenfield Technologies) and France (e.g., SpaceSense). This layer underscores the scalability of the XFARM ecosystem and its capacity to adapt its digital solutions to diverse agroecological and market contexts beyond national borders.

Lastly, the Human Capital Layer addresses the challenges of digital literacy and cultural resistance to technology among farmers. Through the XFARM Academy, this layer promotes digital skills development, awareness-building, and inclusive access to digital agriculture tools, thereby fostering a human-centered approach to innovation.

Together, these interconnected layers demonstrate how XFARM leverages DTs, collaborative governance, and structured knowledge transfer to construct a scalable, sustainability-oriented, and inclusive IE within the agri-food sector.

XFARM's success is fundamentally rooted in its ability to foster collaboration across diverse stakeholder groups, including farmers, research institutions, technology firms, and public organizations. The platform functions as a DIE, facilitating knowledge exchange and enabling co-innovation.

For example, XFARM's collaboration with the University of Bari has resulted in the development of decision support systems (DSS) that assist farmers in optimizing agricultural operations. These joint efforts have produced predictive models aimed at mitigating climate-related risks and improving crop resilience.

Furthermore, XFARM has established synergistic partnerships with agribusinesses and agricultural machinery manufacturers, such as Trimble and Greenfield Technologies, to ensure seamless integration between digital tools and physical equipment. This approach enhances the interoperability of machinery and digital platforms, reinforcing the robustness and effectiveness of the broader ecosystem.

4.3 Challenges and Barriers to Adoption and the Search for Sustainability

Despite XFARM's significant technological advancements, several challenges continue to hinder the widespread adoption of digital tools in agriculture.

A primary barrier is the high initial investment cost, which is particularly burdensome for small and medium-sized farms. Many farmers perceive these technologies as financially inaccessible, thereby slowing their adoption and limiting their impact across the sector.

Another substantial challenge is the digital divide among agricultural operators. While large-scale farms have increasingly adopted digital innovations, smaller agricultural enterprises often lack the technical expertise and financial capacity to implement such tools effectively. In response, XFARM has sought to bridge this gap by developing user-friendly digital interfaces and offering targeted training programs through the XFARM Academy, aiming to facilitate more inclusive adoption of DTs.

Cultural resistance to technological change also remains a significant hurdle. Traditional farmers may be hesitant to shift from conventional methods to data-driven decision-making. To counter this, XFARM has initiated awareness campaigns that showcase the practical benefits of digital agriculture, including increased efficiency, cost savings, and improvements in environmental sustainability.

Sustainability lies at the core of XFARM's innovation strategy. Through the facilitation of data-driven practices, the platform supports more efficient use of agricultural inputs. Farmers employing XFARM's precision farming tools have reported reductions in water consumption, fertilizer application, and pesticide use, which contribute to lower greenhouse gas emissions and promote environmentally sustainable agricultural practices.

Looking ahead, XFARM is focused on expanding its technological capabilities by integrating advanced AI models and blockchain technology to enhance traceability and transparency in agricultural supply chains. Furthermore, the company is exploring new strategic partnerships to extend its IE, with particular emphasis on increasing accessibility to digital farming solutions in developing regions.

5 Conclusion, Limitation and Future Research Avenue

This study has explored the role of XFARM as a DIE within the AgTech sector, illustrating how DTs and ecosystem orchestration can drive agricultural transformation. The findings, while centered on XFARM, offer broader applicability to other AgTech ecosystems, particularly those in fragmented or resource-constrained contexts. Grounded in IE theory [2, 22], the analysis underscores the critical role of stakeholder interdependencies, technological integration, and coordinated value creation in shaping effective IEs.

XFARM's success demonstrates how modular digital platforms—through strategic partnerships, collaborative governance, and sustainability-focused innovation—can generate scalable impact across diverse contexts. Key enabling technologies, such as IoT sensors, AI-based analytics, and satellite monitoring, have improved resource efficiency and decision-making. Partnerships with research institutions, tech providers, and agribusinesses have contributed to the diffusion of innovations and reinforced the ecosystem's robustness. Nonetheless, adoption barriers remain, particularly regarding financial accessibility, cultural resistance, and technological literacy.

This study contributes to the literature on digital agriculture and IEs by revealing how a startup can act as an ecosystem orchestrator. Building on Adner's framework, three pillars emerge as central to XFARM's ecosystem:

- Interdependency Management: XFARM aligns diverse stakeholders and technologies to ensure system-wide compatibility.
- Orchestration and Coordination: The platform mediates knowledge exchange and facilitates co-innovation across the ecosystem.
- Value Creation and Capture: XFARM delivers concrete benefits to all participants, enhancing ecosystem participation and sustainability.

However, significant challenges remain. Cultural resistance among farmers, high upfront costs of digital transformation, and fragmented regulatory environments are barriers to AgTech scalability. These challenges are consistent with previous research [28] (Yoon et al. 2021) highlighting barriers to digital adoption in agriculture.

The paper present several implications. From a theoretical standpoint, this research advances ecosystem theory by showing how digital platforms mitigate ecosystem risks and facilitate open innovation. It extends Adner's and Chesbrough's frameworks into AgTech, portraying startups as strategic orchestrators in ecosystem-driven innovation.

Practically, this study offers insights for managers, policymakers, and investors. AgTech startups should prioritize collaborative engagement and align their technological development with stakeholder needs. For policymakers, promoting digital literacy, reducing the digital divide, and incentivizing technology adoption—especially for small and medium-sized farms—are crucial. Targeted subsidies, public–private partnerships, demonstration farms, and improved rural connectivity are effective strategies for increasing adoption rates. Educators and extension services should co-develop modular, context-specific digital literacy programs to enhance long-term adoption and user confidence.

Like ever study, this work present limitations. Specifically, the single-case methodology, while offering in-depth insights, limits generalizability. Future studies could explore other startups across regions and agricultural systems to identify transferable practices and challenges. Longitudinal analyses may reveal how IEs evolve over time in response to external pressures and internal dynamics.

Incorporating quantitative methods—such as surveys and statistical modeling—could validate and extend the findings, evaluating correlations between ecosystem strategies and adoption outcomes. Mixed-methods approaches would further enhance analytical robustness.

Future research should also examine psychological and behavioral barriers to technology adoption, offering guidance on how to design more user-centered and persuasive digital literacy interventions. Exploring farmer motivations, risk perceptions, and contextual influences can help develop more effective ecosystem engagement strategies.

Ultimately, the XFARM case highlights how startups can catalyze digital transformation in agriculture through strategic orchestration and stakeholder alignment. To unlock the full potential of digital agriculture, coordinated action is needed among startups, institutions, investors, and policymakers. Continued research is essential to support this transition, ensuring that digital ecosystems evolve inclusively, sustainably, and at scale.

References

1. Adner, R.: Ecosystem as structure: an actionable construct for strategy. J. Manag. **43**(1), 39–58 (2017). https://doi.org/10.1177/0149206316678451
2. Adner, R.: Match your innovation strategy to your innovation ecosystem. Harvard Bus. Rev. (2006)
3. Aguinis, H., Solarino, A.M.: Transparency and replicability in qualitative research: THE case of interviews with elite informants. Strateg. Manag. J. **40**(8), 1291–1315 (2019)
4. Arditi, A.B., Camio, M.I., Velazquez, L., Errandosoro, F.: Early adoption of Industry 4.0 technologies in the agricultural sector: a phenomenological analysis. J. Int. Council Small Bus. **4**(3), 230–257 (2023)
5. Autio, E., Thomas, L.D.: Innovation ecosystems. Oxford Handbook of Innovation (2014)
6. Bahn, R.A., Yehya, A.A.K., Zurayk, R.: Digitalization for sustainable agri-food systems: potential, status, and risks for the MENA region. Sustainability **13**(6), 3223 (2021)
7. Baldwin, C.Y., Bogers, M.L.A.M., Kapoor, R., West, J.: Focusing the ecosystem lens on innovation studies. Res. Policy **53**, 104949 (2024)
8. Bambirra Silveira, G., Farina, M.C., Santos, I.C.: High-technology based startup in agribusiness sector: mapping linkages, products, and services. Rev. Econ. Sociol. Rural. **61**(1), e246210 (2023)
9. Barile, D., Secundo, G., Del Vecchio, P.: Food 4.0 for competing during the COVID-19 pandemic: experimenting digitalization in family firms. Eur. J. Innov. Manag. **27**(4), 1381–1402 (2024). https://doi.org/10.1108/EJIM-07-2022-0373
10. Barile, D., Secundo, G., Garzoni, A., Fiorentini, A.: Accelerating corporate innovation ecosystems: the exprivia business incubator case study. J. Eng. Tech. Manag. **74**, 101845 (2024)
11. Bellon-Maurel, V., Piot-Lepetit, I., Lachia, N., Tisseyre, B.: Digital agriculture in Europe and in France: which organisations can boost adoption levels? Crop Pasture Sci. **74**(6), 573–585 (2023)
12. Bereznoy, A., Meissner, D., Scuotto, V.: The intertwining of knowledge sharing and creation in the digital platform-based ecosystem: a conceptual study on the lens of the open innovation approach. J. Knowl. Manag. **25**(8), 2022–2042 (2021)
13. Boehlje, M., Roucan-Kane, M., Bröring, S.: Future agribusiness challenges: strategic uncertainty, innovation and structural change. Int. Food Agribus. Manag. Rev. **14**(5), 53–82 (2011)
14. Brea, E.: A framework for mapping actor roles and their innovation potential in digital ecosystems. Technovation **125**, 102783 (2023). https://doi.org/10.1016/j.technovation.2023.102783
15. Corbin, J., Strauss, A.: Basics of Qualitative Research: Techniques and Procedures for Developing Grounded Theory. Sage Publications (2014)
16. Corvello, V., Felicetti, A.M., Steiber, A., Alänge, S.: Start-up collaboration units as knowledge brokers in corporate innovation ecosystems: a study in the automotive industry. J. Innov. Knowl. **8**, 100303 (2023)
17. Dedehayir, O., Mäkinen, S.J., Ortt, J.R.: Roles during innovation ecosystem genesis: a literature review. Technol. Forecast. Soc. Chang. **136**, 18–29 (2018)
18. Dutia, S.: Agtech: The Emergence of a New Industry Sector. VentureBeat (2014)
19. Eisenhardt, K.M., Graebner, M.E.: Theory building from cases: opportunities and challenges. Acad. Manag. J. **50**(1), 25–32 (2007)

20. Eisenhardt, K.M.: Building theories from case study research. Acad. Manag. Rev. **14**(4), 532–550 (1989)
21. Glaser, B.G., Strauss, A.L.: The Discovery of Grounded Theory Strategies for Qualitative Research. Routledge (1967)
22. Granstrand, O., Holgersson, M.: Innovation ecosystems: a conceptual review and a new definition. Technovation (2020)
23. Gumbi, N., Gumbi, L., Twinomurinzi, H.: Towards sustainable digital agriculture for smallholder farmers: a systematic literature review. Sustainability **15**(16), 12530 (2023)
24. Kamilaris, A., Kartakoullis, A., Prenafeta-Boldú, F.X.: A review on the practice of big data analysis in agriculture. Comput. Electron. Agric. **143**, 23–37 (2017). https://doi.org/10.1016/j.compag.2017.09.037
25. Kapoor, R., Furr, N.R.: Complementarities and competition: unpacking the drivers of entrants' technology choices in the solar photovoltaic industry. Strateg. Manag. J. **36**(3), 416–436 (2015). https://doi.org/10.1002/smj.2223
26. Klerkx, L., Villalobos, P.: Are AgriFoodTech start-ups the new drivers of food systems transformation? An overview of the state of the art and a research agenda. Glob. Food Sec. **40**, 100726 (2024)
27. Lachman, J., López, A.: The nurturing role of the local support ecosystem in the development of the AgTech sector in Argentina. J. Agribus. Dev. Emerg. Econ. **12**(4), 714–729 (2022)
28. Mikhailov, A., Reichert, F.M., Pivoto, D.: Innovation in agribusiness: the case of agricultural technology new ventures. In: Disruptive Innovations: Better Business, Management, Science and Government, IFAMA 28th World Conference, Buenos Aires (2021)
29. Myers, M.D.: Qualitative Research in Business & Management, 2nd edn. Sage, London (2008)
30. Pham, X., Stack, M.: How data analytics is transforming agriculture. Bus. Horiz. **61**(1), 125–133 (2018). https://doi.org/10.1016/j.bushor.2017.09.011
31. Yin, R.K.: Case Study Research: Design and Methods. Sage, CA (2015)
32. de Vasconcelos Gomes, L.A., Facin, A.L.F., Salerno, M.S., Ikenami, R.K.: Unpacking the innovation ecosystem construct: Evolution, gaps and trends. Technol. Forecast. Soc Chang. 136, 30–48 (2018)
33. Autio, E.: Orchestrating ecosystems: a multi-layered framework. Innov. 24(1), 96–109 (2022)
34. França, A., Ferreira, J.: Resources and capabilities through the lens of value (co-) creation: a literature review. Int. J. Innov. Sci. 8(3), 230–253 (2016)
35. Triomphe, B., Rajalahti, R.: From concept to emerging practice: what does an innovation system perspective bring to agricultural and rural development?. In Renewing innovation systems in agriculture and food (pp. 57–76). Wageningen Academic (2012)
36. Del Sarto, N., Ferrigno, G., Parida, V., Di Minin, A.: Do start-ups benefit from coworking spaces? An empirical analysis of accelerators' programs. Rev. Manag. Sci. 17(7), 2471–2502 (2023)
37. Barile, D., Secundo, G., Garzoni, A., Fiorentini, A.: Accelerating corporate innovation ecosystems: The exprivia business incubator case study. J. Eng. Technol. Manag. 74, 101845 (2024)
38. Barile, D., Secundo, G., Magnusson, M. Exploring the digital innovation ecosystem from the perspective of platform-based startups: a case study in the film industry. Eur. J. Innov. Manag. 1–24 (2025)
39. Rambe, P., Khaola, P.: The impact of innovation on agribusiness competitiveness: the mediating role of technology transfer and productivity. Eur. J. Innov. Manag. 25(3), 741–773 (2022)
40. Ramos, P.H.B., Pedroso, M.C.: Classification and categorization of Brazilian agricultural startups (AgTechs). Innov. Manag. Rev. 18(3), 237–257 (2021)
41. Wolfert, S., Ge, L., Verdouw, C., Bogaardt, M.J.: Big data in smart farming—a review. Agric. Syst. 153, 69–80 (2017). https://doi.org/10.1016/j.agsy.2017.01.023
42. Walrave, B., Talmar, M., Podoynitsyna, K.S., Romme, A.G.L., Verbong, G.P.: A multi-level perspective on innovation ecosystems for path-breaking innovation. Technol. Forecast. Soc. Chang. 136, 103–113 (2018). https://doi.org/10.1016/j.techfore.2017.04.011
43. Siggelkow, N.: Persuasion with case studies. Acad. Manag. J. 50(1), 20–24 (2007)

44. Secundo, G., Barile, D., Garzoni, A., Ieva, G.: Orchestrating open innovation ecosystems based on digital technologies: the Lutech case study. In: Digital Innovation Management: People, Process, Platforms and Policy, pp. 21–40. Springer, Cham (2025)
45. Rehman, A.U., Alamoudi, Y., Khalid, H.M., Morchid, A., Muyeen, S.M., Abdelaziz, A.Y.: Smart agriculture technology: an integrated framework of renewable energy resources, IoT-based energy management, and precision robotics. Clean. Energy Syst. 9, 100132 (2024)

Education and Technology in the AI Era

Bridging the Gap Between Visibility and Excellence: A Bibliometric Analysis of Hungarian Research in Business, Management and Accounting

Noémi Hajdú and Péter Sasvári

Abstract Academic visibility and excellence are key indicators in evaluating research performance, especially in higher education. Visibility refers to the inclusion of publications in indexed databases and open access platforms, while excellence is reflected in high citation counts and presence in top-ranked journals. This study examines the publication performance of Hungarian universities in the fields of economics, management, and accounting between 2019 and 2023. Using bibliometric data from Scopus and Web of Science, we compare Hungary's output to that of Austria and the Visegrád countries, focusing on metrics such as Q-ranking distribution, citation frequency, and the presence of highly cited articles. The results reveal that while Hungarian institutions have improved their visibility, they lag significantly behind in excellence, with few high-impact publications. This structural gap suggests the need for targeted institutional and policy-level interventions to enhance research quality and international recognition.

Keywords Visibility · Publication excellence · Scopus indexed journals · Web of Science · SDG9

1 Introduction

In recent years, publication metrics have gained increasing importance in the evaluation of research performance across higher education systems. Visibility and excellence have emerged as two key dimensions: visibility refers to the presence of scholarly outputs in indexed platforms such as Scopus and Web of Science, while excellence is associated with academic recognition, high citation counts, and publication

N. Hajdú (✉) · P. Sasvári
University of Miskolc, Miskolc, Hungary
e-mail: noemi.hajdu1@uni-miskolc.hu; peter.sasvari@uni-miskolc.hu

P. Sasvári
Ludovika University of Public Service, Budapest, Hungary

A. Visvizi et al. (eds.), *Research and Innovation Forum 2025 Vol2*, Springer Proceedings in Complexity, https://doi.org/10.1007/978-3-032-09062-1_18

in top-ranked journals [1, 2]. These indicators are now widely used by universities, funding agencies, and policymakers to assess institutional performance [3, 4].

Although citations are often accepted as proxies for research quality, they are shaped by complex motivations [5], and many publications remain uncited [6]. By contrast, highly cited papers—known as "citation classics" or "gold bars of science"—represent a small yet influential subset of scientific output [7–10]. Both visibility and excellence are influenced by factors such as scientific quality, relevance, methodological rigor, accessibility, and journal ranking [11–13].

In Hungary, these dynamics have been reinforced by structural reforms in higher education. At foundation-run universities ("modellváltó" institutions), performance-based funding mechanisms increasingly link institutional support to publication output. Formal expectations regarding the number of publications—often defined by the maintaining authority—have contributed to a system where quantity is prioritized over quality. This pressure may result in a growing number of indexed publications without a corresponding increase in research excellence, particularly in fields such as Business, Management and Accounting (BMA).

The core problem addressed in this study is the apparent discrepancy between the growing visibility of Hungarian publications and their relatively weak performance in excellence indicators. This raises a broader research gap: while the conceptual difference between visibility and excellence is acknowledged in the literature [5, 6], few empirical studies have examined how these dimensions interact in regional publication patterns, particularly in Central and Eastern Europe. Moreover, there is little insight into how institutional type or publication strategies influence research outcomes across national systems.

In addition to these gaps, recent research has raised concerns that visibility is overemphasised without corresponding excellence. Emerging digital and innovative paradigms such as Industry 5.0 and blockchain applications may increase the output of technical publications, but do not necessarily contribute to long-term academic quality or societal relevance [14]. This perspective emphasises the importance of critically evaluating not only where and how research is published, but also its broader epistemological and policy implications.

To address this gap, our study focuses on the comparative bibliometric performance of Hungarian universities relative to the Visegrád countries (Czech Republic, Poland, Slovakia) and Austria. The V4 countries share historical, economic and academic development trajectories, while Austria provides a relevant benchmark due to its geographical proximity and stronger performance in global rankings.

We aim to answer the following research questions:

1. How do Hungarian universities compare to Austria and the Visegrád countries in terms of visibility and excellence in the BMA fields?
2. What structural publication patterns can help explain Hungary's relative performance?
3. To what extent does visibility (e.g., number of indexed publications) correlate with excellence (e.g., citation impact, Q1 journal placement) in the context of Hungarian and Central European (V4, Austria) research output?

To answer these questions, we conducted a comparative bibliometric analysis using Scopus and Web of Science data from 2019 to 2023. Our sample includes 19 Hungarian higher education institutions, classified by institutional type. Performance indicators include Q-ranking distribution, citation frequency, and the presence of highly cited publications. Based on regional trends and preliminary data, we formulated the following hypotheses:

H1: Hungarian universities have improved their visibility, but still lag in terms of research excellence.
H2: Researchers in the Czech Republic, Poland and Slovakia are more likely to publish in indexed national journals than their Hungarian counterparts.
H3: Austria consistently outperforms the Visegrád countries in producing high-impact publications.

The remainder of this paper is organized as follows. Section 2 presents the methodology and the scope of data used for the comparative bibliometric analysis. Section 3 provides the main findings, with emphasis on both visibility and excellence indicators. Section 4 discusses the results in light of structural factors and institutional strategies. Finally, Sect. 5 concludes with policy implications and directions for future research.

2 Methodology

This study applies a comparative bibliometric analysis based on two major international databases: Scopus and Web of Science (WoS). The aim is to assess and compare the visibility and excellence of research output in the fields of Business, Management, and Accounting (BMA) across Hungary, the Visegrád countries (Czech Republic, Poland, Slovakia), and Austria. The Visegrád countries were selected based on their common regional characteristics and comparable research policy frameworks. The analysis covers a five-year period, from 2019 to 2023. Citation-based metrics such as the H-index and SJR are often used to assess research performance; however, their limitations as indicators of scientific quality are well documented [15]. Furthermore, research productivity is a multi-layered concept that cannot be fully captured by bibliometric indicators alone, and quantitative measures should be interpreted with caution [16]. The use of both databases is justified by their recognized role in research performance evaluation. Web of Science provides detailed excellence indicators (e.g., Highly Cited Papers, Hot Papers, Top 1%), while Scopus offers broader coverage, particularly in applied and regional domains. Their combined use increases the reliability and comprehensiveness of the analysis, as supported by previous bibliometric studies [17, 18].

2.1 Sample Selection and Institutional Classification

The sample includes 19 Hungarian higher education institutions that published at least one article in the selected BMA fields during the study period. Institutions were identified using affiliation data from Scopus, verified through institutional repositories and official websites.

The participating institutions are as follows (with abbreviations used throughout the text after first mention): Budapest Business School—University of Applied Sciences (BGE), Budapest University of Technology and Economics (BME), Corvinus University of Budapest (Corvinus), University of Debrecen (DE), Eszterházy Károly Catholic University (EKKE), Eötvös Loránd University (ELTE), Károli Gáspár University of the Reformed Church in Hungary (KRE), Hungarian University of Agriculture and Life Sciences (MATE), University of Miskolc (ME), Budapest Metropolitan University (METU), John von Neumann University (NJE), University of Public Service—Ludovika (NKE), Óbuda University (ÓE), University of Pannonia (PE), University of Pécs (PTE), Semmelweis University (SE), University of Sopron (SOE), Széchenyi István University (SZE), University of Szeged (SZTE).

These institutions were classified into three types based on their ownership and governance structure:

- Foundation-run (privately governed) universities: 13 institutions (BGE, Corvinus, DE, MATE, NJE, ÓE, PE, PTE, SE, SOE, SZE, SZTE, and METU)
- Ecclesiastical universities: 2 institutions (EKKE, KRE)
- Public (state-maintained) universities: 3 institutions (BME, ELTE, NKE).

The selection criteria for inclusion in the sample were as follows:

1. At least one indexed publication in BMA fields between 2019 and 2023;
2. Clearly stated institutional affiliation in Scopus or WoS;
3. Recognition as a higher education institution in Hungary.

Austria and the Visegrád countries (Czech Republic, Poland, Slovakia) were included for comparative purposes. These countries share similar historical, economic, and academic development paths, while Austria serves as a high-performing regional benchmark due to its strong international publishing track record.

2.2 Subject Categories and Disciplinary Coverage

The study focuses on BMA disciplines as defined by Scopus and WoS:

- Scopus includes 10 subdisciplines: Accounting, Business and International Management, Business, Management and Accounting (general), Industrial Relations, Management Information Systems, Management of Technology and Innovation, Marketing, Organizational Behavior and Human Resource Management, Strategy and Management, and Tourism, Leisure and Hospitality Management.

  Each subdiscipline covers distinct research focuses:

- Accounting relates to financial reporting and taxation.
- Strategy and Management emphasizes long-term business planning.
- Organizational Behavior addresses motivation, culture, and HR practices.
- Etc. (see Appendix for full category descriptions.)
- WoS, while using broader subject categories, includes a Business category that incorporates economic, entrepreneurial and business studies. These classifications guided the filtering of relevant documents.

2.3 Metrics for Visibility and Excellence

We applied the following bibliometric indicators:

- Visibility: Number of indexed publications, share in Scopus and WoS, and Q-ranking distribution (Q1–Q4) based on SJR (Scopus) and Journal Impact Factor (WoS).
- Excellence: Number of Highly Cited Papers (HCP), documents in the Top 1% (WoS), Hot Papers, citation counts, and H-index values.

These indicators allow us to capture both the volume and the scientific impact of publications. Scopus' SJR quartile system was used to identify journal prestige levels across all countries.

2.4 Justification and Contribution

This methodology supports the research questions by offering a robust comparative approach to visibility and excellence. It enables the identification of structural differences between institutional types, facilitates the comparison of national publication strategies (such as the use of national journals), and allows for the assessment of the gap between visibility (i.e., publication output) and excellence (i.e., scientific impact), as formulated in our hypotheses.

Our approach aligns with recent research trends in bibliometrics and academic performance assessment, where combined database usage and cross-country benchmarking are increasingly recommended [19, 20].

3 Results

3.1 Excellence Indicators in the Field of Economics (WoS)

According to data retrieved from Web of Science, the 19 Hungarian universities examined published a total of 61,000 scientific papers and received approximately 497,000 citations between 2019 and 2023. However, only a small fraction of this scholarly output was classified under the economics category: 0.62% of the publications and 0.30% of the citations. This indicates that economics remains a marginal discipline in the overall publication profile of Hungarian universities.

To assess scientific excellence, three Web of Science indicators were used: Highly Cited Papers (HCP), Top 1% of most cited papers, and Hot Papers (HP). As shown in Table 1, only four such publications were produced in the economics category across all Hungarian universities during the five-year period.

Table 1 Indicators of excellence of some Hungarian universities in the category of economics between 2019 and 2023

SN	Abbr. name of the inst.	No. of notifications (pcs)	Highly cited art. (pcs)	Documents in the top 1% (pcs)	Current art. (pcs)
1	Corvinus	93	0	0	0
2	MATE	51	**1**	**1**	0
3	PTE	36	0	0	0
4	ELTE	34	0	**1**	0
5	DE	28	0	0	0
6	SZTE	22	0	0	0
7	BGE	18	0	0	0
8	BME	18	0	0	0
9	SZE	18	0	0	0
10	PE	17	0	**1**	0
11	ME	10	0	0	0
12	NJE	10	0	0	0
13	SOE	6	0	0	0
14	ÓE	6	0	0	0
15	NKE	5	0	0	0
16	METU	4	0	0	0
17	KRE	3	0	0	0
18	EKKE	3	0	0	0
19	SE	1	0	0	0

Source Own editing

Table 1 presents a detailed breakdown of these indicators across institutions. Notably, the MATE produced both a Highly Cited Paper and a Top 1% paper, while ELTE and PE also each had a Top 1% paper. The remaining universities did not register any publications under these excellence criteria.

This extremely low performance highlights a critical gap in the field of economics research excellence. Despite some institutions (e.g., Corvinus, DE, PTE) having relatively high publication counts, none of these outputs were considered highly cited or top-tier based on WoS criteria.

3.2 *Visibility Indicators*

In terms of visibility, the results are similarly modest. Only a limited number of articles were published in Q1 and Q2 journals, with most falling into Q3 and Q4 categories. Additionally, the H-index values for the economics category remained low across institutions (Table 2). This suggests that even when Hungarian universities are present in indexed databases, they are often not publishing in the most influential journals, further reinforcing the visibility–excellence gap.

3.3 *Cross-Country Comparison Based on Scopus/Scimago Data*

According to Scimago data, approximately 2.5% of all indexed publications in the past five years (totaling around 442,000 documents) were categorized under Economics, Management and Accounting (BMA). A comparative breakdown for Austria and the Visegrád countries (Table 3) reveals substantial variation in national output across these disciplines.

Poland leads the region with over 11,000 publications, followed by Austria (6339), the Czech Republic (5220), Slovakia (3232), and Hungary at the bottom with 2697 publications. This is particularly notable given that Hungary and Poland have relatively similar population sizes and higher education capacities. The Hungarian BMA output represents only 24% of Poland's performance during the same period.

Table 3 further disaggregates the national output by subdiscipline. Across all five countries, the two dominant categories are:

- Business and International Management
- Strategy and Management

Together, these fields account for approximately 40–45% of the BMA publications in most countries.

In Hungary, the distribution follows a similar pattern:

- Strategy and Management (26%) and

Table 2 Visibility indicators for some Hungarian universities in the category of economics between 2019 and 2023

SN	Inst. abbr. name	No. of notifications (pcs)	No. of references (pcs)	% of documents cited	H-index	No. of Q1 art (pcs)	No. of Q2 art (pcs)	No. of Q3 art (pcs)	No. of Q4 art (pcs)
1	Corvinus	93	333	63.44	12	6	12	11	2
2	MATE	51	289	70.59	10	4	2	2	3
3	PTE	36	201	55.56	8	3	4	0	0
4	ELTE	34	106	58.82	6	0	4	6	2
5	DE	28	108	67.86	7	1	2	1	9
6	SZTE	22	18	22.73	3	0	3	1	0
7	BGE	18	96	66.67	5	0	4	1	0
8	BME	18	51	55.56	4	0	1	5	2
9	SZE	18	86	55.56	4	1	1	0	4
10	PE	17	89	47.06	4	2	1	1	1
11	ME	10	26	60.00	3	0	0	0	5
12	NJE	10	21	70.00	3	1	3	0	1
13	SOE	6	2	16.67	1	0	0	0	0
14	ÓE	6	12	66.67	2	0	1	0	0
15	NKE	5	9	40.00	1	0	0	0	0
16	METU	4	3	25.00	1	0	0	0	0
17	KRE	3	4	66.67	1	0	0	0	0
18	EKKE	3	12	100.00	2	0	1	0	0
19	SE	1	6	100.00	1	0	0	1	0

Source Own editing

- Business and International Management (21%) dominate,
- suggesting a focus on managerial and operational topics over more technical or financial specializations.

In contrast, subfields such as Accounting and Industrial Relations remain underrepresented in every country, though Hungary reports a relatively higher share in Industrial Relations (5%) compared to Slovakia (0%) or the Czech Republic (1%).

These findings underscore several key points:

- Hungary significantly lags behind regional peers in terms of BMA publication volume.
- The thematic focus across countries is largely consistent, concentrated around strategy, international business, and general management.
- There may be systemic factors (e.g. research funding priorities, institutional incentives, disciplinary traditions) influencing national publication capacities in specific subfields.

Table 3 Publication output of Austria and the Visegrád Group in economics, management and accounting between 2019 and 2023

Science category	Poland		Austria		Czech Republic		Slovakia		Hungary	
	Pcs	Rate (%)	Pcs	Rate (%)	Pcs	Rate (%)	Pcs	Rate (%)	Pcs	Rate (%)
Strategy and management	2257	20	1336	21	1339	26	709	22	694	26
Business and international management	2465	22	1102	17	1208	23	828	26	563	21
Management of technology and innovation	1531	14	981	15	701	13	690	21	273	10
Management information systems	913	8	481	8	479	9	240	7	125	5
Marketing	836	8	492	8	359	7	223	7	170	6
Tourism, leisure and hospitality management	864	8	448	7	336	6	113	3	254	9
Organizational behaviour and human resource management	748	7	542	9	270	5	164	5	190	7
Business, management and accounting	853	8	374	6	305	6	167	5	180	7
Accounting	501	5	461	7	188	4	84	3	116	4
Industrial relations	96	1	122	2	35	1	14	0	132	5
Total	11,064	100	6339	100	5220	100	3232	100	2697	100

Source Own editing

3.4 *Visibility Through Domestic and Controversial Journals (MDPI)*

An additional layer of visibility in the fields of Economics, Management and Accounting (BMA) comes from publications in domestic journals and those published by open-access platforms with questionable academic reputation, such as MDPI. According to Scopus data, between 8 and 24% of all indexed BMA publications in Poland, the Czech Republic, Slovakia, and Hungary were published in either domestic journals or MDPI outlets between 2019 and 2023 (Fig. 1).

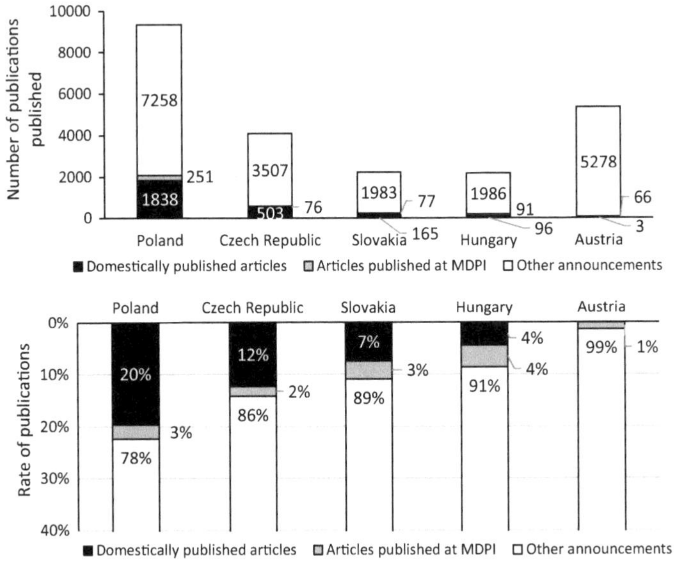

Fig. 1 Number and share of domestic and dubious MDPI articles by country in the field of business, management and accounting between 2019 and 2023 (*source* own editing)

The SJR (Scimago Journal Rank) values of these domestic journals range from 0.103 to 0.987, which may correspond to Q1, Q2, Q3 or even Q4 rankings depending on the field and the specific year. While this formal classification offers visibility in indexing databases, it does not necessarily reflect broad international recognition or scientific impact. In many cases, these journals primarily serve national authors, with limited readership and citation outside their country of origin.

Although such platforms contribute to increased output and technical visibility, they may also reinforce the gap between visibility and excellence if not accompanied by citation impact, international authorship, or scientific relevance. Particularly in Hungary and Slovakia, a notable share of indexed BMA publications falls into this category, raising concerns about the long-term reputation and global integration of national research efforts.

3.5 Challenges Related to National Orientation and MDPI Publication Patterns

The growing share of publications in Scopus-indexed journals that are either published in the author's home country or issued by controversial open-access publishers such as MDPI raises several concerns for scientific integrity and international integration.

First, language and cultural barriers may arise. Journals that primarily publish content written by authors from the same country and in the national language (or style) may hinder access and collaboration with researchers from other academic systems. This can isolate national research from international scholarly discourse.

Second, such journals often suffer from a lack of disciplinary diversity, focusing on domestic issues or perspectives, and neglecting broader debates within the global academic community. This can result in limited scientific cross-pollination, which is essential for interdisciplinary innovation.

Third, these outlets can foster bias and one-sidedness, as editorial preferences and peer review standards may favor national narratives or familiar institutions. This diminishes scientific debate and undermines the credibility of published findings.

Fourth, reliance on domestically oriented or low-reputation international journals reduces the international competitiveness of national research. Researchers may meet formal publication targets without improving their position in global rankings or citation metrics.

Table 4 illustrates the role of MDPI in the BMA field across Austria and the Visegrád countries. The data show that 1–4% of BMA publications appear in seven MDPI journals. While many of these journals are ranked in Q1 or Q2, their credibility and citation practices remain controversial in international scholarly debates.

In Hungary, 90 articles were published in MDPI outlets between 2019 and 2023. The most frequently used MDPI journals were: Risks (26 articles, Q2), Journal of

Table 4 Number of articles published by MDPI by country in the field of economics, management and accounting between 2019 and 2023

Name of the Journal (number of articles, pcs)	SJR 2023	Q 2023	Austria	Czech Republic	Poland	Hungary	Slovakia
Administrative Sciences	0.630	Q2	8	26	38	12	32
Big Data and Cognitive Computing	0.820	Q2	10	0	7	9	4
Journal of Risk and Financial Management	0.485	Q2	14	33	91	31	32
Journal of Theoretical and Applied Electronic Commerce Research	0.892	Q1	3	1	15	1	2
Logistics	0.740	Q2	10	1	6	5	1
Publications	0.793	Q1	7	3	0	6	0
Risks	0.403	Q2	14	12	94	26	6

Source Own editing

Risk and Financial Management (31 articles, Q2), and Administrative Sciences (12 articles, Q2).

Although these publications improve technical visibility in Scopus, their academic credibility and impact remain contested, especially when national publication strategies rely heavily on such platforms.

3.6 National Journal Dominance and Structural Risks of MDPI Publishing

While open-access platforms such as MDPI offer rapid publication and wide accessibility, their rising dominance in certain disciplines has raised several concerns within the scientific community—particularly in Central and Eastern Europe.

Among the most frequently cited issues with MDPI journals are:

- Inconsistent quality control: The peer-review process is often accelerated, and critics argue that editorial standards may not match those of traditional publishers.
- Lower impact metrics: While many MDPI journals are technically indexed in Scopus or WoS, their citation performance and scientific influence often lag behind more established journals.
- High publication costs: Article Processing Charges (APCs) can create barriers for researchers with limited institutional or grant support.
- Volume over curation: The high number of published articles in MDPI journals makes it difficult to discern the scientific relevance of individual contributions.
- Limited editorial transparency: Concerns have been raised about inconsistent peer review practices, rapid acceptance times, and vague ethical frameworks.

These issues may undermine the credibility of articles published in MDPI and reduce their acceptance and perceived value among international scholars.

Beyond MDPI, the prevalence of nationally published journals—i.e. those produced within the country and primarily serving domestic audiences—also affects research visibility and excellence. According to Table 5, the proportion of BMA publications in national journals varies significantly across countries and disciplines.

The most striking example is Hungary, where 73% of Industrial Relations publications were published in a single national journal, Society and Economy. While this journal is indexed in Scopus and provides technical visibility, its limited international reach may restrict the impact and reception of Hungarian research in this field.

Similarly, Poland shows high national publication shares in several categories, including: Tourism, Leisure and Hospitality Management (41%), Organizational Behaviour and Human Resource Management (41%), Management Information Systems (35%).

These figures suggest a regional reliance on domestically oriented publication channels, which—while important for local dissemination—may hinder international integration, citation impact, and disciplinary development.

Table 5 Share of articles published by national publishers by country and scientific category between 2019 and 2023

Category of the science	Austria	Czech Republic	Poland	Hungary	Slovakia	Total
Industrial relations	0%	0%	0%	***73%***	0%	*24%*
Tourism, leisure and hospitality management	0%	*26%*	**41%**	0%	0%	*22%*
Management information systems	0%	*15%*	**35%**	0%	0%	*18%*
Business and international management	0%	*15%*	*28%*	*17%*	*9%*	*17%*
Marketing	0%	*11%*	**31%**	0%	*26%*	*17%*
Organizational behaviour and human resource management	0%	0%	**41%**	0%	0%	*16%*
Business, management and accounting	0%	0%	**34%**	0%	0%	*15%*
Management of technology and innovation	0%	*12%*	**34%**	0%	4%	*15%*
Strategy and management	0%	*11%*	*23%*	*14%*	*4%*	*12%*
Accounting	0%	*19%*	*7%*	0%	0%	*5%*

Source Own editing
Bold Highest value within the row
Italics Second-highest within the row
Bold-italics Tied highest values

The combination of national journal dependence and MDPI overuse can distort publication strategies and reinforce the gap between technical visibility and substantive excellence. These patterns must be considered in future policy recommendations and institutional strategies aiming to elevate regional research performance.

3.7 *Top-Tier Journal Output and Regional Excellence Patterns*

A deeper look at the publication venues of high-impact research provides further insight into the excellence gap observed in the region. Table 6 summarizes the number of top 1% cited articles published between 2019 and 2023 in the field of Economics, Management and Accounting, across Austria, the Czech Republic, Poland, Slovakia and Hungary.

These articles were published in elite international journals, such as: Academy of Management Journal (SJR 8.271), Journal of Finance (SJR 19.139), Review of Financial Studies (SJR 17.654), and Journal of International Business Studies (SJR 4.600), among others.

Table 6 Number of award-winning articles by publishers and country in the field of economics, management and accounting between 2019 and 2023

Name of the journal	Published by	SJR 2023	Austria	Czech Republic	Poland	Hungary	Slovakia
Academy of Management Journal	Academy of Management	8.271	8	0	0	0	0
Academy of Management Review	Academy of Management	10.486	1	0	1	0	0
International Journal of Information Management	Elsevier	5.775	9	4	7	0	0
Journal of Consumer Research	Oxford University Press	5.428	2	0	0	0	0
Journal of Finance	Wiley-Blackwell	19.139	1	0	0	0	0
Journal of Financial Economics	Elsevier	13.655	6	0	1	0	3
Journal of International Business Studies	Palgrave Macmillan	4.600	7	0	0	0	0
Journal of Management	SAGE	7.539	1	0	1	0	0
Journal of Marketing	American Marketing Association	11.799	2	0	0	0	0
Journal of Travel Research	SAGE	3.408	2	0	1	0	0
Marketing Science	INFORMS	5.643	3	0	0	0	0
Organization Science	INFORMS	5.632	5	0	0	0	0
Review of Accounting Studies	Springer	5.481	5	0	0	0	0
Review of Financial Studies	Oxford University Press	17.654	7	0	0	0	0

(continued)

Table 6 (continued)

Name of the journal	Published by	SJR 2023	Austria	Czech Republic	Poland	Hungary	Slovakia
Strategic Management Journal	Wiley-Blackwell	7.82	3	1	0	0	0
Tourism Management	Elsevier	3.352	12	1	9	0	0

Source Own editing

The distribution of such publications clearly highlights Austria's leading position. Austrian researchers contributed to 74 articles across all 10 BMA categories, including 8 in Academy of Management Journal, 9 in International Journal of Information Management, and 12 in Tourism Management.

The Czech Republic produced a total of 6 articles, with contributions to Tourism Management, Strategic Management Journal, and Information Management.

Poland contributed between 6 and 20 articles, with concentrations in Tourism, Financial Economics, and Information Systems—but notably none in Business and International Management or Industrial Relations.

Slovakia had a limited presence, with a small number of high-impact articles in Accounting and Strategy.

Hungary, in contrast, did not produce a single publication in any of the top-tier journals listed in Table 6 during the five-year period.

This total absence underscores the depth of the excellence gap, especially given the country's relatively high number of Scopus-indexed BMA publications. It confirms that technical visibility alone does not translate into academic recognition, and that Hungary remains disconnected from the highest-impact international publication circuits in the BMA field.

The contrast is further illustrated in Tables 6 and 7, which map the disciplinary spread of top publications:

- Austria: articles in all BMA subfields except Industrial Relations.
- Czech Republic: visible in 5 subfields.
- Poland: strong in 8 subfields.
- Slovakia: limited to 2.
- Hungary: absent from all 10 categories.

This not only reveals a quantitative gap, but also a disciplinary vacuum in Hungarian research at the elite level.

Table 7 Change in the SJR in economics, management and accounting between 2019 and 2023 in the countries analyzed

Year	2019		2020		2021		2022		2023	
Country	SJR value	Change (100% = 2019) (%)	SJR value	Switch to visit (%)	SJR value	Switch to visit (%)	SJR value	Switch to visit (%)	SJR value	Switch to visit (%)
Poland	0.567	100	0.607	107	0.751	132	0.766	135	0.885	**156**
Czech Republic	0.592	100	0.611	103	0.671	113	0.790	134	0.883	**149**
Hungary	0.548	100	0.618	113	0.741	135	0.695	127	0.738	135
Austria	1.233	100	1.385	112	1.430	116	1.422	115	1.576	128
Slovakia	0.436	100	0.458	105	0.509	117	0.478	110	0.516	118
Journals average. Weighted SJR	0.914	100	1.119	122	1.214	133	1.213	133	1.276	140

Source Own editing
Bold Marks the two highest percentage values for 2023

3.8 SJR Trends as an Indicator of Qualitative Progress

The Scimago Journal Rank (SJR) reflects journal prestige and is often used to track improvements in publication quality. Table 7 shows the evolution of average weighted SJR values in BMA fields across the analyzed countries between 2019 and 2023.

Austria maintains the highest values throughout the period, with a 28% increase. Poland and the Czech Republic show significant improvement—56% and 49% respectively—exceeding the average growth of journals (40%).

Hungary improved until 2021 but then stagnated, ending the period with a 35% increase, in line with the average but below the regional trend. Slovakia lags furthest behind, with only an 18% growth over five years.

These results suggest that while Poland and the Czech Republic are moving toward excellence by publishing in higher-ranked journals, Hungary and Slovakia remain largely static in journal quality. This underlines the need for strategic shifts in publication targets—not just more output, but better journals.

4 Discussion

This study examined the relationship between academic visibility and research excellence in the field of Business, Management, and Accounting (BMA), focusing on Hungarian universities and comparing their performance to Austria and the Visegrád countries. The analysis was guided by three central research questions:

1. How does Hungary perform regionally in terms of visibility and excellence?
2. What structural publication patterns shape national performance?
3. To what extent does increased visibility correlate with excellence?

The findings support all three hypotheses, although with important nuances.

First, Hungary has made measurable progress in visibility, with an increased presence in Scopus and Web of Science. This supports the hypothesis that international visibility has improved, especially in terms of publication volume. However, this visibility has not translated into equivalent research excellence. The number of highly cited papers, hot papers, or top 1% articles remains extremely low. Institutions such as Corvinus and MATE contribute significantly to output, but their excellence indicators remain limited.

Second, the data confirm regional differences in publication strategy. Researchers in Poland, the Czech Republic, and Slovakia often publish in national journals indexed in Scopus, which support moderate visibility and foster local academic networks. Hungarian authors, in contrast, are less present in domestic journals and tend to publish in international outlets—including those issued by publishers such as MDPI—where reputational and impact levels may vary. This asymmetry may hinder participation in collaborative citation networks and reduce regional research integration.

Third, Austrian universities clearly outperform the Visegrád countries in producing high-impact research. Austrian scholars publish widely in Q1 journals and regularly appear in leading international outlets such as the Strategic Management Journal, Journal of Finance, and Academy of Management Journal. This confirms that sustained excellence is linked to systemic support, established publication cultures, and access to elite scholarly platforms.

Although Hungary's SJR-weighted publication output has increased since 2019, its growth rate lags behind Poland and the Czech Republic. The stagnation in 2022 and the complete absence of Hungarian publications in the top 1% of BMA fields between 2019 and 2023 underscore the persistence of a visibility–excellence gap.

5 Conclusion

This study addressed the gap between research visibility and excellence in the Visegrád countries by providing a comparative bibliometric analysis of institutional performance in the fields of economics, management, and accounting (BMA). While

Hungarian universities have increased their presence in international databases, structural challenges still limit their progress toward research excellence. The results underline the importance of targeted institutional strategies, effective funding frameworks, and meaningful international cooperation.

Based on the findings, several strategic actions are recommended for higher education institutions, research policymakers, and funding agencies. These include: enhancing academic training in scientific writing, peer review, and impact-oriented research design; promoting deep international collaboration through joint projects and researcher mobility; and recruiting internationally recognized scholars to raise standards and support early-career researchers. Aligning institutional incentives with excellence—such as rewarding publications in high-impact journals and recognizing long-term citation performance—can foster a research culture that prioritizes both quality and visibility.

These efforts must be embedded in broader structural reforms. Public funding should be complemented by competitive national and EU-level grants, particularly in light of Hungary's restricted participation in Horizon Europe. Increased engagement with private foundations, policy think tanks, and transnational networks may also diversify financial support and promote innovative research strategies.

Despite its contributions, this study has limitations. It focuses solely on BMA fields and relies exclusively on quantitative bibliometric indicators, which may not fully capture the qualitative dimensions of research performance. In addition, the analysis is limited to indexed outputs and does not consider language, institutional missions, or discipline-specific publication cultures.

Future research should expand on this framework by integrating qualitative approaches, such as interviews or case studies, to explore institutional practices, researcher motivations, and the effects of national science policy. A more holistic view could support the development of a research culture of excellence in Hungary and the broader region.

Acknowledgements Project no. TKP2021-NKTA-51 has been implemented with the support provided by the Ministry of Culture and Innovation of Hungary from the National Research, Development and Innovation Fund, financed under the TKP2021-NKTA funding scheme.

Appendix

Supplementary data (Annex Tables 8 and 9): Available at: https://drive.google.com/file/d/1XhRBEG2UrlJicc7sEH9uTEz0JUpBdoOW/view?usp=sharing.

References

1. Drivas, K., Kremmydas, D.: The Matthew effect of a journal's ranking. Res. Policy **49**(4), 103951 (2020). https://doi.org/10.1016/j.respol.2020.103951
2. Guerrero-Botea, V.P., Moya-Anegón, F.: A further step forward in measuring journals' scientific prestige: the SJR2 indicator. J. Informetrics **6**(2012), 674–688 (2012). https://doi.org/10.1016/j.joi.2012.07.001
3. James, J., Heckman, S.M.: Publishing and promotion in economics: the tyranny of the top five. J. Econ. Lit. **58**(2), 419–470 (2020). https://doi.org/10.1257/jel.20191574
4. Moed, H.F.: Citation Analysis in Research Evaluation, p. 346. Springer, Dordrecht (2005). https://doi.org/10.1007/1-4020-3714-7
5. Brooks, T.A.: Private acts and public objects: an investigation of citer motivations. J. Am. Soc. Inf. Sci. **36**(4), 223–229 (1985). https://doi.org/10.1002/asi.4630360402
6. Derek, R., Smith, P.A., Leggat, A.: Ten citation classics from the New Zealand Medical Journal. New Z. Med. J. **32**(2), 105–106 (2008). https://doi.org/10.1111/j.1753-6405.2008.00183.x
7. Martínez, M.A., Herrera, M., Contreras, E., Ruíz, A., Herrera-Viedma, E.: Characterizing highly cited papers in Social Work through H-Classics. Scientometrics **102**, 1713–1729 (2015). https://doi.org/10.1007/s11192-014-1460-y
8. Garfield, E.: Introducing citation classics. The human side of scientific reports. Curr. Comments **1**, 5–7 (1977). https://garfield.library.upenn.edu/essays/v3p001y1977-78.pdf
9. Stack, S.: Citation classics in suicide and life-threatening behaviour: a research note. Suicide Life-Threat. Behav. **42**(6), 628–639 (2012). https://doi.org/10.1111/j.1943-278X.2012.00117.x
10. Wahid, M., Warraich, N.F., Tahira, M.: Factors influencing scholarly publication productivity: a systematic review. Inf. Discovery Deliv. **50**(1), 22–33 (2022). https://doi.org/10.1108/IDD-04-2020-0036
11. Binswanger, M.: Excellence by nonsense: the competition for publications in modern science. In: Bartling, S., Friesike, S. (eds.) Opening Science. Springer, Charm (2014). https://doi.org/10.1007/978-3-319-00026-8_3
12. Sasvári, P., Fejes, Z.S.: Publication opportunities in Scopus indexed journals in the case of state and law. Sci. Tech. J. **70**(1), 38–58 (2023). https://doi.org/10.3311/tmt.13209
13. Heckman, J., Moktan, S.: Publishing and Promotion in Economics: The Tyranny of the Top Five National Bureau of Economic Research 25093 (2018). https://doi.org/10.3386/w25093
14. Visvizi, A., Malik, R., Guazzo, G.M., Çekani, V.: The Industry 5.0 (I50) paradigm, blockchain-based applications and the smart city. Eur. J. Innov. Manag. **28**(1), 5–26 (2025). https://doi.org/10.1108/EJIM-09-2023-0826
15. Bornmann, L., Marx, W.: How to evaluate individual researchers working in the natural and life sciences meaningfully? Scientometrics **98**(1), 487–509 (2014). https://doi.org/10.1007/s11192-013-1161-y
16. Abramo, G., D'Angelo, C.A.: How do you define and measure research productivity? Scientometrics **101**(2), 1129–1144 (2014). https://doi.org/10.1007/s11192-014-1269-8
17. Moed, H.F.: Citation Analysis in Research Evaluation. Springer (2005)
18. Abramo, G., D'Angelo, C.A., Di Costa, F.: National research assessment exercises: a comparison of peer review and bibliometrics rankings. Scientometrics **89**(3), 929–941 (2011). https://doi.org/10.1007/s11192-011-0452-8
19. Bornmann, L., Leydesdorff, L.: Scientometrics in a changing research landscape. EMBO Rep. **15**(12), 1228–1232 (2014). https://doi.org/10.15252/embr.201439608
20. De Rijcke, S., Wouters, P.F., Rushforth, A.D., Franssen, T.P., Hammarfelt, B.: Evaluation practices and effects of indicator use: a review of literature. Res. Eval. **25**(2), 161–169 (2016). https://doi.org/10.1093/reseval/rvv038

Beyond the Classroom: Private Tutoring in Mathematics in the Baltic States

Zane Varpina and Kata Fredheim

Abstract Private tutoring has emerged as a significant educational phenomenon worldwide, often supplementing or compensating for gaps in formal schooling. In the Baltic States—Latvia, Estonia, and Lithuania—this practice has gained traction, particularly in mathematics, a subject pivotal for both academic success and future opportunities. This paper examines the prevalence, characteristics, and implications of private tutoring in mathematics for primary school students across the region, using data from the 2022 OECD PISA survey. The study addresses two key aspects: (1) the demographic and socio-economic profiles of students engaging in private tutoring and (2) the associations between private tutoring and performance in mathematics. It assesses the potential implications of private tutoring on educational equity and access. A comparative lens is applied to highlight similarities and differences among the three countries, reflecting their distinct yet similar educational systems as Estonia is top ranked among OECD countries in student performance in PISA tests, while Latvia and Lithuania, being above OECD average, lag behind the Northern neighbour.

Keywords Private tutoring · Shadow education · Mathematics · Latvia · Estonia · Lithuania · Baltic countries · SDG 4—quality education

1 Introduction

Private tutoring, often referred to as "shadow education," has emerged as a significant component of modern educational systems worldwide. Shadow education supplements formal schooling by providing additional academic support, particularly in high-stakes subjects such as mathematics. This phenomenon has become increasingly prevalent due to its perceived benefits in enhancing academic outcomes

Z. Varpina (✉) · K. Fredheim
Stockholm School of Economics in Riga, Riga, Latvia
e-mail: zane.varpina@sseriga.edu

K. Fredheim
e-mail: kata.fredheim@sseriga.edu

A. Visvizi et al. (eds.), *Research and Innovation Forum 2025 Vol2*, Springer Proceedings in Complexity, https://doi.org/10.1007/978-3-032-09062-1_19

and preparing students for competitive examinations. The Baltic States—Latvia, Estonia, and Lithuania—have experienced a rise in private tutoring in recent years, driven by a combination of parental aspirations, systemic educational challenges, and socioeconomic diversity. However, research on the prevalence, characteristics, and implications of private tutoring in Baltics remains limited.

The educational contexts of the Baltic States offer a useful lens through which to examine shadow education. Estonia, recognized for its consistently high performance in the Organisation for Economic Co-operation and Development (OECD) Programme for International Student Assessments (PISA), demonstrates the effectiveness of a robust public education system. In comparison, Latvia and Lithuania, while performing above the OECD average in PISA reading and mathematics scores, show room for improvement, which may contribute to a greater reliance on private tutoring. These variations in educational outcomes present an opportunity to explore how private tutoring complements or compensates for formal education in differing national contexts. Moreover, the socioeconomic diversity within the region and within countries highlights issues related to equity and access, as private tutoring may amplify existing educational disparities.

Recent sociological literature has highlighted how families' educational strategies are deeply shaped by class position and access to resources. Building on classical theories of social reproduction, contemporary researchers have shown that private tutoring functions not only as an academic supplement but also as a tool through which middle- and upper-class families consolidate advantage [1–3]. These practices reflect broader dynamics of educational inequality, where access to additional support mechanisms is unevenly distributed and linked to economic and cultural capital. In highly competitive educational systems, such as those in the Baltic States, private tutoring may play a critical role in shaping access to elite secondary schools, reinforcing stratification rather than mitigating it.

This study addresses two core questions: (1) What are the demographic and socioeconomic profiles of students engaging in private tutoring in mathematics in the Baltic States? And (2) What is the relationship between private tutoring and mathematics performance? By leveraging data from the 2022 OECD PISA survey of 15-year-olds, this research aims to provide insights into the issue of shadow education in this region.

This study seeks to contribute to the growing body of literature on private tutoring by exploring its prevalence, characteristics, and implications in the Baltic States. The paper begins by reviewing relevant literature and providing a sociological and empirical background to the study. Further it outlines the methodological approach, including data sources and analytical strategies. The findings are presented in two sections, corresponding to the research questions. The paper concludes with a discussion of the implications for educational equity and policy in the Baltic context.

2 Background

Prevalence and characteristics of private tutoring

Private tutoring has grown into a global phenomenon, reflecting the increasing importance placed on educational achievement. Shadow education—so named because it mirrors the formal curriculum—typically focuses on core academic subjects such as mathematics, which is widely regarded as essential for academic and professional success [4, 5]. In regions such as East Asia, private tutoring is deeply embedded in the educational culture. Countries like South Korea and Japan are reporting some of the highest participation rates globally [6]. In Europe, including the Baltic States, the growth of shadow education is more recent, driven by high academic competition and parental demand for personalized instruction [7, 8].

In the Baltic States, private tutoring in mathematics has become increasingly common, although comprehensive research on its scope and impact is lacking. Estonia, known for its strong public education system, may exhibit a lower dependence on private tutoring compared to Latvia and Lithuania, where systemic challenges could lead to higher demand for supplementary education [9, 10]. Additionally, the urban–rural divide and income disparities within these countries may influence access to and reliance on private tutoring services, potentially exacerbating educational inequities [11].

In Latvia, the education system is decentralized, a lot of autonomy granted to schools and municipalities. Research indicates [12] that educational performance varies notably across Latvian schools, with factors such as school size and teacher salaries playing crucial roles. Larger schools and those offering higher teacher salaries tend to exhibit better student performance. Competition for admission to these schools is hence high, bearing in mind also the relatively small size of the country and non-complicated commute. This disparity has led to discussions about the potential benefits of structural reforms, including school mergers and increased teacher compensation, to enhance educational quality. In this context, private tutoring has emerged as a strategy for students and parents aiming to overcome the limitations of smaller or under-resourced schools, seeking to bridge gaps in educational provision and improve academic outcomes. Estonian education system is recognized for its high performance, as evidenced by its students' strong results in international assessments. The system is decentralized, with clear divisions of responsibility between the state, local governments, and individual schools.

The Estonian Education Strategy 2021–2035 guides long-term developments, emphasizing uniform national curricula and the autonomy of schools to develop their own curricula based on national standards. Despite the overall strength of the public education system, private tutoring exists, though its prevalence appears lower compared to neighbouring countries [13]. The availability of high-quality public education may reduce the demand for supplementary private instruction, positioning private tutoring more as an enrichment tool rather than a compensatory mechanism. In Lithuania, private tutoring has become a notable component of the education system too [14]. Families frequently invest in private teachers to prepare students for critical examinations, reflecting concerns about the adequacy of the formal education system in meeting all students' needs. This trend suggests that private tutoring is often utilized to compensate for perceived deficiencies in the public education system, particularly in preparing for high-stakes assessments such as gymnasium or university admission examinations. The reliance on private tutoring raises questions about educational equity, as access to such services is be limited for students from lower socioeconomic backgrounds, potentially exacerbating existing disparities.

Demographic and socioeconomic profiles of students

Socioeconomic status (SES) significantly influences access to private tutoring, with higher SES families more likely to invest in supplementary education to enhance their children's academic performance. This investment is often driven by the financial capacity and educational awareness of these families, enabling them to seek additional academic support [15, 16]. Conversely, students from lower SES backgrounds frequently encounter barriers to accessing tutoring services, which can perpetuate or exacerbate existing educational inequalities [8, 17].

In high-stakes testing environments, such as those in South Korea and China, private tutoring is prevalent among high-achieving students across all SES levels, reflecting a cultural emphasis on educational attainment [18]. In contrast, European countries exhibit more stratified participation rates, with urban, middle- to upper-class students predominantly engaging in shadow education [7]. This pattern is likely mirrored in the Baltic States, where socioeconomic and regional disparities influence access to supplementary educational resources [19].

Parental education levels and household income are critical determinants of private tutoring participation. Families with higher income levels can more readily afford private tutoring services, thereby providing their children with additional academic support that may not be accessible to lower-income families. This financial capability enables higher SES families to invest in supplementary education, further enhancing their children's academic prospects [20].

Geographical location also plays a significant role in access to private tutoring. Urban areas typically offer a more extensive range of tutoring services compared to rural regions, where such opportunities may be limited or non-existent. This urban–rural divide is known to lead to disparities in educational support, with rural students having less access to supplementary tutoring, potentially impacting their academic performance and perpetuating educational inequalities [21]. At the same time, with online communication tools, especially after Covid-19, geographical location has become less of a concern. Much, if not most, of tutoring takes place remotely even if the tutor and pupils live in the same area.

Gender can influence private tutoring participation. Studies indicate that female students are more likely to engage in private tutoring than male students. This trend may be attributed to various factors, including differences in academic motivation, parental expectations, and perceived academic needs. For instance, research in Poland found that female students were more likely to enrol in private tutoring in mathematics, reflecting a gender-based disparity in supplementary education engagement [20].

Cultural attitudes towards education and parental aspirations significantly impact the demand for private tutoring. In societies that highly value educational achievement and view it as a pathway to social mobility, there is a greater propensity for parents to invest in supplementary education. This cultural emphasis on academic success drives the demand for private tutoring, particularly among families aiming to enhance their children's educational outcomes and future opportunities [21].

Impact on academic outcomes

The relationship between private tutoring and academic outcomes is a focal point of shadow education research. Numerous studies highlight the potential of private tutoring to improve student performance, particularly in mathematics, by addressing individual learning needs [18, 22]. However, the effectiveness of tutoring varies across contexts, influenced by factors such as instructional quality, frequency, and student motivation [7, 17]. In some settings, tutoring serves as a remedial tool for struggling students, while in others, it functions as an enrichment activity for high-achievers.

In the Baltic context, private tutoring may play different roles depending on the national educational landscape. Estonia's strong public education system might reduce the need for private tutoring, whereas Latvia and Lithuania may rely more heavily on tutoring as a compensatory mechanism [10].

Use of OECD PISA data in shadow education research

The Organisation for Economic Co-operation and Development's (OECD) Programme for International Student Assessment (PISA) has been instrumental in advancing research in education. It is the largest single coordinated and comparable education data source. In the year 2000, PISA began with 32 participating countries. By the 2022 assessment cycle, this number had grown to 85 countries. PISA conducts triennial assessments of 15-year-old students' competencies in reading, mathematics, and science across participating countries. Beyond evaluating academic performance, PISA gathers extensive background information through student, parent and school questionnaires, including demographic data, socioeconomic status, and educational practices. This comprehensive dataset enables researchers to investigate various factors influencing student achievement, including the prevalence and impact of private tutoring.

PISA has included questions on additional instructional support in mathematics and other subjects since at least 2009, with refinements over subsequent cycles [23]. In 2012, PISA explicitly incorporated inquiries into students' participation in out-of-school-time lessons, allowing for a more systematic examination of private tutoring trends [23]. Researchers analysing PISA data have explored how participation in shadow education varies by country, socioeconomic background, and school characteristics in several countries. For example, PISA data have been used to compare private tutoring trends in East Asian countries [24], where high-stakes testing drives widespread participation, and in Europe, where private tutoring is more stratified by socioeconomic factors. Similarly, PISA data have been leveraged to examine disparities in access to private tutoring across European nations, highlighting the role of socioeconomic status in shaping participation [7, 8].

PISA data has also been used to assess the effectiveness of private tutoring. Research on Europe, drawing on PISA 2015 data, investigated how student and school characteristics affect tutoring choices and whether supplementary education improves academic performance [25]. Findings suggest that while private tutoring can enhance outcomes, its impact varies based on factors such as the quality of instruction and students' baseline performance levels. This analysis has been extended further by exploring how PISA 2018 data reflect the evolving role of private tutoring in response to changes in national education policies [18].

For the Baltic States, the 2022 PISA survey presents an opportunity to examine the role of private tutoring in mathematics education that has not been explored yet.

Educational equity and access

Private tutoring raises significant concerns regarding educational equity. Higher-income families' ability to afford additional educational support creates disparities that can undermine efforts to ensure equal educational opportunities [8, 21]. These inequities are particularly evident in countries where public education systems struggle to meet diverse student needs, driving demand for private tutoring among affluent households [18].

In the Baltic States, addressing the equity challenges is important. The Estonia's education system has been recognised as comparatively inclusive, but Latvia and Lithuania face greater obstacles in ensuring equitable access to quality education. Policymakers must consider strategies to mitigate the potential inequities associated with private tutoring, such as providing subsidies for low-income families or enhancing the quality of public education to reduce the perceived need for supplementary instruction [9, 17].

Given its' profound effects, the appearance of Generative AI in educational contexts has sparked a debate among scholars, educators, and policymakers. If Covid-19 changed the form of instruction and disrupted the classroom logistics, Generative AI technology, especially large language models with capacity to create human-like text, has more far-reaching effects on the content of learning and the learning outcomes [26].

3 Approach

This study employs a quantitative research design, utilizing secondary data from the OECD PISA 2022 survey. The approach is descriptive and inferential, allowing for the examination of patterns in private tutoring participation and its association with mathematics performance. By employing statistical techniques, the study aims to provide empirical insights into the role of socioeconomic factors in shaping student engagement in private tutoring and its potential impact on academic achievement.

Data

The study utilizes data from the 2022 OECD Programme for International Student Assessment (PISA) survey, which assesses the competencies of 15-year-old students in various domains, including mathematics. The dataset includes all OECD countries, but for the purpose of this study we have selected the observations on the three Baltic States: Latvia, Lithuania, and Estonia, providing a comprehensive cross-national perspective.

The PISA 2022 dataset includes a range of variables related to student demographics, socioeconomic background, and academic performance. This study primarily considers mathematics performance, measured through plausible values that estimate student proficiency based on the PISA test results. Additionally, private tutoring participation is derived from multiple survey items capturing whether students engaged in paid, out-of-school mathematics lessons. Socioeconomic factors such as parental education (PAREDINT), home possessions index (HOMEPOS), books at home (BOOKS), and the Economic, Social, and Cultural Status index (ESCS) are included to assess their role in tutoring participation. Demographic variables such as gender and age are also examined.

Given the standardized methodology of the PISA survey, the dataset ensures comparability across countries and maintains high reliability due to rigorous sampling procedures. The final sample for this study includes a total of 19,022 students, with 7257 students from Lithuania, 5373 from Latvia, and 6392 from Estonia.

Method

The data analysis follows a structured approach to address the two research questions regarding tutoring participation and its relationship with mathematics performance.

To examine the demographic and socioeconomic profiles of students engaging in private tutoring, descriptive statistics are used to compare students who receive tutoring versus those who do not. Frequency distributions and cross-tabulations are conducted for categorical variables such as gender and parental education, while mean comparisons are performed for continuous variables such as socioeconomic indices. Chi-square tests are employed to assess statistical differences in tutoring participation across demographic groups.

To investigate the relationship between private tutoring and mathematics performance, several statistical techniques are applied. First, independent samples t-tests compare the mean mathematics scores of students who receive tutoring versus those who do not. Additionally, correlation analysis is conducted to measure the strength and direction of the relationship between tutoring participation and mathematics scores. It is important to note that the analysis is based on cross-sectional data, which limits the ability to draw causal conclusions. While associations between private tutoring and mathematics performance are examined, the direction and nature of these relationships should be interpreted with caution, as unobserved confounding factors may influence both tutoring participation and academic outcomes.

4 Findings

This section presents the results of the analysis, structured according to the two core research questions. The findings provide a comparative perspective on private tutoring in mathematics among secondary school students in Latvia, Lithuania, and Estonia. The results highlight demographic and socioeconomic characteristics of students engaged in private tutoring and examine its relationship with mathematics performance.

4.1 Demographic and Socioeconomic Profiles of Students Engaged in Private Tutoring

This subsection examines the prevalence and characteristics of students participating in private tutoring across the three Baltic States. The analysis focuses on key demographic indicators (such as gender and age) and socioeconomic factors (including parental education, household wealth, and access to educational resources).

Prevalence of Private Tutoring

The proportion of students engaged in private tutoring in mathematics varies across the three countries. Table 1 provides an overview of the percentage of students who reported participating in private tutoring by country.

The results indicate that private tutoring in mathematics is widespread across the Baltic States, with the highest participation rate observed in Estonia (69.9%) and the lowest in Lithuania (53.0%). Latvia falls in between with 64.6% of students engaging in some form of private tutoring. This suggests that private tutoring plays a significant role in the education systems of all three countries, though with notable variations in participation levels.

Table 1 Percentage of students engaged in private tutoring in mathematics by country (PISA: engagement in additional math instruction)

Country	*Engaged in any form of private tutoring*	One-on-one tutoring with a person	Internet or computer tutoring with a programme	Video-recorded instruction by a person	Small group study or practice (2–7 students)	Large group study or practice (8+ students)
Estonia	*69.9*	32.8	50.5	36.3	20.1	12.7
Latvia	*64.6*	42.3	46.1	23.7	20.3	14.6
Lithuania	*53.0*	25.6	31.9	22.6	15.9	10.5

Source Authors. Calculations from PISA2022 data

The form of tutoring also differs across countries. In Estonia, internet- or computer-based tutoring (50.5%) is the most common form, followed by video-recorded instruction (36.3%) and one-on-one tutoring (32.8%). In contrast, Latvia has the highest proportion of students receiving one-on-one tutoring (42.3%), while Lithuania reports the lowest engagement in all tutoring forms, except for large-group instruction, which remains low across all countries. Interestingly, a substantial proportion of students use multiple forms of tutoring, highlighting a multimodal approach to private instruction. This suggests that students do not rely on a single tutoring method but rather combine different approaches to supplement their mathematics education.

Socioeconomic characteristics of students receiving private tutoring

To better understand the profile of students engaging in private tutoring, we analysed key socioeconomic indicators such as parental education level, household income, and the availability of learning resources at home. Table 2 presents the average scores for these socioeconomic variables among students who receive private tutoring compared to those who do not.

The results suggest that private tutoring participation is linked to higher parental education levels in Latvia and Lithuania, while in Estonia, parental education does not significantly differ between tutoring and non-tutoring students. Household wealth, measured by home possessions, is a significant predictor of tutoring participation in Estonia and Lithuania, but not in Latvia, suggesting that economic factors may drive tutoring demand differently across these countries. Additionally, overall socioeconomic background (ESCS) is significantly higher for students receiving tutoring in Estonia and Latvia, while in Lithuania, ESCS does not show a statistically significant difference. These findings highlight that while private tutoring is more prevalent among students from advantaged socioeconomic backgrounds, the specific factors driving participation vary across the Baltic States.

4.2 Relationship Between Private Tutoring and Mathematics Performance

This subsection explores the association between private tutoring and student performance in mathematics. The analysis includes mean comparisons of mathematics scores, correlation analysis, and regression models controlling for socioeconomic background.

Mean Mathematics Scores by Private Tutoring Participation

A comparison of average mathematics scores between students who receive private tutoring and those who do not is presented in Table 3.

The results indicate that students who participate in private tutoring tend to have higher mathematics scores compared to those who do not across all three Baltic

Table 2 Socioeconomic characteristics of students by private tutoring participation

Socioeconomic Indicator	Private tutoring (yes)	Private tutoring (no)	Significance (p-value)
Estonia			
Index highest parental education	14.6806	14.5421	0.219
Home possessions	0.2054	0.1130	0.019
Index of economic, social and cultural status	0.2070	0.1304	0.005
Latvia			
Index highest parental education	14.4160	14.1562	< 0.001
Home possessions	0.4256	– 0.0689	0.733
Index of economic, social and cultural status	0.0346	– 0.0976	0.025
Lithuania			
Index highest parental education	14.6318	14.3144	< 0.001
Home possessions	0.0081	– 0.1063	0.012
Index of economic, social and cultural status	0.0853	– 0.0244	0.470

Source Authors. Calculations from PISA2022 data
Note Index highest parental education (PAREDINT) represents the highest level of education attained by either parent and is measured in years of schooling, ranging approximately from 0 to 20 years; Home possessions (HOMEPOS) is index is standardized with a mean of 0 and a standard deviation of 1 across OECD countries. HOMEPOS > 0—a student has more home possessions than the OECD average, HOMEPOS < 0—a student has fewer home possessions than the OECD average. HOMEPOS is a continuous variable with range (– 4 to + 4) Index of economic, social and cultural status (ESCS) is a composite index to measure a student's socioeconomic background. ESCS is standardized with a mean of 0 and SD of 1 across OECD countries. ESCS > 0—a higher socioeconomic status (wealthier, better-educated parents, more home resources). ESCS < 0—a lower socioeconomic status. ESCS is a continuous variable with range (– 4 to + 4)

Table 3 Mean mathematics scores by private tutoring participation

Country	Private tutoring (yes) (SD)	Private tutoring (no) (SD)	Statistical significance (*p*-value)
Estonia	523.19 (77.3453)	508.20 (77.3453)	< 0.001
Latvia	489.59 (77.2115)	478.73 (74.5343)	0.019
Lithuania	476.99 (86.7257)	466.70 (79.4632)	< 0.001

Source Authors. Calculations based on PISA2022

Table 4 Correlation between private tutoring participation and mathematics scores

Country	Correlation coefficient	Statistical significance (*p*-value)
Estonia	0.46	< 0.001
Latvia	0.39	< 0.001
Lithuania	0.32	< 0.001

Source Authors. Calculations based on PISA2022

States. The difference is most pronounced in Estonia, where students who receive tutoring score on average 15 points higher than their non-tutored peers ($p < 0.001$). Similarly, in Lithuania, tutoring students score approximately 10 points higher, with the difference being statistically significant ($p < 0.001$). In Latvia, the gap is smaller, with tutoring students scoring about 11 points higher, and the effect is statistically significant at the 0.019 level. These results suggest that private tutoring is associated with improved mathematics performance, though the magnitude of the effect varies by country. The stronger association in Estonia and Lithuania might indicate that tutoring provides a notable academic advantage in these countries, while in Latvia, the difference, though still significant, is relatively smaller.

Correlation Between Private Tutoring and Mathematics Scores

A correlation analysis was conducted to examine the strength and direction of the relationship between private tutoring participation and mathematics performance. Table 4 presents the correlation coefficients for each country.

The correlation analysis reveals a moderate positive relationship between private tutoring participation and mathematics performance in all three Baltic States. The strongest correlation is observed in Estonia ($r = 0.46$, $p < 0.001$), indicating that students who engage in tutoring tend to achieve significantly higher math scores. Latvia follows with a correlation coefficient of 0.39, suggesting a slightly weaker but still substantial association. Lithuania shows the lowest correlation ($r = 0.32$, $p < 0.001$), implying that while tutoring is still linked to better mathematics performance, the effect is somewhat smaller compared to the other two countries. These findings suggest that the impact of private tutoring varies across the region, with Estonia exhibiting the strongest connection between tutoring and math achievement, possibly due to differences in the quality, accessibility, or intensity of tutoring services.

5 Conclusions

This study examined the prevalence, characteristics, and implications of private tutoring in mathematics among secondary school students in the Baltic States using data from the 2022 OECD PISA survey. The findings highlight differences across Estonia, Latvia, and Lithuania in terms of tutoring participation, socioeconomic backgrounds of students engaged in tutoring, and the relationship between tutoring and mathematics performance. Private tutoring was found to be more prevalent among students from higher socioeconomic backgrounds, reinforcing existing inequalities in educational access and opportunities.

The results indicate that students who participate in private tutoring tend to achieve higher mathematics scores than their peers who do not, with the strongest effects observed in Estonia. However, it remains unclear whether these higher scores are a direct result of tutoring or whether students who are already academically inclined are more likely to seek additional support. The correlation between tutoring and performance, while statistically significant, suggests that tutoring provides an academic advantage, but its impact varies by country. This raises important questions about the effectiveness of tutoring and whether it serves as a supplement for struggling students or an enhancement tool for high achievers.

A key concern emerging from this study is that private tutoring reinforces educational inequalities by disproportionately benefiting students from wealthier families. Given the competitive nature of secondary school admissions, particularly to gymnasiums, students from advantaged backgrounds have greater access to tutoring services, increasing their chances of securing spots in elite schools. This dynamic perpetuates social stratification within the education system, as students from lower-income families may lack the financial resources to access similar opportunities, thereby widening the educational gap.

These findings align with previous research on private tutoring and educational inequality. Previous studies [5, 15] highlight that private tutoring, while often perceived as a tool for academic improvement, tends to exacerbate existing disparities by providing additional advantages to already privileged students. Similar patterns have been observed in European countries, where higher-income families are more likely to invest in tutoring as a strategy for securing better educational opportunities [7, 11]. Additionally, research on East Asian countries, where private tutoring is highly institutionalized, suggests that high-stakes transitions between educational levels drive demand for tutoring [25]. This aligns with the Baltic States' context, where 15-year-olds are preparing for competitive gymnasium admissions, leading to an increased reliance on tutoring.

It is also important to recognize that the high prevalence of tutoring among 15-year-olds is likely linked to the upcoming transition to secondary education, where competition for gymnasium placements is intense. Private tutoring might be less common among students at other ages, highlighting the role of high-stakes educational transitions in driving demand for additional instruction.

References

1. Ball, S.J.: Class Strategies and the Education Market: The Middle Classes and Social Advantage. Routledge (2003)
2. Lareau, A.: Unequal childhoods: class, race, and family life. In: Inequality in the 21st Century, pp. 444–451. Routledge (2018)

3. Reay, D.: Miseducation: inequality, education and the working classes. Int. Stud. Sociol. Educ. **27**(4), 453–456 (2018)
4. Bray, M.: The Shadow Education System: Private Tutoring and Its Implications for Planners. UNESCO International Institute for Educational Planning (1999)
5. Zhang, W., Bray, M.: Comparative research on shadow education: achievements, challenges, and the agenda ahead. Eur. J. Educ. **55**(3), 322–341 (2020)
6. Zhang, E., Liu, Y.: Effects of private tutoring intervention on students' academic achievement: a systematic review based on a three-level meta-analysis model and robust variance estimation method. Int. J. Educ. Res. **112**, 101949 (2022)
7. Silva, R.S.: Shadow Schooling: An International Comparative Analysis Using Data from PISA. Católica Porto Business School (2021)
8. Zwier, D., Geven, S., van de Werfhorst, H.G.: Social inequality in shadow education: the role of high-stakes testing. Int. J. Comp. Sociol. **61**(6), 412–440 (2021). https://doi.org/10.1177/0020715220984500
9. Manzon, M.: Shadow educations: mapping the global discourse. Asia Pac. J. Educ. **34**(4), 389–402 (2024). https://doi.org/10.1080/02188791.2014.969194
10. Jansen, M.M., Schroeder, S., Helmke, A.: Private tutoring in mathematics: patterns, effects, and equity concerns. Eur. J. Educ. Res. **12**(4), 565–580 (2023). https://doi.org/10.12973/eu-jer.12.4.565
11. Guill, K., Lüdtke, O., Köller, O.: Academic outcomes of shadow education in Germany: benefits of individual tutoring in mathematics. Educ. Assess. Eval. Account. **32**(2), 1–22 (2020). https://doi.org/10.1007/s11092-020-09330-y
12. Krasnopjorovs, O.: Do changes in student-teacher ratio affect educational outcomes? Evidence from a quasi-natural experiment in Latvia. Econ. Transit. Inst. Change **25**(3), 421–437 (2017). https://doi.org/10.1111/ecot.12227
13. Eurydice: Estonia: Overview. European Commission (n.d.). Retrieved from: https://eurydice.eacea.ec.europa.eu/national-education-systems/estonia/overview
14. OECD: Education in Lithuania. OECD Publishing, Paris (2017). https://doi.org/10.1787/9789264281486-en
15. Bray, M.: Benefits and tensions of shadow education: comparative perspectives. Asia Pac. Educ. Rev. **14**(1), 1–10 (2013). https://doi.org/10.1007/s12564-013-9253-8
16. Buchmann, C., Condron, D.J., Roscigno, V.J.: Shadow education, American style: test preparation, the SAT and college enrollment. Soc. Forces **89**(2), 435–461 (2010). https://doi.org/10.1353/sof.2010.0095
17. Hajar, A.: Private tutoring across different education systems: an international comparison. Educ. Res. Rev. **37**, 100454 (2022). https://doi.org/10.1016/j.edurev.2022.100454
18. Zhang, W.: Extending the comparisons of shadow education and its nexus with schooling. ECNU Rev. Educ. **7**(1), 3–11 (2024). https://doi.org/10.1177/20965311231218015
19. Guill, K., Bos, W.: Does private tutoring increase students' academic performance? Evidence from a large-scale longitudinal study. Learn. Individ. Differ. **71**, 37–48 (2019). https://doi.org/10.1016/j.lindif.2019.03.001
20. Safarzynska, K.: Socio-economic determinants of demand for private tutoring. Eur. Sociol. Rev. **29**(2), 139–154 (2013)
21. Bray, M.: The Challenge of Shadow Education: Private Tutoring and Its Implications for Policy Makers in the European Union. European Commission (2011)
22. Karakus, M., Tlessov, A., Hajar, A.: Illuminating the shadows: the role of private supplementary tutoring on student math performance in PISA 2022. Large-scale Assess. Educ. **12**(42). https://doi.org/10.1186/s40536-024-00228-5
23. OECD: PISA 2012 Technical Report. OECD Publishing, Paris (2012). https://www.oecd.org/pisa/pisaproducts/PISA-2012-technical-report-final.pdf
24. Byun, S., Kim, K.: Shadow education and educational inequality in South Korea: examining effect heterogeneity using propensity score matching. Res. Soc. Stratif. Mob. **55**, 61–74 (2018). https://doi.org/10.1016/j.rssm.2018.03.003

25. Entrich, S.R.: Who is more likely to participate in private tutoring and does it work? Evidence from PISA 2015. Eur. Educ. Res. J. **17**(3), 327–346 (2018). https://doi.org/10.1177/1474904118759249
26. Varpina, Z., Fredheim, K.: Well it seems I'm caught up in a trap. In: Adoption of AI in University Writing: Implications for Critical Thinking and Labour Market Engagement. The International Research & Innovation Forum 2024 (2025, forthcoming)

Knowledge Brokering in Boundary Spanning Activities Between Academia and Industry: Organizational Practices in ICT Sector

Iivari Kunttu and Leena Kunttu

Abstract This paper examines knowledge brokering activities within university–industry relationships (UIRs), with a particular focus on the role of boundary actors in facilitating knowledge transfer, creation, and implementation. Drawing on a qualitative comparative case study of long-term collaborations between global industrial technology firms in the ICT sector and universities in Finland, the study highlights the critical importance of mutual understanding between partners. Specifically, boundary actors on both sides must possess a deep awareness of each other's objectives, priorities, and organizational cultures to effectively support knowledge co-creation and utilization within UIRs. The cases underscore that boundary spanning is widely regarded as one of the most effective mechanisms for enabling knowledge brokering between academic and industrial institutions.

Keywords Knowledge brokering · Boundary spanning · University-industry collaboration · Industry · Infrastructure and innovation

1 Introduction

Industrial companies are increasingly expanding their research, development, and innovation capacities by collaborating and co-developing with external actors who offer valuable expertise, knowledge, and capabilities. This trend is particularly significant in the information and communications technology (ICT) sector, which is inherently knowledge-intensive [2]. As a result, the strategic exploration and integration of

I. Kunttu (✉)
Häme University of Applied Sciences, Hämeenlinna, Finland
e-mail: iivari.kunttu@hamk.fi

L. Kunttu
Finnish Environment Institute, Helsinki, Finland
e-mail: leena.kunttu@syke.fi

A. Visvizi et al. (eds.), *Research and Innovation Forum 2025 Vol2*, Springer Proceedings in Complexity, https://doi.org/10.1007/978-3-032-09062-1_20

external capabilities have driven the growth of collaborative university–industry relationships (UIRs) [1, 10]. These partnerships contribute to the formation of innovation networks characterized by diverse and heterogeneous knowledge bases [9, 13, 14].

Despite the steady growth in research-based collaboration and its clear benefits for industrial actors, university–industry relationships (UIRs) often face challenges in fully capitalizing on the results of such partnerships. In many cases, the outcomes of collaboration do not translate into tangible inventions or market-ready products. One of the primary obstacles to effective utilization lies in the differing and sometimes conflicting objectives, goals, and organizational cultures of the partners involved [4, 8]. While companies typically prioritize short-term research and development efforts that directly impact product and service innovation, universities are primarily focused on advancing open science and generating public knowledge through research and education [4, 8]. These divergent orientations can hinder the seamless integration and application of collaborative outcomes. Previous research has shown that boundary-spanning activities play a crucial role in bridging these institutional divides. Such activities facilitate mutual relational learning and support the integration of new knowledge across organizational boundaries [7, 12]. Boundary actors effectively serve as intermediaries between the "customers" (firms) and "suppliers" (academics), who operate within distinct working environments and cultural frameworks [12]. Recent literature also highlights cases where industrial representatives actively contribute to facilitating collaboration. For instance, individuals with prior academic experience often demonstrate a strong interest in building bridges between their industrial organizations and the academic community [7, 8]. These actors can play a pivotal role in aligning expectations, translating knowledge, and fostering long-term, productive partnerships.

This study focuses on the role of boundary actors as knowledge brokers within UIRs. Knowledge brokers are individuals or organizations that facilitate the movement of knowledge and foster connections between researchers and their various audiences. As such, they play a critical role in enabling the creation, sharing, and application of knowledge across institutional boundaries [6]. The primary aim of this research is to explore the practices and mechanisms through which boundary actors contribute to relational learning processes in UIRs. The central research question guiding this study is: *What kinds of boundary-spanning mechanisms and practices facilitate knowledge brokering in collaborations between industry and academia?* To address this question, the study adopts an inductive, qualitative research approach. This methodology is particularly well-suited for examining complex, context-specific phenomena such as knowledge brokering in UIRs, as it allows for a nuanced understanding of the social and organizational dynamics involved. To this end, the argument in this paper is structured as follows. Section 2 presents the theoretical framework focusing on knowledge brokering and relationship learning. In Sect. 3, the methodological choices of the paper are discussed. Section 4 presents the results of the interviews whereas the results and contribution of the paper is discussed and concluded in Sect. 5.

To support the development of the argument, this paper is organized into five sections. Section 2 presents the theoretical framework, with a particular focus on the concepts of knowledge brokering and relationship learning within UIRs.

Section 3 outlines the methodological approach, including the case study design, data collection methods, and analytical procedures. Section 4 reports the empirical findings derived from the interviews conducted with boundary actors from both academic and industrial organizations. Finally, Sect. 5 discusses the implications of the findings, highlights the study's contributions to the literature, and offers concluding reflections.

2 Theoretical Framework

In collaborative networks between industrial companies and academia, the transfer of information, knowledge, and new ideas often occurs between actors with differing interests, goals, and epistemological orientations [9]. Industrial stakeholders typically emphasize the commercial and application-oriented value of knowledge, while academic actors prioritize the advancement of open science and the generation of publicly accessible knowledge [4, 8]. These fundamental differences can be so pronounced that a dedicated interpretive function is required to bridge the gap between the two domains. [5] refers to this interpretive function as "information brokerage." However, knowledge brokering extends beyond merely connecting actors involved in the knowledge creation process. It also encompasses relational learning and the joint development of knowledge through shared, iterative processes. In this context, brokers play a vital role in translating, aligning, and integrating diverse perspectives, thereby enabling more effective collaboration and innovation across institutional boundaries.

In research collaborations between industry and academia, knowledge brokering is a critical skill for actors who actively facilitate the transfer of academic knowledge into industrial contexts. This activity is commonly referred to as boundary spanning, and the individuals who perform this role are known as boundary actors. These actors serve as interpreters and bridge-builders between organizations that operate in fundamentally different environments, with distinct motives, cultures, and operational logics [6, 12]. Boundary spanning skills are therefore essential capabilities for both academic and industrial participants engaged in knowledge transfer across institutional boundaries [7, 8]. Universities, with their broad knowledge bases and educational competencies, have traditionally developed mechanisms to disseminate and facilitate access to scientific knowledge. In contrast, industrial actors often operate with a narrower, business-specific knowledge base and may adopt a more guarded approach to knowledge sharing, driven by competitive and proprietary concerns.

Building on a practice-based perspective, situated learning plays a key role in shaping organizational practices [15]. This form of learning occurs within the boundaries of organizations, particularly in local communities of practice (CoPs), where members learn through participation in shared social practices. In such settings, knowledge is socially constructed through interactive processes within informal groupings of individuals. In the context of boundary spanning, individuals involved in CoPs often represent different organizations, enabling inter-organizational learning at the individual level. When this learning extends across organizational boundaries, it takes place within what [3] describe as networks of practice (NoPs)—loosely connected individuals engaged in similar practices but dispersed across different locations. These networks facilitate knowledge sharing due to overlapping knowledge bases among participants. In UIRs, learning processes are inherently inter-organizational. Since UIRs are fundamentally relational in nature, they are best understood through the lens of relationship learning. According to Selnes and Sallis ([11], p. 80), relationship learning is a joint activity in which organizations share information, interpret it collaboratively, and integrate it into a relationship-specific memory. This process unfolds in three key phases: Knowledge sharing, (2) Joint sensemaking and (3) Knowledge integration. Throughout these phases, the role of individuals who facilitate the learning process—such as boundary actors—is essential [8]. These facilitators help bridge organizational divides, enabling the co-creation and internalization of knowledge within collaborative partnerships.

In this article, we seek answers to our main research question by using these three phases as theoretical lenses. For this, we have divided our research question into three subquestions: *(1) How to facilitate knowledge transfer in an efficient manner, (2) how to support joint knowledge creation between partners, and (3) how to support utilization and integration the jointly created knowledge?* These sub-questions are related to the three above mentioned phases of relationship learning, respectively.

3 Methodology

This study investigates the role of boundary actors as knowledge brokers in eight university–industry relationships (UIRs) in Finland, using a qualitative case study approach. The cases were selected purposively, based on the concept of information-rich cases [16], rather than through random sampling. The primary selection criterion was the presence of long-term, productive collaboration between a university or research institute and an industrial company operating in the information and communications technology (ICT) sector—including electronics, mobile communications, and software. To explore the practices associated with boundary spanning and knowledge brokering, data were collected through semi-structured interviews. The interview protocol was designed around three key phases of the knowledge process: (1) Knowledge Transfer and Relationship Building—Interviewees were asked how academic and industrial actors establish effective working relationships that enable knowledge sharing, (2) Knowledge Creation—Questions focused on practices that support the joint development of new knowledge, and (3) Knowledge Integration and Utilization—This phase explored how collaboratively developed knowledge is embedded and applied within the organizations. All interviews were recorded and transcribed for analysis. To maintain confidentiality, interviewees are identified only by their organizational affiliation: UNIV (university) or IND (industry).

4 Results

Following the research question, the case analysis focuses on boundary spanning and knowledge brokering activities within university–industry relationships (UIRs), with particular attention to the role of boundary actors on both sides of the collaboration. The analysis is structured according to the process outlined in Table 1, and is guided by three sub-questions derived from this framework. These sub-questions are explored across three thematic subsections: (1) Knowledge Sharing, (2) Knowledge Creation and (3) Knowledge Utilization. Each subsection examines how boundary actors contribute to the respective phase of the knowledge process, highlighting the mechanisms and practices that facilitate effective collaboration between academic and industrial partners.

Table 1 Phases of the learning process in three levels

Phase	Situated learning (individual level)	Relationship learning (organizational level	Knowledge brokering
1. Knowledge sharing	Development of knowledge sharing practices between individuals having overlapping knowledge bases	Development of knowledge sharing practices in the shared process between organizations	Linking the partners and opening and facilitating knowledge sharing between them in trusted atmosphere
2. Knowledge creation	Learning by participating in social practice that enables inter-organizational learning on individual level	Combining and making sense of the previous knowledge from participating organizations and creating new knowledge in a shared process between individuals representing different organizations	Helping the partners to find a common language to discuss and solve complex technical problems, and jointly set common targets for the collaboration
3. Knowledge utilization	Interaction and learning between individuals result in incremental changes in organizational practice	Integrating the new, relationship-specific knowledge into the organizational memories	Facilitating the utilization of the jointly created knowledge on both sides of the relationship

Source Authors

4.1 Knowledge Sharing

As UIRs are typically much more driven by individuals than organizations, relationship learning taking place in UIRs is typically based on personal-level relations between the key persons in the relationships, in which information exchange, knowledge sharing and learning can take place [4, 8]. This ensures efficient knowledge transfer and sharing, which is a necessary part of the mutual learning process requiring commitment and engagement from the collaboration parties [10]. All the academic interviewees agreed that long-term personal contacts are a necessary condition for knowledge sharing and transfer:

> The co-creation activities practically take place in close interaction between the university researchers and industrial developers (Univ E).
>
> Personal contacts and relationships between the industrial partner's R&D staff and our researchers really facilitate close collaboration and open communication in this relationship (Univ D).

As indicated in the previous UIR research, knowledge transfer in UIRs is particularly important for industrial firms to absorb new and valuable information and knowledge that may be useful to their R&D activities [12]. Whereas creation of open science and knowledge is a central principle in universities, the industrial firms are interested in the economic value facilitated by the knowledge [8]. In this manner, various institutional norms and restrictions concerning public and private knowledge may cause barriers to effective knowledge sharing in UIRs [4], and the mainstream of the UIR literature sees the sharing of knowledge as particularly problematic for industrial actors. Scholars have also indicated that the key enabler for effective knowledge transfer is mutual trust developed in long-term collaboration [4, 12]. For this, boundary actors' role is essential [12] Based on this, a central theme in the interview was to understand the mechanisms behind the trust building in UIRs. The interviewed industrial managers emphasized the meaning of previous collaboration experience and personal-level contacts to the key contacts on the university side. Most of the interviewed industrial managers actually had had prior contacts to their academic partners already before starting the collaboration, which helped building the bridge between the partners:

> I had known the research team leaders from the past, when I was conducting research work in the academic world (Ind B).
>
> We had been working in the same research projects with them [university partner] during my previous work in R&D (Ind D).

This indicates that the industrial actors with a background in academic research are more likely to build trust and adapt to the collaboration with universities than those managers who do not know the academic world well. This was supported also in the views of university actors:

> Many of the researchers and PhD students have continued their career at the industrial partner after their graduation. This certainly facilitates our collaboration (Univ F).

> Some of the interviewees also emphasized the reputation of the research team as indicator of a trusted partner:
>
> My colleagues had good collaboration experiences with this research team (Ind E).

When considering the boundary spanning activities, it is interesting to understand how individuals act in the relationship when they facilitate the knowledge transfer between the partners.

> I try to share as much information as possible with the university partner. High levels of trust is of course mandatory in this (Ind B).
>
> I encourage our developers and university researchers to team-work with close contacts and frequent interactions. Remote meeting tools make this easier nowadays (Ind A).
>
> University side can take an active role in facilitating information sharing between the partners. Making close personal contacts between the researchers and industrial developers is a key issue there (Univ C).

Based on the interview data, it seems that a crucial task for a boundary actor is to help the partners to make such close contacts between each other that the university and industry representatives can jointly form virtual teams, in which the project work is carried out.

4.2 Knowledge Creation

Joint sensemaking between the industrial and academic partners means that the partners combine and make sense of the previous knowledge that they may have from their experience, and in the joint learning process create new knowledge. This process is shared between individuals representing different organizations. In joint sensemaking, partners work together in order make sense of the problem to be solved, and together find potential solutions to it. Since the partners' knowledge and capabilities are typically heterogeneous in nature, the partners are making efforts together to combine their previously obtained knowledge that may be both tacit and formal in nature. In the process they create new and unique knowledge to solve the problem to be studied [11]. In the interviews, the role of the boundary actors was investigated from the viewpoint of knowledge creation. The interviewees mentioned that finding a common language is a typical problem between industrial and academic world, especially in the beginning of the collaborative relationship:

> In many cases, the industrial and academic people have difficulties to understand each other. Especially in the beginning, there must be someone who can understand both parties (Ind F).
>
> Industry people need someone who "translates" the scientific language to them (Ind C).

Thus, the boundary actors may greatly facilitate the joint sensemaking between the partners by acting as a translator between them. This process is not only helping the partners to communicate, but also to facilitate the joint process in finding common goals and targets to the joint knowledge creation activities:

> Finding common goals that are meaningful for both parties is important in the beginning (Ind D).
>
> Even if the researchers are experts in the very same area as the industrial partner, it is not straightforward that they can easily find a common goal that is meaningful for both parties (Univ B).

Thus, the role of the boundary actor is central in setting suitable goals for the joint activities. For this, experience from both industrial and academic work may be greatly helpful:

> Some members of our research team have working experience from industry. This experience helps them to understand the goals and targets of our industrial partners (Univ D).
>
> I have been working as university researcher in the past, so I understand also the academic partners' interests related to the project work (Ind H).

This supports the earlier indications on the importance of boundary crossing capabilities as an important skill for knowledge brokers who act on the boundary of industry and academia. In this manner, the personnel mobility over this boundary seems to develop the boundary spanning skills on both sides. Many of the interviewees also emphasized the meaning of long-term collaboration, since the knowledge accumulates in the relationships and the members of the participating organizations during a longer time period. Also in this, the boundary actors have a central role:

> Collaboration with good partners should not be seen as short projects, but rather as a long-term interaction. In this manner, we have time to develop new and valuable knowledge, and also to utilize it (Univ C).
>
> Collaboration tends to continue with good university partners. An active role of the key contacts is essential in this (Ind H).
>
> Working in the project gets more and more effective as the people learn to know each other better (Ind C).

In a well-facilitated collaboration, the partners do not only learn from each other to find new solutions to the jointly defined problems, but also to learn how to collaborate between each other [2]. The interview data is consistent with the earlier research that indicates that the bridging mechanisms between industry and academia are being developed over time, as the collaboration skills, attitudes and opinions of the key persons of the collaboration are developed. This happens often through positive experience and interaction on personal-level interactions [2].

4.3 Knowledge Integration

The integration of the jointly developed knowledge into the organizational practices may be involved with the relationship structures and practices, routines, experiences, competences and capabilities developed in the learning relationship between partners. In the UIR relationships, the collaboration partners may also develop the results of their joint learning process as concrete implementations, such as prototypes that can be steps for the industrial development process [10]. The industrial and academic partners may have contradictory interests related to the utilization and implementation of the results of collaborative research [8]. This is because the industrial partners usually prefer to utilize the results in the development of new products and services, which often requires the research results to be kept confidential. However, the university actors usually wish to publish the results obtained in the research collaboration [4]. The role of boundary actors is central in finding consensus between these two views:

> It is really important to understand the goals and targets of both parties to agree on how the results will be utilized (Univ B).

> We often have negotiated how we can utilize the outcomes of our project in such a way that it is satisfying for both parties (Univ D).

The interviewees were asked about the practices that facilitate the integration of the knowledge in the effective manner. The industrial interviewees emphasized the role of the university researchers in helping the developers in the industrial side to implement the results of the research into concrete outcomes:

> In my experience, the university people should be very active in helping our R&D staff in developing the results of the research into prototypes that can be tested in a real-world environment (Ind F).

> It is a real benefit if the academic people can work and collaborate in the industrial environment to further develop the inventions and other outcomes made in the collaboration (Ind C).

The interview data showed that the role of the university-side boundary actors is central in making the knowledge integration effective also on the industrial side. The university actors must have capabilities to understand the industrial environments and their goals and targets to effectively lead the researchers in a way that allows efficient industrial utilization of the research:

> In my group, the researchers routinely work very closely with industrial partners, and implement the results with them (Univ D).

> During the years of collaboration, we have found a working model where our researchers interact closely with the industrial developers through the project. Also the results have been implemented together (Univ H).

On the university side, knowledge integration means the accumulation of the new practically oriented knowledge obtained in the industrial collaboration. The interview data shows that the university actors can use this knowledge in many different manners. The knowledge can, for example, be used in practical way to give the researchers understanding on the real-world applications of their research topics. This is particularly beneficial for academic researchers, since it provides insights into industrial way of working, and gives them valuable boundary spanning skills. Thus, the practical knowledge obtained from the university-industry collaboration can lower the barrier between industry and academia, and facilitates personnel mobility:

> Many of my students have also been recruited by the industrial partner after their graduation. They have been able to use the knowledge developed in the joint research project in their new work at the industrial partner (Univ H).

The interviewees also mentioned the value of the practical knowledge in the education. Many of the boundary actors at the university side used the new knowledge created in the industrial project as practical material in their teaching:

> Using the case examples obtained from our industrial projects makes also theoretically oriented teaching more practical, and provides the students with real-world application examples (Univ E).

5 Discussion and Conclusions

This study contributes to the literature on inter-organizational learning by examining the role of boundary actors as knowledge brokers in university–industry relationships (UIRs). By identifying specific knowledge brokering practices, the study makes two key contributions. First, it extends existing research by providing empirical evidence on how boundary actors facilitate knowledge brokering in long-term UIRs. Second, it highlights a set of collaborative practices that help academic and industrial partners overcome barriers to effective learning and foster productive collaboration. The study is based on a comparative case analysis of eight UIRs in Finland. In each case, key boundary actors from both academia and industry were interviewed to uncover the practices that support successful knowledge brokering across the three phases of relationship learning: knowledge sharing, knowledge creation, and knowledge integration. In the first phase, information sharing, the most critical capabilities of boundary actors involve building trust and cultivating close personal relationships across organizational boundaries. These relationships enable open knowledge exchange, which is a prerequisite for effective collaboration. A deep understanding of the partner's goals, expectations, and organizational culture—often gained through active boundary-spanning experiences—is essential for bridging institutional divides. In the second phase, joint knowledge creation, partners combine their existing knowledge, experiences, and practical insights to co-create new knowledge that is often specific to the relationship and its objectives. Tacit knowledge plays a central role in this process. Boundary actors facilitate joint knowledge creation by helping partners establish a shared language and align their goals. Prior experience in both academic and industrial settings enhances their ability to mediate and support this collaborative process. The third phase, knowledge integration and utilization, focuses on implementing the outcomes of joint knowledge creation. Due to differing organizational goals and norms, conflicts may arise—such as disagreements over the publication of research findings. Boundary actors who understand both parties' preferences are instrumental in negotiating compromises and ensuring that the jointly created knowledge is effectively utilized. They also play a key role in translating scientific knowledge into forms that are actionable and commercially viable. Effective brokering practices in this phase include supporting industrial implementation and prototyping, often facilitated by academic partners, as well as promoting personnel mobility between academia and industry.

In conclusion, this study demonstrates that knowledge brokering is a critical capability for boundary actors operating within university–industry relationships (UIRs). As these individuals navigate the boundary between academia and industry, they must also overcome a range of barriers stemming from differing organizational cultures, goals, and norms. The mechanisms that enable effective bridging are often cultivated over time through sustained collaboration, during which mutual trust and individual collaboration skills are gradually developed. In successful knowledge co-creation processes, UIR partners not only work together to address shared challenges but also co-develop the very practices and frameworks that make their collaboration effective. In this context, knowledge brokering is not merely a support function—it is a central component of the collaborative process. However, generating new knowledge is only part of the equation. Equally important is the ability to translate and adapt that knowledge into forms that are meaningful and actionable for both academic and industrial stakeholders. This aspect of brokering requires a deep understanding of both domains and the ability to mediate between them. One effective way to build such capabilities is through personnel mobility across institutional boundaries, allowing individuals to gain firsthand experience in both academic and industrial environments.

This paper makes also a practical contribution, since boundary spanning in university–industry relationships is arguably one of the most effective mechanisms for facilitating knowledge brokering between these institutions. Boundary actors play a pivotal role in transferring newly created knowledge to relevant stakeholders both within and beyond their organizations. Programs that promote personnel and researcher mobility between academia and industry are valuable tools in supporting this function [7]. However, to fully leverage the potential of boundary actors, policy-makers and industrial leaders should explore additional avenues for enabling academic researchers to gain industrial experience, and for industry professionals to contribute to academic work as part of their career development. Recognizing and supporting the role of boundary actors is essential for both universities and companies, as these individuals are not only facilitators of collaboration but also transmitters of tacit knowledge, which is often critical to innovation. To this end, both sectors should actively promote boundary-spanning roles and encourage personnel to engage in such activities. These initiatives not only enhance knowledge brokering but also help lower institutional and cultural barriers between academia and industry. This study underscores the significant contribution of boundary actors to the innovation process, highlighting the need for both academic and industrial stakeholders to fully appreciate and invest in boundary spanning and knowledge brokering as strategic capabilities.

References

1. Ankrah, S., AL-Tabbaa, O.: Universities-industry collaboration: a systematic review. Scand. J. Manag. **31**(3), 387–408. https://doi.org/10.1016/j.scaman.2015.02.003
2. Bellini, E., Piroli, G., Pennacchio, L.: Collaborative know-how and trust in university–industry collaborations: empirical evidence from ICT firms. J. Technol. Transfer. **44**(6), 1939–1963 (2019). https://doi.org/10.1007/s10961-018-9655-7
3. Brown, J.S., Duguid, P.: Knowledge and organization: a social-practice perspective. Organ. Sci. **12**(2), 198–213 (2001)
4. Bruneel, J., D'Este, P., Salter, A.: Investigating the factors that diminish the barriers to university-industry collaboration. Res. Policy **39**(7), 858–868 (2010). https://doi.org/10.1016/j.respol.2010.03.006
5. Burt, R.S.: Structural holes and good ideas. Am. J. Sociol. **110**(2), 349–399 (2004)
6. Haas, A.: Crowding at the frontier: boundary spanners, gatekeepers and knowledge brokers. J. Knowl. Manag. **19**(5), 1029–1047 (2015). https://doi.org/10.1108/JKM-01-2015-0036
7. Kunttu, L., Huttu, E., Neuvo, Y.: How doctoral students and graduates can facilitate boundary spanning between academia and industry. Technol. Innov. Manag. Rev. **8**(6) (2018). https://doi.org/10.22215/timreview/1164
8. Kunttu, L., Neuvo, Y.: Balancing learning and knowledge protection in university-industry collaborations. Learn. Organ. **26**(2) (2019). https://doi.org/10.1108/TLO-06-2018-0103
9. Mao, C., Yu, X., Zhou, Q., Harms, R., Fang, G.: Knowledge growth in university-industry innovation networks—results from a simulation study. Technol. Forecast. Soc. Chang. **151**, 119746 (2020). https://doi.org/10.1016/j.techfore.2019.119746
10. Perkmann, M., Tartari, V., McKelvey, M., Autio, E., Broström, A., D'Este, P., Fini, R., Geuna, A., Grimaldi, R., Hughes, A., Krabel, S., Kitson, M., Llerena, P., Lissoni, F., Salter, A., Sobrero, M.: Academic engagement and commercialisation: a review of the literature on university-industry relations. Res. Policy **42**(2), 423–442 (2013). https://doi.org/10.1016/j.respol.2012.09.007
11. Selnes, F., Sallis, J.: Promoting relationship learning. J. Mark. **67**(3), 80–95 (2003). https://doi.org/10.1509/jmkg.67.3.80.18656
12. Siegel, D.S., Waldman, D.A., Atwater, L.E., Link, A.N.: Toward a model of the effective transfer of scientific knowledge from academicians to practitioners: qualitative evidence from the commercialization of university technologies. J. Eng. Technol. Manag. JET-M **21**(1–2), 115–142 (2004). https://doi.org/10.1016/j.jengtecman.2003.12.006
13. Steiber, A., Alvarez, D.: How culture and technology go hand-in-hand in digital transformations into an ecosystem business. In: Visvizi, A., Troisi, O., Corvello, V. (eds.) Research and Innovation Forum 2023. RIIFORUM 2023. Springer Proceedings in Complexity. Springer, Cham (2024). https://doi.org/10.1007/978-3-031-44721-1_1
14. Stemberkova, R., Maresova, P., David, O.O., Adeoye, F.: Knowledge management model for effective technology transfer at universities. Ind. High. Educ. **35**(6), 638–649 (2021). https://doi.org/10.1177/0950422220978046
15. Weckowska, D.M.: Learning in university technology transfer offices: transactions-focused and relations-focused approaches to commercialization of academic research. Technovation **41**, 62–74 (2015). https://doi.org/10.1016/j.technovation.2014.11.003
16. Welsh, R., Glenna, L., Lacy, W., Biscotti, D.: Close enough but not too far: assessing the effects of university-industry research relationships and the rise of academic capitalism. Res. Policy **37**(10), 1854–1864 (2008)

Information and Communication Technologies (ICTs) as Tools in Teaching and Learning in Physical Education Classes for Children

Plácido Rogério Pinheiro, Herleson Paiva Pontes, and João Batista Furlan Duarte

Abstract The present case study involves assistive technologies based on motion capture as immersive and interactive pedagogical tools in Physical Education classes to enhance and empower students' teaching and learning processes. Based on relevant theoretical frameworks, an educational structure was developed that incorporates lesson plans promoting the integration of these devices into Physical Education classes, using digital games that require students to perform physical movements to execute actions and events during virtual matches. Through an applied, qualitative, and exploratory methodology, this proposal is characterised by its low cost and accessibility, making it feasible for implementation in any basic educational setting. Additionally, the lesson plans were developed to address all thematic units of Physical Education as defined by the National Common Curricular Base (BNCC) established by Brazil's Ministry of Education for various educational levels. Among the results obtained, the creation of a strategic guideline for the use of ICTs in Physical Education by teachers and students stands out, as well as the transformation of immersive and interactive experiences into learning moments through the meaning-making of these experiences. Moreover, the gap between the classroom and the world "beyond the school walls" was reduced. While the use of digital games as pedagogical tools is widely discussed in scientific research, this study contributes by structuring lesson plans that incorporate motion capture sensors in physical activities, promoting a meaningful learning environment and connecting the classroom to the digital reality of contemporary students.

H. P. Pontes (✉)
Federal Institute of Ceará (IFCE), Jaguaribe, Brazil
e-mail: herleson@gmail.com

P. R. Pinheiro · J. B. F. Duarte
Graduate Program in Applied Informatics, University of Fortaleza (UNIFOR), Fortaleza, Brazil
e-mail: placido@unifor.br

J. B. F. Duarte
e-mail: furlan@unifor.br

A. Visvizi et al. (eds.), *Research and Innovation Forum 2025 Vol2*, Springer Proceedings in Complexity, https://doi.org/10.1007/978-3-032-09062-1_21

Keywords Educational informatics (SDG 4) · Assistive technologies (SDG 4 · SDG 10) · Physical education (SDG 3 · SDG 4) · Natural user interfaces (SDG 9) · Motion capture (SDG 9)

1 Introduction

In the context of the evolution of Information and Communication Technology (ICT) in contemporary society and its transformative impact on how people communicate, work, and learn, the teaching and learning process has undergone significant changes due to the adoption and use of ICT in educational settings [1]. Students experience the pervasive presence of technology in their daily lives through digital devices such as smartphones, tablets, and computers, as well as through connectivity and rapid access to information enabled by the Internet.

One of the major challenges arising from the pervasive integration of technology into daily life is the efficient and pedagogically appropriate incorporation of these tools by schools and teachers to enhance the teaching and learning process [2]. Technology holds considerable potential to improve students' learning experiences by fostering knowledge construction in a more interactive and engaging manner. Examples of such technological applications in the classroom include initiatives involving educational games, videos, mobile applications, and virtual and/or augmented reality devices [3].

As a knowledge-based discipline, Physical Education promotes students' physical health, motor skills, teamwork, and lifelong wellness habits through structured movement and sports activities. It could use technology to offer various benefits, such as increasing student engagement and participation in activities, as well as supporting the development of skills and competencies [1].

However, the integration of technology into the classroom also presents numerous challenges for teachers, students, and schools [3]. Educators must possess technological skills to effectively utilize these tools, as well as the time to structure and implement lessons incorporating them [4]. From the students' perspective, not all have equal access to technology at home, given the socioeconomic disparities within society. The lack of adequate infrastructure for conducting activities, along with the absence of necessary equipment and Internet access, also constitutes significant obstacles to the use of ICT in schools [1]. This gap is notorious in Physical Education, since the use of ICT in classrooms to support the development of students' motor, cognitive, and socio-emotional skills is scarce. This prompted the following research question:

RQ: How can technology engage students in Physical Education classes?

This study presents structured lesson plans and initiatives that incorporate ICT into Physical Education, fostering student motivation and enriching knowledge construction through novel technological experiences. More specifically, the proposed lesson plans employ motion capture devices and sensors as tools for developing gross motor coordination, promoting engagement and entertainment within the pedagogical context. These educational experiences are designed using accessible devices, with costs below two minimum wages, and are easily replicable.

To this end, the argument in this paper is structured as follows. Initially, Sect. 2 will present a literature review on ICT in Education, Meaningful Learning and Natural User Interfaces. Section 3 will detail the structuring and planning of the lessons, including the methodology applied, besides the pedagogical and technical specifications employed to the lessons, and describing the experiments conducted. Following this, Sect. 4 will discuss the results obtained and their implications for improving pedagogical practice. Finally, Sect. 5 will conclude the paper, summarizing the main contributions and pointing out directions for future research.

2 Literature Review

2.1 *ICT in Education*

Nowadays, Information and Communication Technologies (ICT) represent a crucial area for human development, becoming increasingly present at various stages of human evolution [5, 6]. Although ICT has become an integral part of people's daily lives by facilitating, streamlining, and automating tasks, significant efforts are still required to develop and implement innovative resources that enhance learning in education [7].

It is crucial for various educational stakeholders to understand the impact of ICT in school environments based on the specific context in which these solutions are implemented and the target audience of such initiatives, as these factors significantly influence students' knowledge construction [8–10]. Furthermore, ICT contributes to student development only when it serves a clearly defined educational purpose.

Moreover, contemporary students, immersed in digital technologies from birth, exhibit advanced skills in online activities like downloading, content creation, and sharing, suggesting accelerated learning [2].

Given students' intense interaction with technology, fostering active behaviors like interactivity, immersion, and protagonism, the school environment must adopt new teaching methodologies [4]. This aims to harness and direct this potential to optimize the learning process, adapting to how these students process and interact with information.

2.2 *Meaningful Learning*

The acquire of new knowledge in the human mind occurs when the learning experience carries meaning, relevance, and personal significance for the student [11]. From an educational perspective, learning a particular subject takes place when the student, as the protagonist of the process, actively and voluntarily engages in the classroom activity.

Meaningful Learning encompasses a set of processes, techniques, and tools aimed at engaging students in the construction of both individual and social knowledge [11]. Its central driving force lies in intentional, voluntary, and purposeful interactions between the student and the environment, with its representations and models. A pedagogical activity is considered meaningful when it exhibits the following five characteristics: Active, Cooperative, Intentional, Constructive and Authentic.

2.3 Natural User Interfaces

In recent years, Natural User Interfaces (NUIs) have experienced remarkable growth, as they aim to make interactions more intuitive and immersive for humans [12]. Applications that base their interaction processes on touch, motion, and voice recognition are examples of NUIs. All natural interfaces adhere to two fundamental design principles: the interface should be virtually imperceptible to the user and the user's interaction with the interface should be based on natural elements or real-world experiences.

Typically, natural user interfaces are implemented through the design of physical input/output devices that incorporate these fundamental principles. Examples of devices that enable natural interaction include the Kinect [13] and the Switch [14]. These assistive devices facilitate the development of gesture-based control applications with satisfactory levels of capture precision and relatively low acquisition costs, driving their use in scientific research, commercial solutions, and digital gaming.

3 Structuring and Planning of Lessons

3.1 Methodology

This is an applied, qualitative and exploratory research, which took place in an elementary school in Jaguaribe (Ceará, Brazil). It was developed through the authors' experiences and observations, based on four classroom practices with one teacher and thirty-five students (Age 7; 21M, 14F). Figure 1 illustrates all the steps carried out.

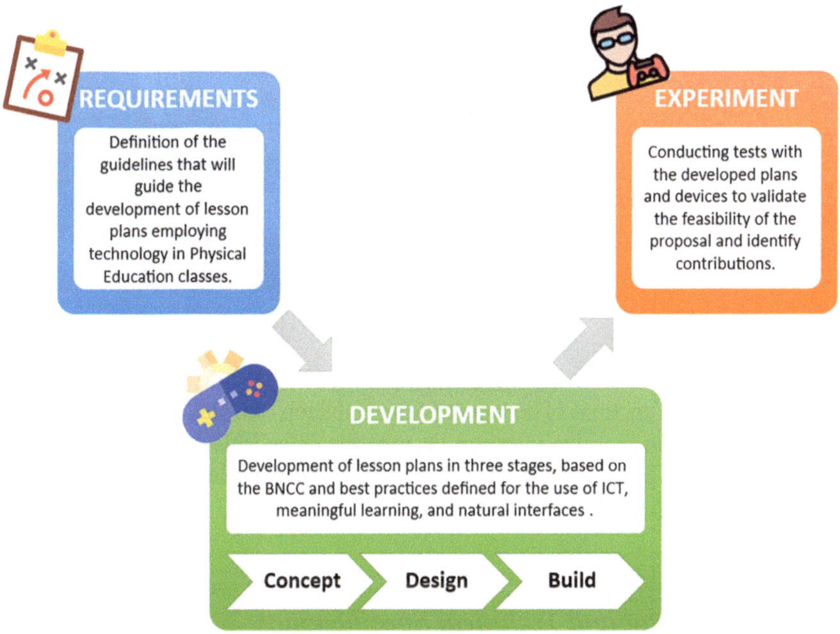

Fig. 1 Research roadmap

3.2 *Assistive Technologies Employed*

In educational environments, some technologies play a role in immersion and interaction, as they enhance a person's sensory perceptions through the generation of specific stimuli tailored to the target audience of the tool.

In this context, assistive technologies support the user in performing certain tasks, as well as enabling the inclusion of individuals by promoting bodily movement in an unrestricted manner. Two of them offer good capture precision with low costs: Microsoft Kinect [13] and Nintendo Switch [14].

Initially developed for Microsoft's Xbox360 gaming console in 2010, Kinect is a motion capture sensor that promotes interaction with computational systems through the detection of spatial position, movements, and voice commands from users [13]. Figure 2 shows this device.

Fig. 2 Microsoft Kinect

Fig. 3 Nintendo Switch

Released by Nintendo in 2017, the Nintendo Switch is a hybrid video game console that combines the functions of both portable and tabletop consoles [14]. The Nintendo Switch provides an intuitive interface through the detachable Joy-Con controllers, which can be used in various ways: attached to the console, used separately, or combined in a grip to form a conventional controller. Figure 3 illustrates the device.

3.3 Pedagogical and Technical Specifications

The proposed lesson plans and their activities are aimed at children and adolescents of both genders, literate, across all educational levels, within the age range of 7–17 years. Besides, any educational space in schools with sufficient room for participants' movement and access to power outlets can be utilized for lessons involving ICTs. The application of technology-assisted activities can be carried out by teachers in Physical Education and Information Technology. Additionally, it is possible to optionally involve teachers from other subjects, tutors, and occupational therapists from the school.

It is important to highlight that the teacher does not need both NUI devices to carry out the activities: one of them is sufficient to implement various immersive and interactive pedagogical proposals in Physical Education lessons. The low acquisition cost, compared to the price of computers and laptops, as well as the simplicity involved in setting up and using the games, emphasize the inclusive appeal of using technologies in the daily academic activities of institutions. Additionally, these devices can be utilized in initiatives from other subjects, such as History, Geography, Mathematics, and Physics.

The list of materials and services involved in the delivery of proposed activities is in Table 1.

Table 1 List of materials and services involved in the delivery of proposed activities

Material/service	Price
Xbox360 + Kinect Sensor	US$ 50.00
Nintendo Switch	US$ 200.00
3 (three) Xbox360 games	US$ 15.00
3 (three) Nintendo Switch games	US$ 45.00
Total	US$ 310.00

3.4 Digital Game Selection

Our approach to using ICT in Physical Education lessons is based on the employment of digital games that promote player movement to perform in-game actions, which are captured by the Kinect sensor or the Switch Joy-Con controllers.

To achieve this, games from both platforms were selected that encourage player interaction through movement and that could be linked to themes present in the Physical Education curriculum, considering the three fundamental principles and six thematic axes established by Brazilian's National Common Curricular Base (BNCC) [15].

- Kinect Adventures (Xbox360)
  - *Description*: An action and adventure game that uses the Kinect sensor to allow players to control the game through body movements, offering a series of mini games based on physical and interactive activities.
  - *Topics*: Play and Games, Adventure Physical Practices
  - *Suggested Activities*: The game features a series of competitive and cooperative mini-games that require players to perform full-body movements, including jumps and squats, making it suitable for use in relay races. Additionally, it presents forests and oceans as environments, providing contextualization for adventure practices.
- *Kinect Sports (Xbox360)*
  - *Description*: An interactive sports game that uses the Kinect sensor to allow players to participate in a variety of sports, such as soccer, bowling, and boxing, through body movements.
  - *Topics*: Sports, Combat Sports, Gymnastics
  - *Suggested Activities*: The game offers a series of sports, including alternative modalities, that can be experienced by players through the reproduction of movements identical to those in the respective sports, such as serves, kicks, and throws. The game includes Soccer, Bowling, Boxing, Table Tennis, Athletics, Beach Volleyball, American Football, Golf, Tennis, Darts, Skiing, and Baseball. Additionally, it is possible to work on actions and rules without the need for additional equipment, promoting immersion at a low cost.
- *Just Dance 2019 (Xbox360)*

  - *Description*: An interactive dance game that uses the Kinect sensor to allow players to imitate choreographies of popular songs through body movements.
  - *Topics*: Dance
  - *Suggested Activities*: The list and styles of songs are extensive and diverse, allowing players to perform various rhythms. It is also possible to improve motor coordination through cooperative dances, in which there is a single choreography performed by a team of players.

- *Nintendo Switch Sports (Nintendo Switch)*

  - *Description*: An interactive game that uses the Joy-Con controllers to allow players to participate in various sports, such as bowling, tennis, and soccer, through realistic body movements.
  - *Topics*: Sports, Combat Sports
  - *Suggested Activities*: The game offers a series of sports, including alternative modalities, that can be experienced by players through the reproduction of movements identical to those in the respective sports, such as serves, kicks, and throws. The game includes Soccer, Bowling, Chambara, Tennis, Volleyball, Badminton, and Tennis. Additionally, it is possible to work on actions and rules without the need for additional equipment, promoting immersion at a low cost.

- *Everybody 1–2-Switch (Nintendo Switch)*

  - *Description*: A game that offers a collection of mini-games where players compete and interact with each other using Joy-Con controllers and mobile phones.
  - *Topics*: Play and Games, Gymnastics
  - *Suggested Activities*: The game features a series of competitive and cooperative mini-games that require players to perform full-body movements, including jumps and squats, making it suitable for use in relay races. One of the highlights is the simultaneous support for up to 100 players using mobile phones.

- *Ring Fit Adventure (Nintendo Switch)*

  - *Description*: An adventure game for Nintendo Switch that combines physical exercises with RPG elements, where players use the Ring-Con and Leg Strap to explore a virtual world and defeat enemies through physical activities.
  - *Topics*: Play and Games, Adventure Physical Practices
  - *Suggested Activities*: The proposal to combine the execution of movements with the enhancement of an RPG character and the existence of a story as contextualization allows for greater player engagement.

Figure 4 presents the games proposed for the Xbox 360 with Kinect sensor, while Fig. 5 illustrates the games for the Nintendo Switch with Joy-Con controllers.

Fig. 4 Microsoft Kinect games selected

Fig. 5 Nintendo Switch games selected

3.5 *Development of Lessons*

Our proposal for constructing Physical Education lessons had three stages: requirements' definition of the guidelines that would guide the development of lesson plans employing technology in Physical Education classes, development of lesson plans based on the BNCC and best practices defined for the use of ICT, meaningful learning, and natural interfaces, and testing the developed plans and devices to validate the feasibility of the proposal and identify contributions, as illustrated in Fig. 1.

Considering the games presented and their suggested use, we conducted a study to align the resources provided by the digital game with the learning objectives outlined in the lesson plan. Additionally, we defined how the structural variables (Space, Time, and Tactics) should work during the learning process, generating outcomes such as movement variety, difficulty levels, and appropriateness for the target audience.

The lessons were structured to integrate the selected games and activities with the specific learning objectives of the Physical Education curriculum, including an introduction that contextualizes the topic/theme to be addressed, along with the concepts and skills to be worked on. Individual or team activities, cooperative or competitive modes, comparative analyses of rules and their applications, are examples of the potential of devices like Kinect and Joy-Con controllers in encouraging the active participation of all students.

It is important to highlight that the teacher must play an active role in monitoring and providing guidance to students during the activities, ensuring that movements are executed correctly and safely. Moreover, it is the teacher's responsibility to validate the construction of knowledge through formative assessments that identify students' progress, as well as to allow for adjustments to be made to future activities as needed.

3.6 Validation Experiment

With the goal of validating the feasibility of the developed lesson plans, the experiment took place during four classroom practices with one teacher and thirty-five students in an elementary school. The authors' perceptions and learning gained during the conduction of classes were extracted to validate the lesson plans and identify potential contributions.

4 Results and Discussion

4.1 Results

During the planning and structuring of the proposal, the experiences inherent in using technologies as a pedagogical tool during supervised stages were observed and recorded. As a result, some important observations suggest the positive potential of this approach in Physical Education lessons.

From the student's perspective, the implementation of lessons mediated by digital games draws attention and stimulates engagement. By bringing the technological world into the classroom, the gap between "school" and "outside" is reduced, making the learning meaningful.

The features related to the modes and levels of the games used allow students to experience different contexts in controlled environments. They have the freedom to experiment, act, and take risks without harm to their health, in scenarios whose real-world reproduction would be quite complex or impossible; behavior observed

Fig. 6 A class using Microsoft Kinect games

at several moments when students mimicked the movements of their peers who were playing.

Moreover, the cooperation, competition, and challenge promoted by the games significantly engaged the students, creating different feelings among the participants, such as competitiveness, joy, motivation, and moments of social interaction among the players.

Additionally, the components of the graphical interface capture the students' attention, assisting in their immersion and interaction with the game and its proposed activities/actions.

From the teacher's point of view, considerable ease was observed in preparing the environment for the lessons, conducting activities with the students, and dismantling the equipment after the lessons ended, both closed and open spaces. Figure 6 displays photos of some classes based on the lesson plans developed.

4.2 Discussion

The observed results suggest potential benefits for the learning process when utilizing technologies such as the Microsoft Kinect and Nintendo Switch as tools that catalyze the construction of knowledge through student experimentation, as expected by [12].

An important contribution lies in the fact that it brings suggestions for strategies to integrate ICTs into Physical Education lesson plans. As presented in this study for both teachers and students, it is evident that the use of digital games requiring student movement enables the promotion of experiences that combine entertainment and learning, assisting the student in the construction of new knowledge as described by [7].

A digital game differentiates itself by offering motivation as an element connected to an experience, and by converting this experience into learning through the significance of that moment: the combined use of the senses to perform the game actions, captured by movement sensors, is active, promotes cooperation, is intentional, builds new knowledge, and is authentic, consolidating all the characteristics of meaningful learning as proposed by [11].

Another important feature of the proposal for adopting ICTs in Physical Education is the broad scope of themes covered by digital games with motion sensors. As described in BNCC [15], the games presented encompass all six thematic units for Physical Education, allowing for work across the vast curriculum defined for this subject. In many cases, a single game addresses multiple curricular axes, offering flexibility in how the teacher can employ this tool. Additionally, it is possible to create interdisciplinary activities involving other areas of knowledge, such as Physics, Arts, and History, expanding the reach of ICTs within the school environment, like described in [4].

Furthermore, an issue raised by the integration of technology in Physical Education is the reduction of the distance between the classroom and the world "beyond the school walls", a challenge described in [2]. The use of ICTs in schools reduces the archaic feeling present in traditional education, aligning the learning process with the dynamism, speed, and connectivity of the contemporary digital world of this generation.

Lastly, in the context of the applied educational memory game, the following themes raised during the application of this activity in school environments: *Competitiveness*, *Frustration/Sadness/Anger*, *Motivation*, *Deafness* and *Joy*. These topics must be managed by the teacher during gameplay, as suggested in studies [1, 3].

4.3 Limitations

Although digital games and motion capture devices demonstrate potential for use in classrooms, offering good gameplay, easy learning, appropriate levels of challenge, and motivating players to learn Physical Education content, this approach has some limitations.

Conducting games with large groups requires a considerable amount of time. Even if the school provides several consoles with the games, the total number of students is quite high, which increases the complexity of managing player rotation. In addition to the mechanics inherent to the game itself, the fact that students need time to learn the necessary movements for success in the game increases the total time of each match.

Finally, there was no quantification of the impact of the proposal on school environments due to limited time and the large scope of this study. Future research is planned to expand the understanding of the use of devices such as the Microsoft Kinect and Nintendo Switch in classrooms, as well as to address issues related to inclusion in educational settings.

5 Conclusion

This work described the proposal for adopting the assistive devices Microsoft Kinect and Nintendo Switch as immersive and interactive pedagogical tools during Physical Education lessons. A structure was developed for the use of assistive devices in Physical Education classes through digital games that require motions to trigger events.

One important finding of this research is that digital games with low-cost motion sensors foster knowledge construction through active, multisensory, and meaningful learning experiences. Also, ICT integration in lesson plans promotes student motivation, engagement, and authentic learning through movement-based gameplay. Moreover, technologies like games and NUI devices narrow the gap between school and contemporary digital life. Lastly, emotional and behavioral aspects emerge during gameplay and require active teacher mediation.

Although the use of digital games as pedagogical tools is a recurring topic, this work contributes by structuring the use of motion sensors for physical activities at the lesson planning level, promoting a meaningful learning environment, and connecting the classroom to the digital reality of contemporary students.

Finally, future directions can be proposed based on the findings of this work. One future direction involves implementing this proposal in educational environments and collecting data for quantitative analysis to identify its influence on student learning. Another future endeavor could be constructing an interdisciplinary curriculum that combines Physical Education with other areas of knowledge, using ICTs as connecting elements. Additionally, further analysis of other motion sensor-based games can be conducted to assess their viability for use in Physical Education lessons.

References

1. Ohuruogu, B., Ikechukwu, U.J., Mong, E., Farias, A.R.C.: Application of Information and Communication Technology (ICT) in administration of physical education and sports. J. Educ. Pract. **10**(24), 1–5 (2019). https://doi.org/10.7176/JEP/10-24-01
2. Prensky, M.: Digital Game-Based Learning. Translated by Eric Yamagute. Senac São Paulo, São Paulo (2012). https://doi.org/10.1145/950566.950596
3. Akram, M.S., et al.: Teachers' perceptions of ICT integration in teaching-learning practices: a systematic review. Front. Psychol. **13**, 920317 (2022). https://doi.org/10.3389/fpsyg.2022.920317
4. Prensky, M.: "Don't Bother Me, Mom—I'm Learning!": How Video Games Are Preparing Our Children for Success in the 21st Century—and How You Can Help! Translated by Lívia Bergo. Phorte, São Paulo (2010)
5. Hawkridge, D.: New Information Technology in Education. Taylor & Francis (2022)
6. Vidal, E.M., Maia, J.E.B.: Introduction to Distance Education and Basic Informatics, p. 2. Universidade Aberta do Brasil (2015)
7. Raja, M., Priya, G.: Conceptual origins, technological advancements, and impacts of using virtual reality technology in education. Webology **18**(2), 116–134 (2021). https://doi.org/10.14704/WEB/V18I2/WEB18311

8. Stoffová, V., Katyetova A., Obonya, J.: Informatics in the first level of primary school. In: EDULEARN23 Proceedings, pp. 4155–4164 (2023). https://doi.org/10.21125/edulearn.2023.1110
9. Moraes, M.C.: Emerging Educational Paradigm. Papirus Editora (2011)
10. Dogan, S., Dogan, N.A., Celik, I.: Teachers' skills to integrate technology in education: two path models explaining instructional and application software use. Educ. Inf. Technol. **26**, 1311–1332 (2021). https://doi.org/10.1007/s10639-020-10310-4
11. Howland, J., Jonassen, D., Marra, R.: Meaningful Learning with Technology. Pearson Education (2013)
12. Preece, J., Rogers, Y., Sharp, H.: Interaction Design. Bookman (2005)
13. Microsoft Kinect. https://learn.microsoft.com/en-us/archive/msdn-magazine/2012/june/kinect-starting-to-develop-with-kinect. Accessed 01 Oct 2024
14. Nintendo Switch. https://www.nintendo.com/pt-br/switch/. Accessed 01 Oct 2024
15. Brazil. Ministry of Education. National Common Curricular Base. Brasília (2018)

Hackathons as an Approach for Developing New Products for Food Industry Micro-enterprises—Design Science Research of Side Stream Hackathon

Jari Jussila, Anne-Mari Järvenpää, and Kristiina Paju

Abstract Hackathons have been recognized as innovative platforms for developing new solutions to challenges in the food industry, particularly in the context of utilizing side streams—by-products or waste materials generated during food production and processing. Food enterprises in EU and especially in Finland are, however, small and have very limited resources for product development. Hence, design science was used as a methodology to develop an innovation practice applicable to food industry micro-enterprises. As objectives of the solution to be developed in a hackathon format, we defined: how mash from a brewery by-product, the ham stumps from a butcher's shop or the leftover dough from a bakery could be utilized to create new products. Then a side stream hackathon was designed to involve relevant stakeholders for the problem. Student teams demonstrated new product ideas from the side streams, which were taken into production by the micro-enterprises. The major finding was that successful innovation required the commitment and capability of the micro-enterprise to implement the idea in practice. Furthermore, university-industry innovation collaboration was found beneficial in facilitating of the necessary collaboration between micro-enterprises that is required for agri-food innovations. From a design perspective it is essential to understand the needs and limitations of the micro-enterprises and adopt the innovation practice to be accessible and efficient to the target group.

Keywords Side streams · Hackathon · Design science research · SDG12

J. Jussila (✉) · A.-M. Järvenpää
Häme University of Applied Sciences, Hämeenlinna, Finland
e-mail: jari.jussila@hamk.fi

K. Paju
Forssa Vocational Institute, Forssa, Finland

A. Visvizi et al. (eds.), *Research and Innovation Forum 2025 Vol2*, Springer Proceedings in Complexity, https://doi.org/10.1007/978-3-032-09062-1_22

1 Introduction

It has been recognized that in the future, food enterprises can play a role in addressing societal challenges, both in the fight against climate change, for example through the development of food chains, and in promoting nutrition as a driver of product development [1]. Disruptions in food supply chains, such as the conflict in Ukraine, have exposed the vulnerability of the global agri-food system, causing severe shocks to the world's food supply [2] and thus alerting to the importance of the food industry in the security of supply [3]. The current agri-food supply chains are challenged, for instance, by food loss and waste generation along the supply chain [2]. On the other hand, an enormous amount of side stream and food by-products are generated that can end up as waste if left unutilized [4, 5]. On the other side, direct conversion of side streams into food products or soil improvement would save huge amounts of energy and resources [4, 6]. Soil improvement is an example of regenerative development practice in agriculture that has the potential to contribute positively to the environment and leave the environment in a better state than before [7, 8]. Food enterprises in EU and especially in Finland are, however, small and have very limited resources for product development [9]. There is an urgent need to support food industry micro-enterprises innovation practices in tackling challenges and taking advantage of opportunities in food production.

Education can be considered as ecosystem that helps to bring together technologies, knowledge and social innovation by active engagement of users (e.g., customers and students) in co-creation [10]. University-industry innovation collaboration [11] using hackathons has been found as a feasible innovation practice for fast iteration of innovations that can lead to new ideas and concepts to be further developed by enterprises [12]. In this study, we aimed to develop an accessible and efficient innovation practice for food industry micro-enterprises in a hackathon format. Hackathon can be defined as "*one type of organized, goal-driven innovation contest, a short time-bounded event with a challenge to be solved creatively in coopetition and collocation of teams, whose results are presented and recognized in a ceremony at the end of the event.*" [13]. To address the research problem, this paper is structured as follows: in methods and materials we introduce design science research methodology and data collected via workshops and interviews. In results we introduce the developed hackathon concept and its evaluation. In discussion and conclusions, we draw together the study's findings and propose future research direction.

2 Methods and Materials

Design science research was used as a methodology [14] to develop an innovation practice applicable to food industry micro enterprises. The research process was guided by the design science research process (DSRP) model that includes the nominal process sequence of (1) Problem identification and motivation, (2) Objectives of a solution, (3) Design and development, (4) Demonstration, (5) Evaluation and (6) Communication [14].

The design science research process started with problem identification and motivation in a REACT-EU funded project, Data and Innovation to Support Green Transition in Food (DIVA), where the aim was to help SMEs and micro-enterprises to develop new uses for food industry by-products or side streams and to create synergies between the food industry SMEs and micro-enterprises. The project conducted a survey to map food industry SMEs by-product practices and willingness to develop new products and services based on circular economy principles. Based on survey results potential enterprises were identified and invited to one-on-one goal-setting workshops. The problem with food industry SMEs and micro-enterprises is that while by-products or side streams are generated in the production, there is a lack of knowledge on how to collaborate across organizational boundaries and make profitable use of by-products and side streams.

The objectives of the solution were defined in the goal-setting workshops with the micro-enterprises. It was agreed that the objective was to develop an innovation practice that enabled collaboration between food industry micro-enterprises and allowed them to make use of ideas of food experts outside the enterprise and potential users, while ensuring that the result should be something that has value to the enterprises, e.g., new products that can be sold in the market.

It was agreed that a hackathon could be a suitable format for innovation practice, where by-products and side streams from two or more food industry micro-enterprises would be used as inputs for new product innovations. The design and development of a hackathon concept suitable for food industry micro-enterprises took as a baseline previously developed hackathon concepts: educational hackathon [15], virtual hackathon [12]; hybrid hackathon [16], and 'general' hackathon concept [17] and typology [18], while recognizing the need to adapt the innovation practice to meet the resource constraints (e.g., time constraints, human resources and digital competence of the entrepreneurs) and outcomes expectations of the food industry micro-enterprises (a tangible new product that can be sold in the market). The micro-enterprises involved in this study both employed two persons (the entrepreneurs), who could not be away from their daily business operations for a full day hackathon, which is common for many small farms, breweries, and bakeries, for instance, that employ 1–2 people. Additionally, the need for support in creating introductory video materials of the challenge and enabling virtual and hybrid collaboration with hackathon participants was identified in the design and planning processes of the hackathon. Furthermore, the hackathon participants were invited for a pre-hackathon information day, where they got the challenges beforehand and were also allowed to reach out to the micro-enterprises with any questions they might have had regarding the challenge before the hackathon day (Table 1).

Table 1 Description of the side stream hackathon design

Attribute	Description
Organization and stakeholders	External organizer (DIVA project) External facilitator (DIVA project) Micro-enterprises
Innovation process	Goal-setting workshop, pre-hackathon information day, 1 day hackathon event, post-hackathon seminar
Setting	Radical collocation, virtual collocation, hybrid collocation
Challenge	Side stream challenges provided by micro-enterprises accessible as video presentations to the hackathon participants
Focal interest and goal	Create new tangible products from food industry side streams and by-products with two distinct beneficiaries
Teams and participants	8 teams, 4 teams working for each micro-enterprise challenge 3rd year Bachelor students from Degree Programme in Biotechnology and Food Engineering 4–5 participants per team, 37 students in total
Collaboration	Participants collaborated inside the team and with food industry specialists and teachers coaching and mentoring in the event
Co-creation and ceremony process	Facilitators, researchers and teachers supported the co-creation process and provided recognition to the teams, one team was chosen and awarded as the winner for each challenge

Source Authors

3 Results

The demonstration of the developed side stream hackathon was carried out in three phases. The first phase included the goal-setting workshop together with the food industry micro-enterprises, and then a pre-hackathon information day, where participants of the hackathon were familiarized with the challenges and could present questions for the food industry micro-enterprises. DIVA project organized a recording of the side stream challenges of the food industry micro-enterprises that were made available as videos for the hackathon participants. Participants were recruited from the Häme University of Applied Sciences Degree Programme in Biotechnology and Food Engineering and included both full-time and part-time students. The DIVA project researchers organized and facilitated the hackathon event and recruited also biotechnology and food engineering experts to serve as mentors and judges at the event. During the hackathon, the eight teams presented their most feasible side stream innovation idea to the judging panel during a pitching session. Out of those, two ideas were announced as the winners—one for the brewery's challenge and one for the farm's challenge.

The winning idea of the brewery's side stream challenge was the healthy "Mashy" snack bar concept—the main ingredient being the beer-making side stream called mash which consists of the "still" nutritious malted grains e.g. barley, wheat, or rye—that emphasized the importance of combining nutrition, sustainability and innovativeness in the product. As the winner of the farm side stream challenge, the judging

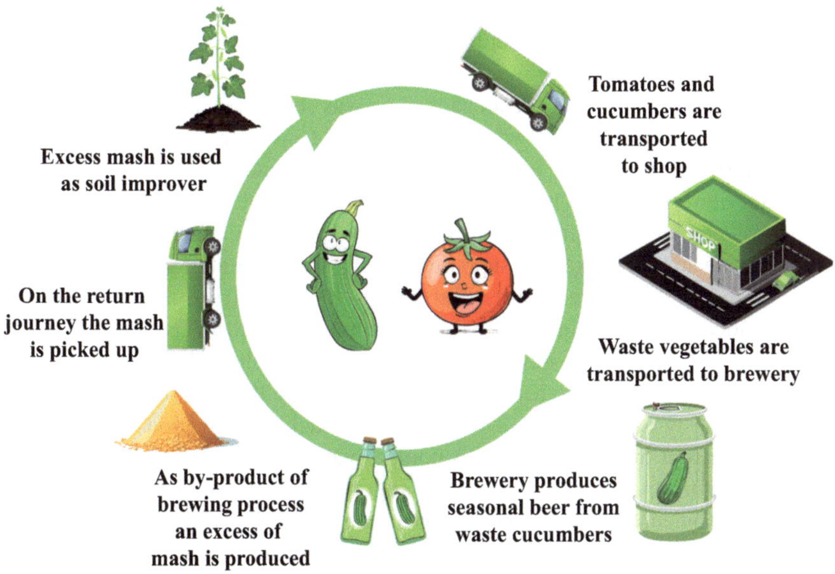

Fig. 1 Cucumber and tomato beer ecosystem. *Source* Authors

panel selected the concept of a Cucumber and Tomato Beer ecosystem (Fig. 1) that combined in its solution side streams and by-products from both participating food industry micro-enterprises.

The winning solution presented an ecosystem, where waste tomatoes and cucumbers are used to create seasonal beers by a local brewery. As a by-product of the brewing process excess mash is produced, which is used by the cucumber and tomato farm as a soil improver. Transportation and production are optimized to reuse waste, by-products, and side streams generated in the whole food system.

Both micro-enterprises received the hackathon results (the pitched "most feasible ideas" as well as all other ideas that were jotted down during the ideation process) after the event as well as an invitation to a joint lunch meeting with the hackathon facilitators to go through the hackathon's results in an informal setting. Surprisingly, both enterprises were inspired by the Cucumber and Tomato Beer ecosystem-concept, the winning idea chosen for the farm. As a result of this joint after-hackathon get-together, the brewery and the farm teamed up to jointly produce the cucumber beer. It was discovered that cucumbers that do not meet the shop specifications but are otherwise of good quality served as excellent raw material for the brewery. After a speedy collaborative experimentation, the brewery launched cucumber beer to the market.

The results of this innovation process were made available for public tasting during the DIVA project's final seminar, where the product got great feedback for its freshness from both local and international (e.g., England, Romania) guests. Alongside the already canned and branded cucumber beer, other products from the food industry side streams were offered for tasting such as sausages made with mash from the brewery, cakes and other sweet products made from bakery side streams among others.

The evaluation of the side stream hackathon was conducted in three phases. First, a joint lunch meeting was held with the hackathon facilitators and the participating micro-enterprises to capture immediate feedback. Secondly, the micro-enterprises were interviewed in a post-hackathon seminar. Finally, the micro-enterprises were interviewed three years after the hackathon took place to evaluate how well the designed side stream hackathon supported a solution to the micro-enterprises challenge. Reflecting three years after, Talvisillan Tähkä ja Trekooli responded that student's ideas except the cucumber beer have not materialized concretely, the idea of a cucumber gel came to mind (cosmetics) and it is still in the process of being developed (maybe it will be tested one day), otherwise the side stream cucumbers will go to charity for a big part. The side stream hackathon method was perceived as a good concept from the company's point of view, an easy and effortless way for entrepreneurs to get new ideas, especially important that the students were involved in the development with so called "open-minded lenses". They also stated that if something similar is organized in the future, they would be willing to participate again. OlutMylly reflected that as a result of the hackathon they got an interesting sales product (the cucumber beer), especially for (sale in) events, and because it is a local thing, it is a liked a beer, which became a standard product produced seasonally. Especially events need something were interesting for the audience, and this product fits well with that category. Only a small amount of cucumber is needed in the production, and the cucumber is surprisingly tasteful. Other benefits include networking, strengthening of collaboration, increased interest in the company, and "that we are making something out of someone else's waste". As a concept, the side stream hackathon "went easily, well", regardless that "we did not (even) see the presentations", "it was fun, (let's have) more of these". As communication of this case study, Research & Innovation Forum 2025 was selected as a scholarly forum to report the study.

4 Discussion and Conclusions

A major finding was that a successful innovation outcome from the side stream hackathon required corporate entrepreneurship [19] including both commitment and capability of the micro-enterprise to implement the idea in practice. The capacity to absorb partner resources (cf. [11]) was necessary to create a new product from food chain side stream and the hackathon supported finding the right partner to do that. The introduced side stream hackathon concept supported combining heterogeneous partners (students, food industry specialists, food industry micro-enterprises) and heterogeneous knowledge ("from open-minded lenses") to create new product concepts. This highlights the importance of involving partners that have complementary resources and knowledge to contribute to the innovation activity.

In a previous study, it was argued that hackathons in university-industry innovation collaboration can mainly support the fast iteration of innovations (fuzzy front end), yielding new ideas and concepts that can be further developed into new products in the slow iteration of innovations (new product development), spanning e.g., 1–3 years [12]. This study shows that it is possible to design and implement hackathons that serve as the fuzzy front end of the innovation process and support faster new product development process [20] leading to a platform product (next generation product and incremental change in production process) based on the typology of new products [21]. Furthermore, it has been demonstrated that hackathons can lead to successful products for micro-enterprises.

Regarding the design science research process [14], the wider objective of the research project to provide innovation know-how for companies and new uses/ products for side streams, and the focal goal of the hackathon to create new tangible products for food industry side streams and by-products, was achieved. Interestingly, the innovative concept developed in the hackathon, Cucumber and Tomato Beer ecosystem (Fig. 1), went beyond the original goals defined in the design science research process. If the concept will be fully implemented in the future, it will go beyond a new circular economy product to a regenerative product innovation [7, 22] that has the potential to provide net positive outcomes to nature via soil improvement of the excess mash generated in production process. University-industry innovation collaboration supporting regenerative product innovations was identified as a promising new research area.

Acknowledgements This research was supported by the European Regional Development Project Green Smart Services in Developing Circular Economy SMEs (A77472) and Optimization of Circular Economy Material Flow Logistics—OPTIMA (A80571)

References

1. Erkkola, M., Fogelholm, M., Silvasti, T.: Ilmastokriisi ja kansanterveys—on tekojen aika. Sos. Aikakauslehti. **56** (2019)
2. Deák, Z.: Circular economy solutions for the agri-food business. Presented at the MEB—22nd International Conference on Management, Enterprise, Benchmarking, Budapest (2024)
3. Glencross, A.: The geopolitics of supply chains: EU efforts to ensure security of supply. Glob. Policy **15**, 729–739 (2024). https://doi.org/10.1111/1758-5899.13388
4. Stuebler, A.S., Heinz, V., Aganovic, K.: Development of food products (2020). https://doi.org/10.1016/j.cogsc.2020.100356
5. Socas-Rodríguez, B., Álvarez-Rivera, G., Ibáñez, E., Cifuentes, A.: Food by-products and food wastes: are they safe enough for their valorization? (2021)
6. Jyske, T., Rasa, K., Korkalo, P., Kohl, J.: Cascade Vision: Regionally Adaptive Circular Bioeconomy—Added Value, Wellbeing and Re-source Wisdom with Cascade Processing. Natural Resources Institute Finland, Helsinki (2023)
7. Järvenpää, A.M., Jussila, J., Henttonen, K., Helander, N., Kunttu, I.: Contrasting restorative economy and regenerative economy in circular economy context. In: Proceedings of XXXIV ISPIM Innovation Conference. International Society for Professional Innovation Management ISPIM, Ljubljana (2023)

8. Hahn, T., Tampe, M.: Strategies for regenerative business. Strateg. Organ. **19**, 456–477 (2021)
9. Järvenpää, A.M., Jussila, J., Kunttu, I.: Barriers and practical challenges for data-driven decision-making in circular economy SMEs. In: Visvizi, A., Troisi, O., Grimaldi, M. (eds.) Big Data and Decision-Making: Applications and Uses in the Public and Private Sector, pp. 163–179. Emerald Publishing Limited, Leeds (2023)
10. Troisi, O., Visvizi, A., Grimaldi, M.: Rethinking innovation through industry and society 5.0 paradigms: a multileveled approach for management and policy-making. Eur. J. Innov. Manag. **27**, 22–51. https://doi.org/10.1108/EJIM-08-2023-0659
11. Rajalo, S., Vadi, M.: University-industry innovation collaboration: reconceptualization. Technovation **62**, 42–45 (2017)
12. Jussila, J., Raitanen, J., Suominen, A.H., Järvenpää, A.M.: Virtual hackathons—a novel approach for university-industry collaboration. In: Visvizi, A., Lytras, M.D., Aljohani, N.R. (eds.) Proceedings in Complexity. Springer, Cham (2021). https://doi.org/10.1007/978-3-030-62066-0_19
13. Halvari, S., Suominen, A., Jussila, J., Jonsson, V., Bäckman, J.: Conceptualization of hackathon for innovation management. In I. Bitran, S. Conn, C. Gernreich, M. Heber, K. R. E. Huizingh, O. Kokshagina, M. Torkkeli, & M. Tynnhammer (Eds.), The proceedings of The XXX ISPIM INNOVATION CONFERENCE - Celebrating Innovation - 500 Years Since Da Vinci The International Society for Professional Innovation Management: 16-19 June 2019 - Florence, Italy (Vol. 93). The International Society for Professional Innovation Management (ISPIM) (2019). https://urn.fi/URN:NBN:fi:tuni-202112078957
14. Peffers, K., Tuunanen, T., Rothenberger, M.A., Chatterjee, S.: A design science research methodology for information systems research. J. Manag. Inf. Syst. **24**, 45–77 (2007)
15. Suominen, A.H., Jussila, J., Lundell, T., Mikkola, M., Aramo-Immonen, H.: Educational hackathon: innovation contest for innovation pedagogy. In: Bitrain, I., Conn, S., Huizingh, K.R.E., Kokshagina, O., Torkkeli, M., Tynnhammar, M. (eds.) Proceedings of the ISPIM Innovation Conference : Innovation, The Name of The Game. Lappeenranta University of Technology; ISPIM, Stockholm (2018)
16. Jonsson, V., Suominen, A., Bäckman, J., Fogelberg, J., Eriksson, E.: Digital competences for virtual/hybrid hackathons: DigComp 2.1 in municipal organization. In: Proceedings of the XXXIII ISPIM Innovation Conference. LUT Scientific and Expertise Publications, Copenhagen (2022)
17. Halvari, S., Suominen, A., Jussila, J., Jonsson, V., Bäckman, J.: Conceptualization refinement of hackathon for innovation management. In: XXXI ISPIM Innovation Conference: Innovating in Times of Crisis. LUT Scientific and Expertise Publications, Copenhagen (2020)
18. Halvari, S., Suominen, A.H., Jussila, J.: A typology of hackathon-use. In: Proceedings of the XXXII ISPIM Innovation Conference: Innovating Our Common Future. Lappeenrannan teknillinen yliopisto, Berlin (2021)
19. Celenta, R., Cucino, V., Feola, R., Parente, R.: Towards Innovation 5.0: the role of corporate entrepreneurship. In: Visvizi, A., Troisi, O., and Corvello, V (eds.) Springer Proceedings in Complexity. Springer, Cham (2024). https://doi.org/10.1007/978-3-031-44721-1_34
20. Koen, P.A., Ajamian, G.M., Boyce, S., Clamen, A., Fisher, E., Fountoulakis, S., Johnson, A., Puri, P., Seibert, R.: Fuzzy front end: effective methods, tools, and techniques. In: The PDMA Toolbook for New Product Development, pp. 5–35. Wiley, New York (2002)
21. Wheelwright, S.C., Clark, K.B.: Revolutionizing Product Development: Quantum Leaps in Speed, Efficiency, and Quality. Simon and Schuster, New York (1992)
22. Kuenkel, P.: Collective stewardship towards life economies. In: Transformation Literacy: Pathways to Regenerative Civilizations. Springer, Cham (2002)

Exploring the Characteristics of Fashion Branding and Architecture Through Storytelling

Zenah Mahmoud and Asmaa Ibrahim

Abstract Fashion and architecture have increasingly intersected, culminating in branded environments that engage the senses and communicate compelling narratives. This paper investigates how architecture serves as a medium for fashion branding, focusing on the translation of brand identity into spatial, sensory, and symbolic experiences. Two internationally recognized projects—the Prada Transformer by OMA in Seoul and the Chanel Mobile Art Pavilion by Zaha Hadid Architects—are examined as case studies. Through thematic analysis of their spatial design, sensory engagement, and narrative strategies, this study identifies the key characteristics that enable architecture and fashion to converge for powerful branding and storytelling. The findings offer insights into how such environments foster brand resonance and visitor engagement, with implications for both temporary and permanent commercial contexts.

Keywords High fashion · Architecture branding · Storytelling architecture · Fusion · Sensory · Cultural aesthetics · Spatial experience

1 Introduction and Background

Fashion and architecture, while distinct in their primary functions, share intriguing similarities in their provision of functional and aesthetically driven services [1]. These fields mutually influence each other through materials, designs, and other characteristics, evolving over time through cultural and artistic sources to create spaces that blend functionality with narrative and sensory engagement. Historically, architectural grandeur inspired fashion, and textile artistry found its way into architectural

Z. Mahmoud
Effat College of Architecture and Design, Effat University, Jeddah, Saudi Arabia

A. Ibrahim (✉)
Urban Design Lab, College of Architecture and Design, Effat University, Jeddah, Saudi Arabia
e-mail: asibrahim@effatuniversity.edu.sa

A. Visvizi et al. (eds.), *Research and Innovation Forum 2025 Vol2*, Springer Proceedings in Complexity, https://doi.org/10.1007/978-3-032-09062-1_23

details. The twentieth century witnessed the explicit fusion of fashion and architecture for branding, reflecting artistic innovation, as seen in collaborations like Zaha Hadid's work with Chanel and Rem Koolhaas's Prada Transformer Pavilion [2, 3]. This paper seeks to explore the crucial characteristics or elements that strengthen the understanding of these intersections, focusing on sensory engagement, movement, and emotional responses in how these branded spaces impact fashion perception [4]. Through the detailed examination of case studies, specifically the Prada Transformer [5] and the Chanel Mobile Art Pavilion [6, 7], this study aims to uncover the strategies that make fashion branding through architecture engaging and memorable by analyzing their architectural design and narrative approaches. The fusion of architecture and fashion in these contemporary examples highlights the growing importance of creating emotionally and culturally resonant spaces that offer immersive brand encounters [3, 8]. The paper will proceed with a literature review, followed by an in-depth analysis of the selected case studies, and conclude with a discussion of the findings and their implications.

2 Literature Review

Fashion and architecture, while traditionally distinct disciplines, are profoundly interwoven in their capacity to shape human experience, identity, and perception through design. Both fields manipulate form, materiality, and space to create environments—whether garments or buildings—that mediate between function and sensory engagement, influencing how individuals perceive, move through, and emotionally connect with their surroundings [1, 3]. This reciprocal relationship has evolved over time, with each field borrowing from and inspiring the other, as seen in the translation of architectural motifs into garment design and the adoption of textile aesthetics in architectural detailing [9].

To systematically analyze these interrelationships, Table 1 presents a framework of key categories and criteria: design elements, aesthetic styles, symbolism and identity, interactivity, lighting and ambiance, sustainability and innovation, user experience, and historical/cultural influences. These categories, grounded in the literature, clarify how architecture influences fashion, how spatial design impacts fashion (and vice versa), and how storytelling is enacted through architecture in branded spaces.

Design elements such as material usage, texture, pattern, and form are central to both disciplines. Reserachers demonstrate how the Mashrabiyya screen, a traditional architectural element, is reinterpreted in contemporary garment design, showing direct material and formal dialogue. Architectural materials and forms inspire textile choices, while spatial forms reflect fashion trends and evoke brand narratives through innovative material applications [8].

Aesthetic styles—including minimalism, organic forms, geometric shapes, and intricate details—act as visual languages that guide both fashion and architectural storytelling. The adoption of specific styles can create fluidity in design, as seen in the organic approach of Zaha Hadid's work, or express heritage and modernity

Table 1 Interrelationships between architecture, fashion, and storytelling in branded spaces

Category	Criteria	Architecture's influence on fashion	Spatial influence on fashion (and vice versa)	Storytelling through architecture
Design elements	Material usage, texture, pattern, form	The inspiration for textile choices on the architectural materials	Spatial forms displaying fashion trends	Implementing materials that aim to evoke brand narratives
Aesthetic styles	Minimalism, organic forms, geometric, intricate shapes	The styles affecting the fashion sense create a fluid design following an organic architectural approach	Fashion designs impact spatial design, creating flowy or direct spaces	The styles guide the visual storytelling of spaces, talking about heritage through ornate details, or about modernity through clean, minimal lines
Symbolism and identity	Cultural references, symbolism, and brand identity	The brand identity, a reflection of the architecture	To express fashion brand values through spatial configurations, using space as a medium	The architecture details, ornaments, structures, and materials symbolize the brand essence
Interactivity	Sensory and physical engagement, touching, moving, and spatial interaction	Interactive fashion experience through architectural forms	Fashion pieces that encourage interactivity and movement in space	Experiencing a story through space interaction
Lighting and ambiance	Experience through the emotional sense, lighting, and mood creation	The visual appeal of fashion elements, highlighting significance and/or details	Enhance the experience through setting the mood and vibe for the space	Using lighting to enhance storytelling
Sustainability and innovation	Sustainable materials, adaptability, reuse, and recycling	Architectural sustainable designs inspire fashion	Inspiring sustainable designs, influencing spatial design	Sustainable practices enhancing brand narratives

(continued)

Table 1 (continued)

Category	Criteria	Architecture's influence on fashion	Spatial influence on fashion (and vice versa)	Storytelling through architecture
User experience	Layout flow, accessibility, sensory experience	Architectural layout encouraging fashion accessibility	Fashion collections influence spatial navigation	Storytelling elements enhancing user immersion
Historical/ cultural influences	Historical design references, cultural elements	Architecture borrowing from fashion history	Fashion capturing spatial cultural aspects	Narrative built around historical or cultural roots

Source Authors based on literature review

through ornate or minimal lines [3, 8]. These styles not only shape the visual identity of spaces but also influence the narrative conveyed by the environment.

Symbolism and identity are constructed through cultural references, brand symbolism, and spatial configurations. Researchers highlight how architecture can serve as a medium for expressing brand values, with details, ornaments, and materials symbolizing the essence of the brand [10, 11]. Fashion brands leverage architectural design to amplify their ethos, creating immersive environments that communicate identity and heritage.

Interactivity is facilitated by sensory and physical engagement, encouraging users to touch, move, and interact with space. Researchers emphasize the importance of multisensory experience, where both architecture and fashion foster user participation and immersion. Interactive fashion experiences can be enhanced by architectural forms that invite exploration, while fashion pieces can encourage movement within space, making the narrative of the brand physically accessible [4, 12].

Lighting and ambiance are critical for shaping emotional responses and setting the mood of a space. Researchers discuss how lighting and atmospheric design influence perception, highlight significant details, and enhance the storytelling dimension of branded environments. The strategic use of light can create dramatic effects, reinforce brand narratives, and guide visitors through curated experiences [13, 14].

Sustainability and innovation are increasingly important in both fields. Researchers note how sustainable materials, adaptability, and the reuse of resources are reflected in both architectural and fashion design. Sustainable practices not only address environmental concerns but also enhance brand narratives by aligning with consumer values [15, 16].

User experience is shaped by layout flow, accessibility, and sensory engagement. Researchers show how architectural layout encourages accessibility and influences how fashion collections are navigated and experienced. Storytelling elements embedded in spatial design enhance user immersion, making the brand encounter memorable and meaningful [17, 18].

Historical and cultural influences are evident in the borrowing of design references and cultural elements. Researchers discuss how architecture and fashion capture and reinterpret historical and cultural aspects, building narratives around roots and heritage [19, 20].

Storytelling emerges as a pivotal and unifying link between the two disciplines. Both fashion and architecture utilize narrative strategies to forge connections with users, guiding them through sequences of experiences and emotions [19, 20]. In the digital age, storytelling has expanded beyond traditional media, with fashion brands leveraging digital platforms and branded content to engage audiences interactively [21]. The effectiveness of brand storytelling is shaped by thematic choices and the development of strong brand personas, which are translated into physical space through visual merchandising and spatial design [11, 18, 22].

Luxury fashion brands have particularly embraced architectural innovation as a means of expressing identity and creating memorable experiences. Researchers analyze how retail spaces function as tools for brand expression and cultural communication, with projects such as the Chanel Mobile Art Pavilion by Zaha Hadid Architects and the Prada Transformer by OMA serving as both architectural statements and narrative devices [5–7, 23–25]. These structures exemplify how architecture can be harnessed to reflect brand values, adaptability, and innovation.

This integration responds to shifting consumer expectations, as outlined in several references, who note a growing demand for experiences that resonate emotionally, align with personal values, and offer opportunities for self-expression [4, 12, 15]. Contemporary branded spaces are thus evolving into sophisticated cultural hubs and experiential destinations, combining fashion, architecture, and narrative into holistic, multisensory environments.

Accordingly, the literature and the analytical framework in Table 1 reveal a dynamic and reciprocal relationship between fashion and architecture, grounded in shared principles of design, sensory engagement, and storytelling. By translating brand narratives into physical and digital spaces, fashion brands can foster deeper connections with consumers, creating environments that are both memorable and meaningful in an increasingly competitive landscape.

3 Materials and Methods

A qualitative case study methodology was adopted to investigate the convergence of fashion and architecture in the creation of branded environments. This approach was selected for its ability to facilitate in-depth exploration of complex, context-dependent phenomena and to provide nuanced insights into how specific, innovative examples succeed in their real-world settings.

Two case studies were purposively selected based on several criteria: (1) international recognition as branded environments by both architectural and fashion scholarship, (2) explicit use of narrative, sensory, and symbolic design strategies, (3) availability of comprehensive documentation—including architectural plans, critical reviews, and visitor accounts—and (4) the capacity to exemplify the dynamic interplay between architecture, fashion, and storytelling. The selected cases are: Prada

Transformer: Designed by the Office for Metropolitan Architecture (OMA), located in Seoul, South Korea. OMA is internationally recognized for its experimental and transformative approach to spatial design [5]. The second one is Chanel Mobile Art Pavilion: Designed by Zaha Hadid Architects, exhibited in cities including Hong Kong, Tokyo, and New York. Zaha Hadid Architects are renowned for their innovative, fluid architectural forms [6]. Both projects have received critical acclaim for advancing the discourse on branded environments and were chosen for their ability to exemplify the cutting edge of the fashion-architecture fusion.

A thematic analysis was conducted on primary and secondary sources, focusing on four key dimensions: spatial sequencing (the progression of visitor experience through designed space), sensory engagement, symbolism, and emotional resonance. Triangulation was achieved by cross-referencing multiple data sources, including published interviews, critical reviews, architectural documentation, and, where available, direct observation through virtual tours and video walkthroughs. To broaden the scope of the conclusions, findings from these pavilions were conceptually compared with established permanent boutiques, drawing on relevant literature and available documentation.

Case Study 1: OMA's Prada Transformer—Seoul, South Korea: The Prada Transformer, designed by OMA and constructed in Seoul, is a dynamic, narrative-driven structure. Its tetrahedral design allows the entire building to rotate, presenting four distinct layouts for different events—fashion, art, cinema, and culture—mirroring the ever-evolving, dramatic world of fashion. Each transformation is intended to surprise and captivate visitors, revealing new possibilities and capturing the constant flux that defines the fashion industry. The tactile and visual elements are central: a soft, elastic membrane envelops the structure, inviting touch and exploration, while shifting light and shadow create a sensory-rich environment. The building's transformative nature symbolizes Prada's avant-garde identity and adaptability, reinforcing the brand's ethos of innovation and unpredictability [5]. The act of physical transformation generates anticipation and emotional engagement, deepening the visitor's connection to the brand and amplifying its high-fashion identity.

Case Study 2: Zaha Hadid Architects' Chanel Mobile Art Pavilion—Hong Kong, Tokyo, New York: The Chanel Mobile Art Pavilion, designed by Zaha Hadid Architects and exhibited in global cities such as Hong Kong, Tokyo, and New York, exemplifies a curvilinear, flowing spatial narrative. The pavilion's fluid plan and sculptural surfaces create a sense of wholeness and timelessness, reflecting Chanel's enduring style and modernity. As visitors move through the pavilion, they embark on an intimate journey exploring different facets of the brand, seamlessly integrating art and culture. Materiality and lighting are crucial: the translucent façade and smooth forms diffuse light, creating an ethereal atmosphere. Adjustable lighting enhances ambiance for each global location, while tactile surfaces invite exploration. The pavilion's mobile nature symbolizes Chanel's global reach and versatility, reinforcing its legacy of craftsmanship and cultural innovation. The design's elegance and grandeur evoke introspection about Chanel's heritage, while the immersive spatial experience fosters a deep emotional connection with the brand [6].

These two case studies—OMA's Prada Transformer in Seoul [5] and Zaha Hadid Architects' Chanel Mobile Art Pavilion in Hong Kong, Tokyo, and New York [6]—offer compelling examples of how groundbreaking architectural design can be strategically used by fashion brands. Both projects demonstrate how narrative, sensory engagement, symbolism, and spatial innovation are harnessed to craft unique consumer experiences and extend brand identity beyond traditional retail or exhibition spaces. By focusing on these exemplary cases, the paper provides detailed insights into the nuanced ways architecture and fashion converge to create branded environments that are immersive, memorable, and emotionally resonant.

4 Discussion of Findings

Examining the Prada Transformer in Seoul and the Chanel Mobile Art Pavilion in cities like Hong Kong, Tokyo, and New York shows how architecture can become a key part of fashion branding. These buildings are more than just places to show clothes—they help tell each brand's story, create memorable experiences, and leave a strong impression on visitors.

The Prada Transformer, designed by OMA, is a great example of how a building can reflect a brand's personality. Its unique tetrahedral shape means the whole structure can rotate and change its layout for different events. This constant change isn't just practical—it's a symbol of Prada's focus on innovation and always staying ahead in fashion [5]. Features like the soft, flexible exterior and the way light moves through the space make the building feel alive and interactive. Visitors aren't just looking at fashion; they're part of an experience that matches Prada's unpredictable and creative spirit. The building's ability to surprise people and adapt to new uses helps visitors feel a real connection to the brand.

The Chanel Mobile Art Pavilion, created by Zaha Hadid Architects, takes a different approach but is just as effective. Its smooth, flowing lines echo Chanel's famous elegant style, and the use of light and materials creates a soft, dreamlike feeling inside [6, 7]. Because the pavilion travels to different cities, it also represents Chanel's global influence and ability to reach people everywhere. The space feels both grand and personal, inviting visitors to reflect on Chanel's history and values. The design blends art, architecture, and fashion in a way that makes the brand's identity come alive.

Comparing these two projects, we see two different but equally strong ways to use architecture in fashion branding. The Prada Transformer is all about movement, change, and excitement, while the Chanel Pavilion focuses on elegance, timelessness, and global appeal. Both use sensory details—like touch, light, and space—to make visitors feel emotionally connected to the brand.

These findings match what experts have said about "brandscapes," where architecture is used to shape how people experience a brand, not just display products [2]. Both case studies also support the idea that architecture can tell a story and help communicate brand values in a way people remember [4, 20].

Overall, the success of these projects shows that architecture is becoming a powerful tool for fashion brands. By investing in creative, innovative buildings, brands like Prada and Chanel can offer visitors more than just a shopping trip—they create unique, emotional experiences that help people remember and connect with the brand long after they leave.

5 Conclusion

This study sets out to answer how architectural design strategies in high-fashion branded environments can translate brand identity into immersive, emotionally resonant experiences for visitors, and to identify the key elements that make this possible. By closely examining the Prada Transformer and the Chanel Mobile Art Pavilion, we found that architecture plays a vital role in telling a brand's story, engaging the senses, and building emotional connections with audiences.

The Prada Transformer demonstrated how adaptable, ever-changing architecture can mirror the dynamic nature of fashion itself. Its rotating tetrahedral form, flexible materials, and dramatic use of light created a space that was not only functional for different events but also became a living symbol of Prada's innovative and constantly evolving identity [5]. Visitors were invited to explore, interact, and experience surprise—making the brand's values tangible and memorable.

The Chanel Mobile Art Pavilion took a different approach, using Zaha Hadid's signature flowing curves and careful lighting to echo Chanel's timeless elegance and global outlook [6, 7]. The pavilion's mobile nature reinforced Chanel's worldwide influence, while its soft, immersive atmosphere drew visitors into a deeper emotional connection with the brand. Here, architecture, art, and fashion blended seamlessly to create a unique, reflective experience.

Comparing these two cases, we see that there is no single formula for success. Prada's space focused on dynamism and change, while Chanel's emphasized elegance and continuity. Both, however, used sensory engagement, symbolism, and narrative design to make the brand experience more meaningful and memorable for visitors. These findings reinforce the idea that architecture can be a powerful storytelling tool in fashion branding, shaping how people feel, interact, and remember a brand [2, 4, 20]. Smart architectural choices can turn a branded space into an emotional journey, not just a backdrop for products.

It is important to note, however, that both the Prada Transformer and the Chanel Mobile Art Pavilion were temporary, high-budget projects. Their scale and ambition may not be easily replicated by all brands or in permanent retail settings. Future papers should explore how these principles can be adapted for more everyday, long-term, or resource-limited environments. It would also be valuable to study the impact of architectural storytelling on consumer behavior in typical retail spaces and to consider the sustainability of such ambitious designs.

Accordingly, this paper shows that when fashion and architecture work together thoughtfully, the result is a branded environment that strengthens brand identity, creates lasting memories, and sets a new standard for immersive consumer experiences.

References

1. Abbasi, S.: Fashion and architecture: Interdisciplinary dialogues. Int. J. Des. **16**(2), 45–59 (2022)
2. Klingmann, A.: Brandscapes: Architecture in the Experience Economy. MIT Press (2007)
3. Quinn, B.: Fashion and Architecture. Laurence King Publishing (2013)
4. Pallasmaa, J.: The Eyes of the Skin: Architecture and the Senses (3rd ed.). Wiley, New Jersey (2012)
5. Viva, C.: OMA's Prada transformer: reconfigurable architecture and brand narrative. Des. J. **24**(3), 215–229 (2021). https://doi.org/10.1080/14606925.2021.189485
6. Minner, J.: Chanel mobile art pavilion: a case study in experiential branding. J. Arch. Des. **28**(1), 55–67 (2024)
7. Architects, Z.H., Bertrand, J.: The channel mobile art pavilion: architecture as brand storytelling. Arch. Digest **81**(2), 98–107 (2011)
8. Boehn, S., Deuber-Mankowsky, A.: Fashion spaces: architecture and the display of style. Fash. Theory **20**(5), 563–580 (2016). https://doi.org/10.1080/1362704X.2016.1173344
9. Akçay, B., Alothman, S.: Mashrabiyya: from architectural element to contemporary garment design. Int. J. Fash. Des. Technol. Educ. **11**(2), 239–248 (2018). https://doi.org/10.1080/17543266.2018.1445056
10. Almila, A.-M.: Fashion, space and law: interdisciplinary perspectives on the spatiality of fashion. Fash. Theory **25**(1), 9–27 (2021). https://doi.org/10.1080/1362704X.2020.1850192
11. Rachbini, W., Yulianti, F., Suryani, T.: Brand storytelling and consumer engagement: the role of narrative in fashion branding. J. Fash. Mark. Manage **27**(1), 1–18 (2023). https://doi.org/10.1108/JFMM-09-2022-0222
12. Hekkert, P., Schifferstein, H.N.J.: Introducing product experience. In: Schifferstein, H.N.J., Hekkert, P. (eds) Product Experience. Elsevier, pp. 1–8 (2008). https://doi.org/10.1016/B978-008045089-6.50003-2
13. Gifford, R.: Environmental Psychology: Principles and Practice (4th ed.). Optimal Books (2007)
14. Böhme, G.: The art of the stage set as a paradigm for an aesthetics of atmospheres. Ambiances **1**, 1–14 (2013). https://doi.org/10.4000/ambiances.315
15. Manzini, E.: Design, When Everybody Designs: An Introduction to Design for Social Innovation. MIT Press (2015)
16. Farahat, A.: Fashion's influence on architectural design: a study of contemporary interiors. J. Inter. Des. **39**(3), 23–39 (2014). https://doi.org/10.1111/joid.1203
17. Copercini, S.: The influence of architectural spaces on creativity in fashion design. Des. Stud. **41**, 89–103 (2015). https://doi.org/10.1016/j.destud.2015.08.001
18. Park, J., Stoel, L., Lennon, S.J.: Cognitive, affective and conative responses to visual merchandising: the role of involvement. Int. J. Retail. Distrib. Manage **42**(9), 714–729 (2014). https://doi.org/10.1108/IJRDM-02-2013-0039
19. Jankoska, A., Stevkovska-Stojanovska, D.: Translating architectural forms into wearable narratives: the fashion-architecture interface. Text.: Cloth Cult. **18**(4), 412–427 (2020). https://doi.org/10.1080/14759756.2020.1754793
20. Trisno, R., Wijaya, A., Santosa, P.I.: Storytelling in fashion and architecture: creating immersive branded environments. Int. J. Arch. Urban Stud. **10**(2), 101–115 (2022). https://doi.org/10.21608/ijaus.2022.123456

21. Rees-Roberts, N.: Fashion film and the digital turn: moving images, branded content, and the new fashion media. Film Fash. Consum. **9**(2), 109–126 (2020). https://doi.org/10.1386/ffc_00020_1
22. Herskovitz, S., Crystal, M.: The essential brand persona: storytelling and branding. J. Bus. Strategy **31**(3), 21–28 (2010). https://doi.org/10.1108/02756661011036673
23. Rocamora, A.: Mediatization and digital media in the field of fashion. Fash. Theory **20**(1), 1–20 (2016). https://doi.org/10.1080/1362704X.2015.1082291
24. Rocamora, A.: Luxury fashion brands and the architectural turn: the case of channel. Fash. Theory **21**(1), 67–90 (2017). https://doi.org/10.1080/1362704X.2016.1173344
25. Koolhaas, R.: Prada transformer: architecture in motion. Archit. Rec. **197**(6), 72–77 (2009)

Sustainability, Resilience, and Inclusion in the Emerging Global Order

Social Responsibility in Smart Cities: A Review of Community Involvement in Urban Technological Choices

Valentina Cicalese, Maria Rosaria Sessa, Roberto de Luca, and Ornella Malandrino

Abstract Smart cities are designed as a model of urban development based on information technology and networks, which implements advanced digital technologies and improves infrastructure and public services with efficiency in resource allocation and environmental sustainability: it aims to take cities to the next level, creating a more efficient and pleasant environment for citizens and businesses. Concurrently, when these kinds of technological solutions take place without the participation of local communities, they can cause social exclusion and a stigmatizing view of innovation. This article analyses the crucial role of social responsibility in promoting more inclusive smart cities, focusing on the involvement of communities in decision-making processes. The methodological approach combines a conceptual overview with the analysis of symbolic case studies. The review explores the concepts of participatory governance, inclusion, and urban sustainability, providing a conceptual framework for understanding how cities can incorporate social responsibility into their technological strategies. The analysis of case studies from Barcelona, Amsterdam, and Helsinki illustrates innovative community engagement practices, such as using digital platforms for direct participation, co-designing public spaces, and shared monitoring through participatory sensors. The results show the benefits of a participatory approach: involved communities show a greater sense of belonging and trust in institutions, while administrations improve the effectiveness of urban policies, reduce operational costs, and promote wider adoption of technologies. However, challenges remain, including technological accessibility and the need to overcome cultural and digital barriers.

V. Cicalese (✉) · M. R. Sessa · R. de Luca · O. Malandrino
Department of Management and Innovation Systems, University of Salerno, Salerno, Italy
e-mail: vcicalese@unisa.it

M. R. Sessa
e-mail: masessa@unisa.it

R. de Luca
e-mail: rodeluca@unisa.it

O. Malandrino
e-mail: ornellam@unisa.it

A. Visvizi et al. (eds.), *Research and Innovation Forum 2025 Vol2*, Springer Proceedings in Complexity, https://doi.org/10.1007/978-3-032-09062-1_24

Keywords Smart cities · Social responsibility · Urban development · Digital technologies · Urban sustainability · Environmental sustainability · SDG 11

1 Introduction

In recent years, the smart city model has really taken off as a leading approach to urban innovation. It encourages the use of digital technologies, big data, and the Internet of Things (IoT) to enhance infrastructure, public services, and environmental sustainability. However, when these technological solutions are crafted and rolled out without actively involving citizens, they can lead to social exclusion, a lack of trust in institutions, and limited transparency in public decision-making.

This article dives into the growing tension between technological efficiency and social responsibility, questioning how smart cities can also foster inclusivity, participation, and social justice. In particular, it addresses two central research questions:

RQ 1: How can digital technologies foster citizens' participation in urban decisions?

RQ 2: What good practices exist to involve local communities in a smart city?

The theoretical discussion is grounded in the belief that technology isn't just a neutral tool; it's an actant [1]—a force that actively shapes social relationships. As Aristotle [2] pointed out with the concept of tekhnè, and later Heidegger [3] reflected on regarding technology, humans are intertwined with the tools they create: technology influences how we engage with the world and each other. Therefore, urban technological governance should be viewed not just through the lens of innovation and efficiency, but also as a space for ethical, cultural, and social negotiation.

From this perspective, the smart city is seen as a socio-technical ecosystem, where infrastructures and data are deeply connected to human dynamics and civic engagement. As Degli Esposti [4] emphasizes, the smart city concept today needs to strike a balance between its "hardware" components—technology, infrastructure, economy—and its "software" elements—social relations, inclusion, and quality of life. This article examines how digital technologies can be harnessed to boost civic participation and weave social responsibility into smart urban strategies. It highlights the experiences of Barcelona, Amsterdam, and Helsinki, cities that have implemented tangible tools to engage local communities in urban transformation.

The following sections introduce the conceptual foundations of smart cities and social responsibility, then examine a selection of emblematic case studies, and finally reflect on the opportunities and limits of civic engagement in digital urban governance.

2 Methodology

This study adopts a qualitative and integrative approach, blending conceptual analysis with the examination of notable case studies. The first phase involves the development of a theoretical framework focused on key themes such as participatory governance, social inclusion, and urban sustainability. Conceptual analysis is employed to critically assess the underlying normative and strategic assumptions of smart city initiatives.

The second phase consists of a comparative analysis of three significant case studies—Barcelona, Amsterdam, and Helsinki—selected for their international relevance in the use of digital tools to foster civic engagement. These cities exemplify diverse yet coherent approaches to participatory technologies, including platforms such as Decidim, Smart Citizen Kit, and Maptionnaire.

Case selection was guided by criteria of relevance, diversity of participatory tools, and data accessibility, aiming to ensure both analytical depth and comparability. The timeframe considered, 2012–2022, reflects a pivotal decade in the evolution and scaling of civic technology initiatives.

The selected cities were not only chosen for their international relevance in digital urban governance, but also for their distinctive and complementary approaches to citizen participation through technology.

Barcelona was selected for its pioneering role in developing Decidim, a modular, open-source platform for participatory democracy, which has become a global benchmark adopted by institutions and communities beyond Spain.

Amsterdam exemplifies grassroots environmental engagement through the use of Smart Citizen Kits, highlighting a bottom-up approach to data collection and civic empowerment in sustainability efforts.

Helsinki offers a unique model of institutionalized feedback systems, such as Maptionnaire and the City Feedback App, which integrate citizen input directly into spatial planning and service delivery.

Data collection was based on triangulation from a range of credible sources, including academic databases (e.g., Scopus, Google Scholar), institutional repositories (e.g., UN-Habitat, European Commission), and open-access platforms related to the tools under study (e.g., Decidim.org, hel.fi, smartcitizen.me). Peer-reviewed literature and public performance indicators were prioritized wherever available, to ensure methodological rigor, data reliability, and transparency.

3 Theoretical Background

This section sets the stage for understanding how smart cities intersect with social responsibility. It begins by examining how the definition of smart cities has evolved, particularly in relation to sustainability goals. Then, it dives into the ethical, inclusive, and participatory principles that are essential for developing smart urban environments that prioritize social responsibility.

a. Defining the Smart City: A Multifaceted Perspective

The idea of a Smart City has changed considerably over the years, shifting from focusing solely on technology to a broader, more human-centered viewpoint. While earlier definitions often highlighted the role of ICTs in enhancing urban efficiency and service delivery [5], modern interpretations stress the need to blend technology with social, economic, and environmental factors to foster sustainable urban growth.

A Smart City can be described as an urban area that utilizes digital technologies, data analysis, and innovative strategies to improve the quality of life for its residents, enhance the efficiency of urban operations and services, and encourage sustainable development [6].

However, focusing solely on technology does not fully capture what makes a city smart. Therefore, a genuine Smart City should place utmost importance on citizen engagement, participatory governance, and social inclusion [7]. This means active citizen engagement in designing, implementing, and monitoring projects within the Smart City while satisfying their needs and aspirations. Further, the aims of a smart city encompass reducing inequalities, promoting social cohesion, and ensuring that every individual will profit from technological progress.

b. The Role of Smart Cities in Achieving the 2030 Agenda

The idea of a Smart City aligns closely with the United Nations' 2030 Agenda for Sustainable Development, especially Goal 11, which seeks to "make cities and human settlements inclusive, safe, resilient and sustainable." [8].

Smart Cities have a huge potential to drive sustainable urban development. By implementing various initiatives, these cities work to reduce their environmental footprint through energy efficiency, the use of renewable energy sources, better waste management, and sustainable transportation systems.

Smart cities leverage data and advanced technologies to maximize resource efficiency—encompassing water, energy, and raw materials. This approach minimizes waste and actively supports the principles of a circular economy.

Additionally, smart technologies improve the quality and delivery of public services such as transportation, healthcare, education, and public safety, making them more efficient, accessible, and responsive to citizens' needs.

On top of that, smart cities boost economic growth by creating a nurturing environment for innovation and entrepreneurship, attracting skilled workers and investments, and opening up new avenues for development.

Finally, they play a vital role in enhancing social inclusion by tackling inequalities, encouraging citizen engagement, ensuring accessibility, and promoting fair access to services for everyone in the community.

c. **Social Responsibility: The Ethical Heart of Smart City Development**

As Smart Cities increasingly depend upon data-driven decision-making and advanced technologies, it is crucial to consider innovative social consequences and ethics. Smart cities' social responsibility is tied to integrating transparency, inclusivity, and sustainability throughout the smart city design pipeline [9]. Those principles are fundamental if cities are to foster innovation and urban justice.

People must understand how data related to them is being collected, used, and shared through Smart City initiatives. The data practices must be transparent so that citizens are informed about the technologies used within this initiative. This builds trust and encourages informed contributions to city life. Key elements of transparency consist of:

- Open data policies that should be established by smart cities so that they can render all data collected and used by urban technologies available for the common good. This allows the citizens to understand how the decisions are made, thus enabling their active participation to shape their city [10].
- Algorithm Transparency: When an algorithm acts in a decision-making process-that is, in traffic management or service allocation—the way it works must be very clear and understandable. Citizens should understand how these algorithms work, why they were used, and their criteria to ensure accountability and fairness [11].
- Data Privacy: The collection and use of personal data must not violate citizens' privacy rights. Smart cities must adopt complete security in the collating and using data while ensuring that data are collected for the specific proposed use; otherwise, misuse of the data could compromise the rights of particular individuals [12].

Inclusivity in smart cities refers to the creation of an environment that accommodates all urban residents—regardless of age, ability, economic status, or digital literacy—ensuring that technologies and services are designed to be accessible and beneficial to everyone. This concept emphasizes that all individuals, irrespective of their background or circumstances, should have the opportunity to engage with and derive value from the offerings of a smart city. Achieving this goal requires a commitment to universal accessibility, where technological solutions and urban services are developed to meet the diverse needs of the entire population, with particular attention to those who are most vulnerable [13]. Equally important is the principle of meaningful participation, which entails actively involving citizens in the processes of value creation and decision-making related to smart city initiatives. This necessitates the establishment of inclusive platforms through which individuals can express their perspectives and influence priority-setting [14]. Ultimately, smart cities must ensure that the advantages of technological innovation are distributed equitably among all residents, thereby preventing the reinforcement or exacerbation of existing social inequalities [15]. Sustainability in smart cities places a strong emphasis on preserving the integrity of natural ecosystems by prioritizing the reduction of emissions, the conservation of resources, and the adoption of circular economy principles. A key aspect of this approach involves enhancing energy efficiency and transitioning towards renewable energy sources, thereby reducing overall consumption and minimizing environmental impact [16]. In parallel, the promotion of sustainable mobility solutions is essential, encouraging a shift away from private car usage in favor of accessible and efficient alternatives such as public transportation, cycling infrastructure, and car-sharing programs [17]. Furthermore, the integration of circular economy strategies plays a critical role in advancing urban sustainability. By focusing on waste reduction and the continuous reuse and recycling of materials, smart cities can move towards more regenerative and resource-responsible systems [18].

According to the literature, the realization of the strong collaborative bonding between the citizens, public administrators, and businesses is the foundation in support of responsible Smart City development [19]. Citizens should have a voice in shaping the technological landscape of their cities, with public administrators working to guarantee that technology benefits the public and not private interests. Businesses should be encouraged to operate within a framework of ethical responsibility as they contribute to the development of a socially responsible smart city ecosystem. This involves the creation of open platforms for dialogue and engagement, where all stakeholders—including citizens, businesses, and public institutions—can articulate their needs and collaborate in identifying and implementing appropriate solutions [20]. In this context, the role of public–private partnerships is particularly significant. While businesses provide essential expertise and resources, these partnerships must be conducted with high degree of transparency to ensure that private interests do not override the broader public good [21]. Moreover, active citizen engagement remains a foundational element of socially responsible smart city development. Citizens must be integrally involved in setting priorities, evaluating initiatives, and monitoring outcomes, thus reinforcing accountability and aligning projects with the genuine needs of the community [22].

This examination of Smart Cities explores their technological foundations, various application areas, and the challenges and opportunities they present. It highlights how Smart Cities reshape urban living and emphasizes the need for a holistic and integrated approach. The smart, equitable, and livable cities that significantly contribute to the 2030 Agenda should be built on this premise, with technology aligned to robust citizen engagement, social responsibility, and the core principles of sustainable development.

Technological tools for participation and good practices: Barcelona, Amsterdam and Helsinki

Having outlined the conceptual and ethical foundations of smart cities, the following section turns to real-world examples. It explores how digital tools have been implemented in three selected cities to promote civic participation and social responsibility.

In today's world, where digitalization is on the rise, technological tools for citizen participation are becoming more critical than ever. They represent a significant shift in how citizens engage with institutions and participate in decision-making. Using digital platforms, participatory sensors, and social media, citizens can air their opinions, propose ideas, monitor their environment, and communicate directly with their government. This section reviews those tools and analyses their promise, several challenges, and their implication while focusing on some considerable case studies.

- Digital participation platforms, crowdsourcing applications, and real-time urban monitoring dashboards.

Digital participation platforms significantly shift how citizens engage with institutions and participate in decision-making. These platforms, often built on open-source software, create a structured online space where individuals can share their views, suggest ideas, vote on proposals, and collaborate on public policy development [23].

Crowdsourcing applications enable the efficient and organized collection of citizens' thoughts and opinions on specific issues or projects. By incorporating voting, commenting, and discussion features, these platforms encourage the development of innovative ideas and co-creating solutions that benefit the community [24]. This approach can enhance citizen engagement, elevate the quality of public decisions, and support transparent decision-making processes. However, it is crucial to ensure that these platforms are accessible to everyone and that the opinions shared online are genuinely considered by institutions [25].

Real-time urban monitoring dashboards provide immediate data on various city aspects, including traffic, pollution, energy use, and service quality. These tools equip citizens with insights to grasp urban dynamics, track city performance, and report issues [26]. Urban dashboards can enhance citizens' understanding of city challenges, foster better communication between the public and institutions, and encourage citizen involvement in urban policymaking. However, it is essential to ensure that the data presented is accurate, current, and easily comprehensible for all citizens [27].

Barcelona's 'Decidim' platform is a prime example of a digital tool for citizen engagement. Decidim is an open-source software that empowers citizens to participate in decision-making processes, propose initiatives, vote on suggestions, and collaborate in shaping public policies [28]. The platform has been effectively utilized in various contexts, including developing the Sustainable Urban Mobility Plan and designing public spaces. Decidim stands out due to its modular architecture, which allows customization to meet diverse needs and contexts. Additionally, Decidim provides sophisticated tools for moderating discussions, managing participatory processes, and evaluating the impact of citizen engagement [29].

The Barcelona City Council introduced the platform in 2016 and has experienced notable growth in citizen engagement. By December 2018, it had attracted over 30,000 registered users, accumulating more than 1.5 million pages views and over 300,000 visitors. During this time, 35 participatory processes were initiated, 1141 public meetings were held through the platform, and 13,297 proposals were submitted. Over 9196 were turned into public policies that citizens could monitor for their implementation status [28].

Decidim's growth was not confined to Barcelona; by 2022, there were over 400 instances of the platform globally, with significant uptake by Catalan administrations.

- Participatory sensors.

Participatory sensors enable citizens to gather environmental data or report issues using smartphones or other devices. This information is then shared with the community and relevant authorities, enhancing collective awareness and improving city management [30].

These sensors can report various problems, such as street potholes, malfunctioning streetlights, noise pollution, or other disturbances. They also collect environmental data, including air quality, noise levels, and temperature.

While participatory sensors offer valuable insights for environmental monitoring, identifying urban challenges, and informing data-driven public policies, it is crucial to ensure the quality of the data collected, protect citizens' privacy, and ensure that institutions interpret the data correctly [31].

An example of participatory sensors in action is the 'Smart Citizen Kit' project, which equips citizens with tools to gather, understand, and share environmental data. Since its inception in 2012, various versions of the kit have been created to actively engage citizens in monitoring their urban surroundings [32].

In Amsterdam, a notable experiment began in 2014 using the Smart Citizen Kit, allowing residents to measure environmental conditions in their area with affordable electronics and sensors, sharing their findings online in real-time. The goal was to enhance participants' air quality awareness and assess how effective citizen involvement could be in environmental monitoring [33].

- Social media and messaging apps.

Social media and messaging apps provide direct communication channels between citizens and their local administrations. These platforms allow citizens to ask questions, report issues, give feedback, and engage in online discussions. In turn, administrations can leverage social media and apps to share important public information, promote initiatives, and gather input from the community [34].

Social media and messaging apps can enhance communication between citizens and institutions, boost citizen engagement, and foster transparent decision-making processes. Nevertheless, ensuring these tools are used responsibly, with clear and concise communication [35] is essential.

The City of Helsinki initiated several activities in developing public spaces where citizens participated, using social media and messaging applications. For example, through social media, citizens can express their opinions about urban redevelopment design projects or report problems related to public spaces [36]. Through active social media, Helsinki communicates with citizens, collects their opinions about urban projects, and builds their engagement regarding citizen matters. Additionally, the city has created specific mobile apps that enable citizens to report problems, provide feedback, and access public services.

Helsinki has introduced the 'City Feedback App', enabling residents to report issues like overflowing rubbish bins, obstacles on bike paths, or damaged playground equipment. These reports go straight to the appropriate authorities and can be seen by other app users [37].

As part of the D-CENT project, Helsinki has created a platform named 'Decisions Helsinki' to enhance citizen involvement in the city's decision-making. This platform provides open-source tools for tracking and influencing administrative choices [38].

Helsinki has also implemented Maptionnaire, a map-based tool that allows for creating questionnaires and discussion forums, connecting citizen feedback to specific locations within the community. This tool was utilized, for instance, to gather residents' ideas about the city's future urban development [39].

These initiatives highlight Helsinki's dedication to leveraging digital tools to foster two-way communication between citizens and the government, encouraging more inclusive and participatory governance.

4 Discussion

The analysis of digital participation platforms highlights both the potential and the limitations of civic technologies. Based on these findings, this section reflects on broader issues of inclusion, trust, and governance in smart cities.

a. **Barriers to Participation in Smart Cities: Challenges and Failures**

While smart cities hold great promise for fostering inclusion and innovation, ensuring effective citizen participation remains challenging. Various obstacles hinder the most vulnerable groups from engaging with digital platforms and influencing urban decision-making.

- Digital Divide and Accessibility: Not everyone has access to the necessary technology or a reliable internet connection. The ITU (2022) reports that 37% of the global population still lacks internet access, with even higher rates among older and low-income individuals [40]. To tackle this issue, cities like Toronto and San Francisco have launched free connectivity initiatives in underserved areas and digital literacy programs for seniors and minority groups [41].
- Lack of Representativeness in Decision-Making: Even when participatory technologies are in place, decision-making can often be skewed towards those who are more tech-savvy, leaving out individuals who are less familiar with digital tools [42].
- Distrust and Lack of Trust in Institutions: Certain smart city initiatives have faced backlash due to their unclear data usage and lack of transparency in decision-making. In response to these issues, Amsterdam has established a 'Data Commons' that empowers citizens to have greater control over the data collected by smart technologies [10].

To ensure that smart cities truly involve everyone, several strategies can be implemented. One of the main ones is to focus on inclusion, creating user-friendly digital interfaces while offering traditional options such as paper forms and community meetings. In this way, all citizens can be reached, regardless of their familiarity with technology.

Another key approach is participatory digital budgeting. For example, platforms such as Decidim in Barcelona allow residents to decide how local funds are spent, increasing their sense of ownership and involvement in community choices. It is essential to ensure fair representation by monitoring the socio-demographic composition of participants. In this way, digital platforms will reflect the diversity of the community and will not only appeal to the tech-savvy.

b. **Impacts of Community Involvement: A Synergetic Approach for Smart Cities**

Citizens' active involvement in urban technological decisions is crucial for the success of smart cities. Participation is not a democratic formality; instead, it directly benefits the community and the administration, producing a positive reverberation of trust and cooperation [14].

The sense of belonging and trust in institutions will be, therefore, primarily heightened among citizens. As a contributor, he feels himself an essential part of the decision-making process, being able to influence decisions that impinge on the quality of life; therefore, he feels he has a measure of control over the affairs of the community [22]. Participation clarifies technological decisions, making them more transparent and understandable [10].

From the administrative perspective, collaborating with citizens to design technological solutions increases the likelihood of these solutions being accepted and integrated into daily life. When technologies align with the community's actual needs, thanks to active participation from the early stages of design and implementation, it reduces resistance to change and encourages broader adoption [25]. Another excellent way to lessen errors in urban policymaking and its execution is getting regular feedback from citizens. Real-time monitoring of the effects of policies allows for instant responses from the administrations, facilitating timely adjustments to ensure their initiatives work in effect [26].

In a wider framework, involving communities, most notably in decision-making regarding urban technology, guarantees a more participatory approach. The gains of this participative interest strengthen the link between the citizenry and institutions and consequently lead to more effective and sustainable public policies. Insights gained through participation help align technologies with community needs, fostering acceptance.

5 Conclusions and Limitations

Drawing from the theoretical discussion and empirical observations, the final section summarizes the study's key contributions, acknowledges its limitations, and outlines directions for future research.

Building smart cities of the future is more than advancing technology—it requires putting humans at the center of innovation. The analysis of the cases of Barcelona, Amsterdam, and Helsinki has shown how civic technologies such as Decidim, the Smart Citizen Kit, and urban feedback platforms can actively foster collaborative decision-making and more inclusive urban governance.

This study contributes to the ongoing debate on smart city development by highlighting three main aspects:

First, it emphasizes the importance of integrating social responsibility into digital urban strategies.

Second, it shows that civic tech tools are not neutral but reflect different models of engagement—top-down, bottom-up, or hybrid—depending on the political and institutional context.

Third, it offers a comparative perspective that helps identify both replicable practices and context-specific challenges in digital participation.

However, these benefits are still constrained by structural barriers such as the digital divide, mistrust, and the underrepresentation of marginalized groups. Moreover, the ways in which data is handled and interpreted do not always create inclusive or empowering environments for all citizens.

In terms of limitations, this study is based primarily on secondary sources—including academic literature, public reports, and open-access platforms—which, while diverse and credible, may lack the depth and granularity of direct empirical fieldwork. Additionally, the focus on European cities may limit the generalizability of the findings to other socio-political contexts.

Future research should therefore explore the long-term impacts of participatory technologies and extend the analysis to non-European or resource-constrained urban environments, where digital infrastructures and civic cultures differ substantially.

Ultimately, building truly smart cities requires not only technological advancement, but also political will, cultural openness, and a shared commitment to equity and transparency.

References

1. Aru, P., Santangelo: RGI Report on Smart Cities (n.d.). Retrieved from iris.unito.it
2. Gurashi, A.: Smart Cities and Governance. RTSA, 1, (2021). Retrieved from rtsa.eu
3. Heidegger, M.: The question concerning technology. In: The Question Concerning Technology and Other Essays (W. Lovitt, Trans). Harper & Row, New York. (Original work published 1954), pp. 3–35 (1977)
4. Degli Esposti, P.: I confini ibridi della smart city nel processo di urbanizzazione della tecnologia. Sociologia urbana e rurale **118**, 161–177 (2019)
5. de Paes, V.C., Pessoa, C.H.M., Pagliusi, R.P. et al.: Analyzing the challenges for future smart and sustainable cities. Sustainability **15**(10), 7996 (2023). https://doi.org/10.3390/su15107996
6. Albino, V., Berardi, U., Dangelico, R.M.: Smart cities: definitions, dimensions, performance, and initiatives. J. Urban Technol. **22**(1), 3–21 (2015). https://doi.org/10.1080/10630732.2014.942092

7. Townsend, A.M.: Smart Cities: Big Data, Civic Hackers, and the Quest for a New Utopia. W.W. Norton & Company (2013)
8. United Nations: Goal 11: Make Cities and Human Settlements Inclusive, Safe, Resilient and Sustainable. United Nations Department of Economic and Social Affairs (2015). Retrieved from https://sdgs.un.org/goals/goal11
9. Kitchin, R.: The ethics of smart cities and urban science. Philos. Trans. R. Soc. A Math. Phys. Eng. Sci. **374**(2083), 20160115 (2016). https://doi.org/10.1098/rsta.2016.0115
10. Janssen, M., Charalabidis, Y., Zuiderwijk, A.: Benefits, adoption barriers and myths of open data and open government. Inf. Syst. Manage. **29**(4), 258–268 (2012). https://doi.org/10.1080/10580530.2012.716740
11. O'Neil, C.: Weapons of Math Destruction: How Big Data Increases Inequality and Threatens Democracy (2016)
12. Zuboff, S.: The Age of Surveillance Capitalism: The Fight for a Human Future at the New Frontier of Power (2019)
13. UN-Habitat: World Cities Report 2016: Urbanization and Development—Emerging Futures. United Nations Human Settlements Programme (2016). https://unhabitat.org/world-cities-report
14. Cardullo, P., Kitchin, R.: Being a "citizen" in the smart city: up and down the scaffold of smart citizen participation. GeoJournal **84**(1), 1–13 (2019). https://doi.org/10.1007/s10708-018-9845-8
15. Sassen, S.: Cities in a World Economy (5ª ed.). SAGE Publications (2018)
16. European Environment Agency.: Sustainability of Europe's Mobility Systems. EEA Report No 18/2024 (2024). https://www.eea.europa.eu/en/analysis/publications/sustainability-of-europes-mobility-systems
17. European Commission: White Paper: Roadmap to a Single European Transport Area—Towards a Competitive and Resource Efficient Transport system (2011)
18. Ellen MacArthur Foundation: Towards a Circular Economy: Business Rationale for an Accelerated Transition (2013)
19. Vanolo, A.: Smartmentality: the smart city as disciplinary strategy. Urban Stud. **51**(5), 883–898 (2014). https://doi.org/10.1177/0042098013494427
20. Healey, P.: Collaborative Planning: Shaping Places in Fragmented Societies (1997)
21. Grimsey, D., Lewis, M.K.: Public Private Partnerships: The Worldwide Revolution in Infrastructure Provision and Project Finance. Edward Elgar Publishing (2004)
22. Susha, I., Janssen, M., Verhulst, S.: Data collaboratives as a new frontier of cross-sector partnerships in the age of open data: taxonomy development. In: Proceedings of the 50th Hawaii International Conference on System Sciences (HICSS), 2691–2700 (2017). https://hdl.handle.net/10125/41404
23. Parycek, P., Sachs, M., Skoric, M.M.: E-participation: a new form of participation in the digital age. In: Proceedings of the 17th Annual International Conference on Digital Government Research (dg.o '16), pp 312–314 (2016). https://doi.org/10.1145/2912160.2912196
24. Nam, T.: Smart city from a systems perspective: smart governance, citizen participation, and urban resilience. Gov. Inf. Q. **34**(3), 383–390 (2017). https://doi.org/10.1016/j.giq.2017.07.002
25. Batty, M., Townsed, A., Cohen, R.: Smart Cities: Big Data, Civic Hackers, Ans the Quest for a New Utopia. Yale University Press (2012)
26. Barandiaran, X.E., Calleja-López, A., Monterde, A., Romero, C.: Decidim, a Technopolitical network for participatory democracy: philosophy, practice and autonomy of a collective platform in the age of digital intelligence. In: Decidim, a Technopolitical Network for Participatory Democracy. Springer, pp. 1–34 (2024). https://doi.org/10.1007/978-3-031-50784-7_1
27. Decidim: Decidim: The Platform for Citizen Participation (2023). https://decidim.org/
28. Nikiforova, A., Lněnička, M., Milić, P. et al.: From the evolution of public data ecosystems to the evolving horizons of the forward-looking intelligent public data ecosystem empowered by emerging technologies ResearchGate (2024). https://www.researchgate.net/publication/383296056

29. Bonney, R., Phillips, T.B., Ballard, H.L., Enck, J.W.: Can citizen science enhance public understanding of science? Pub. Underst. Sci. **25**(1), 2–16 (2016). pubmed.ncbi.nlm.nih.gov
30. Gabrys, J.: How To Do Things with Sensors. University of Minnesota Press (2019). https://doi.org/10.5749/9781452964096
31. Smart Citizen Project: A Citizen-Driven Environmental Monitoring Initiative that Provides Open-Source Sensor Kits for Air Quality, Noise Pollution, and Urban Environment Tracking. Retrieved from smartcitizen.me
32. Waag Society—Smart Citizen Kit Findings—Research and Analysis on Citizen Participation in Environmental Monitoring Through the Smart Citizen Kit, Highlighting Lessons Learned and Impact on Urban Policy. Retrieved from waag.org
33. Mergel, I., Edelmann, N., Haug, N.: Defining digital transformation: results from expert interviews. Government Inf. Q. **36**, Article ID: 101385 (2019). https://doi.org/10.1016/j.giq.2019.06.002
34. Gil de Zúñiga, H., Jung, N., Valenzuela, S.: Social media use for news and individuals' social capital, civic engagement and political participation. J. Comput.-Mediat. Commun. **17**(3), 319–336 (2012)
35. European Commission—Digital Inclusion and Smart Cities—An Official EU Report Analyzing Initiatives for Bridging the Digital Divide and Fostering Inclusive Smart City Policies Across European Regions. Retrieved from ec.europa.eu
36. City of Helsinki—Citizen Participation—Official Portal Detailing Ways for Residents to Engage in Urban Decision-Making, Including Participatory Budgeting, Public Consultations, and Digital Tools for Governance. Retrieved from hel.fi
37. City of Helsinki—City Feedback App—A Mobile Application That Allows Residents to Report Urban Issues, such as Waste Management, Road Conditions, and Public Space Maintenance, Directly to Municipal Authorities. Retrieved from hel.fi
38. Forum Virium Helsinki.: City Feedback App: Enhancing Citizen Participation and Urban Environment Monitoring in Helsinki (2022). Retrieved from https://forumvirium.fi/en/first-test-period-to-begin-feedback-on-the-urban-environment-in-helsinki/
39. GeoAwesome—Citizen Engagement in Helsinki—An Article Exploring Innovative Ways Helsinki Integrates Geospatial Technology and Citizen Input into Urban Planning for a more Livable and Sustainable City. Retrieved from geoawesome.com
40. Kahila-Tani, M., Fagerholm, N., Hausner, V. H., Broberg, A.: Maptionnaire—a community engagement platform for map-based participatory planning. In: Evaluating Participatory Mapping Software. Springer, pp. 71–91 (2023)
41. International Telecommunication Union (ITU)—Measuring Digital Development: A Global Report Providing Key Statistics and Insights into Internet Connectivity, Digital Inclusion, and Smart City Infrastructure Worldwide (2022). Retrieved from itu.int
42. Kolotouchkina, O., Ripoll González, L., Belabas, W.: Smart cities, digital inequalities, and the challenge of inclusion. Smart Cities **7**(6), 3355–3370 (2024). https://doi.org/10.3390/smartcities7060130

A Conceptual Framework for Digital Transformation of Business Models: Advancing Towards Industry 5.0

Antonia Terán-Bustamante, Antonieta Martínez-Velasco, and Ana Beatriz Hernández-Lara

Abstract Digital transformation is progressing unevenly across industries, with varying levels of success influenced by organizational and sector-specific factors. Understanding where to focus investments and what type of transformation to adopt has become a crucial challenge for companies seeking competitiveness and market relevance in the digital era. This paper aims to analyze companies' strategic decision making to foster digital transformation, conducting a literature review, and proposing a conceptual framework for digital transformation of business models. The study identifies key drivers of successful digital transformation, including digital strategy, human capital, scalability, customer focus, security and risk management. Integrating these factors, the proposed model emphasizes the strategic alignment of digital initiatives with organizational goals, fostering a culture of continuous innovation and adaptability. The findings contribute to a deeper understanding of the mechanisms and prerequisites for effective digital transformation, offering insights for organizations navigating the shift toward Industry 5.0.

Keywords Industry 5.0 · Digital transformation · Business models · SDG9 · Innovation · Industry

A. Terán-Bustamante
Facultad de Ciencias Económicas y Empresariales, Universidad Panamericana, Ciudad de México, México
e-mail: ateran@up.edu.mx

A. Martínez-Velasco
Facultad de Ingeniería, Universidad Panamericana, Ciudad de México, Mexico
e-mail: amartinezv@up.edu.mx

A. B. Hernández-Lara (✉)
Department of Business Management, Universitat Rovira i Virgili, Reus, Spain
e-mail: anabeatriz.hernandez@urv.cat

A. Visvizi et al. (eds.), *Research and Innovation Forum 2025 Vol2*, Springer Proceedings in Complexity, https://doi.org/10.1007/978-3-032-09062-1_25

1 Introduction

The rapid evolution of digital technologies has become a transformative force reshaping industry worldwide. Companies are under increasing pressure to adapt to the demands of digital transformation, which not only entails technological upgrades but also requires rethinking business strategies, processes, and models [1–3]. This necessity is especially pronounced in the context of Industry 5.0, where innovation goes beyond automation, emphasizing collaboration between humans, technology, and machines, moving towards more human-centered, resilient, and sustainable industries [4].

Despite the proliferation of digital initiatives, many organizations struggle with fragmented approaches, often lacking a clear direction for investment and implementation. In addition, different industries face unique challenges in navigating digital transformation, influenced by sector-specific requirements, resource availability, and organizational readiness [1]. These variations underscore the critical need for a conceptual framework of business model to guide companies in their digital transformation journey, ensuring that investments lead to meaningful and sustained innovation [5].

This research aims to address the current gap in guiding organizations through digital transformation by synthesizing the existing literature to propose a comprehensive conceptual framework for the digital transformation of business models. The study seeks to bridge theoretical gaps by identifying critical factors that optimize transformation strategies and support their effective implementation. To achieve this, the research explores three core questions: (1) What is the conceptual and thematic structure of the current state of the art in digital transformation of business models? (2) What are the key variables and dimensions that contribute to a successful digital transformation? and (3) What are the managerial implications of how different industries uniquely approach and navigate the challenges of digital transformation?

To this end, the argument in this paper is structured as follows: the next section reviews the relevant literature, followed by the methodology. The findings are then presented and discussed, and the paper concludes with a summary of the main insights.

2 Literature Review

2.1 Defining Digital Transformation

The concept of digital transformation has evolved significantly over the past two decades, shifting from basic digitization efforts to comprehensive strategies that integrate technology into every aspect of an organization.

Digital transformation is a process where digital technologies disrupt traditional value-creation mechanisms [6]. It operates at multiple levels, involving the leverage of digital technologies to enhance existing processes while simultaneously driving digital innovation to revolutionize business models [7].

Digital transformation can be also considered as a process of reconversion and adaptation, where digital technologies seamlessly integrate into daily routines, motivating organizations to adopt innovative practices to meet client demands by delivering novel products and services that address evolving needs [8]. The digital transformation of business models is framed as a crucial connection between innovation and strategic management, aiming to create and capture value [1].

In summary, digital transformation represents the continuous reconfiguration of firms to create value through large-scale implementation of technology [9]. This transformation builds competitive advantages by improving customer experiences, reducing operational costs, and generating sustained value [10].

2.2 Key Competencies and Challenges of Digital Transformation

Digital transformation, being a multifaceted phenomenon, necessitates diverse competencies depending on its drivers and objectives. Developing digital competencies has become essential, as organizational competencies differentiate companies by defining their abilities and approaches.

Key competencies required for digital transformation include "business competencies" involving the development of a vision and strategies, understanding the environment, and identifying opportunities; "technical competencies" which focus on expertise in working with hardware, data, and software; "digital competencies" centered on understanding emerging technologies and possessing technical expertise; and "leadership competencies" which involve managing relationships, fostering communication, and driving collaboration effectively as leaders [11].

Digital transformation entails different phases [12]—identification, definition, design, execution, evaluation, and communication—each requiring specific competencies. The identification phase involves analyzing technological trends and aligning with business strategy, while the definition phase emphasizes vision, motivation, and communication skills. Design requires ICT integration, business process design, and project management, whereas execution focuses on leadership and change management. Evaluation relies on data analysis, risk assessment, and impact measurement, and communication involves effectively sharing goals and results.

Digital transformation also requires coordinated leadership and action across these competencies, encompassing five key elements—customers, competition, added value, data, and innovation—that businesses must continuously adapt to within the dynamic digital landscape [13].

However, digital transformation also presents challenges that may impede the adoption of digital innovation [14, 15]. These challenges often include organizational resistance, where employees and management may resist change due to uncertainty or fear of redundancy; resource constraints, as enterprises, particularly small and medium-sized businesses (SMEs), frequently lack the financial and technical resources needed to implement comprehensive digital strategies; and misalignment with strategy, where digital initiatives may fail if they are not aligned with broader organizational goals or lack strategic coherence.

2.3 The Digital Transformation of Business Models

A business model outlines how a company creates, delivers, and captures value [16]. To remain competitive in the complex digital business environment, companies must integrate and adapt their business models to digital transformation. Digitalization drives the emergence of innovative forms of value creation, enhancing productivity, fostering innovation, and strengthening competitiveness. It also contributes to improving social well-being and advancing environmental sustainability. This value creation is underpinned by the insights derived from digital data, which are captured and analyzed through intelligent systems [17], enabling the development of novel business management models, promoting innovation, and expanding access to new markets [17].

The impact of digitalization extends across multiple levels of business models, requiring companies to navigate and adapt to this evolving landscape strategically [18]. The digital transformation of business models is a complex process driven by disruptive digital technologies like the Internet of Things (IoT), artificial intelligence (AI), and big data [18]. These technologies blur physical-digital boundaries, enabling flexibility and scalability [19], while reshaping workplaces, production, markets, and customer behavior [20, 21]. Their generative and malleable nature demands organizations rethink value creation, capture, and adaptation strategies [22, 23].

The review of existing literature reveals the absence of a unified framework that addresses the multidimensional nature of digital transformation. This study proposes a conceptual framework for digital transformation of business model that synthesizes the main variables and dimensions for a successful digital transformation, offering a holistic perspective for organizations navigating the complexities of this process.

3 Methodology

To achieve this objective, we conducted a literature review of articles published since 2010, covering the last two decades during which digital transformation has undergone significant evolution. Our search was carried out using the Web of Science

(WoS) database with specific keywords searched in the title, and applying logical combinations as shown in the following query:

> "TITLE (*digital transformat** AND *business model*) OR (*digit** AND *in* AND *business model*) OR TITLE (*innovat** AND *in* AND *business model*) OR TITLE (*technolog** AND *adoption* AND *in* AND *business model*)"

The initial search yielded 3303 documents. After excluding reports, conference proceedings, and other non-article formats, the dataset was narrowed down to 1990 articles. Of these, 1557 were classified within the Business and Economics research areas, and 1373 belonged to the Web of Science Core Collection and were written in English. The authors intended to conduct a manual content analysis of the full text of each paper, which required excluding documents that only tangentially addressed digital transformation and business models or lacked a clear focus necessary to contribute to the development of a conceptual framework for digital business model transformation. Through an iterative screening process, the three authors independently reviewed the abstracts and keywords of the 1373 papers. Ultimately, 49 articles were selected for full-text analysis and formed the final sample for this study.

This number was determined based on theoretical saturation—a point at which no new relevant themes or insights emerged from additional articles. Each selected paper was read in full, and its content was carefully analyzed to extract themes and codes relevant to the construction of our conceptual framework for the digital transformation of business models.

The analysis was conducted in two phases. First, we performed a descriptive bibliometric review using RStudio (version 2023.09.1+494) and the bibliometrix package. This initial phase offered a preliminary understanding of the conceptual and thematic structure of the topic, primarily through keyword analysis.

In the second phase, we conducted a content analysis, delving deeply into the dataset of articles to develop the proposed conceptual framework for the digital transformation of business models. The coding process was conducted iteratively by the three authors, who worked independently to identify key themes and patterns related to the digital transformation of business models. Rather than relying on a predefined coding scheme, the codes emerged inductively through close reading of the full texts, guided by the central dimensions of Industry 5.0—particularly the interplay between technological development (being the main themes related to technology and digital transformation), human-centric aspects (with themes like human capital and customer focus), sustainability and organizational resilience (focused on managerial and legal aspects, change management and competitiveness). The independent coding efforts were then compared and consolidated through joint discussions, ensuring consistency and alignment across the themes. This approach allowed us to capture both the technological and socio-organizational aspects that shape digital business model transformation today.

4 Findings and Discussion

4.1 Descriptive Bibliometric Analysis

To gain a deeper understanding of the conceptual and thematic structure of the topic, a comprehensive bibliometric analysis was conducted. This involved three distinct but interconnected analyses: frequency analysis of key terms, mapping co-occurrence networks, and trend topic analysis.

Table 1 presents the keywords most commonly assigned by the authors in their articles, along with their respective percentages. The identification of the most frequent words appearing across the body of literature provided an initial, broad overview of the main concepts and themes emerging within the field.

The results reveal that, in addition to the primary keywords used in our search, other significant terms emerged, such as "information," "technology," and "leadership." These findings underscore the strong relationship between digital transformation and information and technology, while also emphasizing the pivotal role of leadership in driving the digital transformation of business models—an observation that aligns with insights from previous studies [12, 13].

Figures 1 and 2 compare authors' keywords with broader topics derived from cited references. These figures illustrate the relative prominence of the keywords, with the size of each word indicating its frequency of occurrence in the dataset. The analysis highlights that research on digital transformation and business models often aligns with themes like dynamic capabilities, innovation, technology, management, and value creation. These findings underscore the importance of competencies—particularly dynamic capabilities—in driving digital transformation [11, 12].

The mapping of co-occurrence network (Fig. 3) captures the interdependencies between the papers in our dataset, for a better understanding and discovery of patterns in the research topic. The co-occurrence network visualizes how often specific terms appear together within the same context, suggesting potential associations between them.

Figure 3 illustrates three primary clusters that highlight the significance of innovation, dynamic capabilities, and technology. The innovation cluster links to change,

Table 1 Most frequent keywords

Keywords	Occurrences	%
Digital transformation	14	33.33%
Business models	4	9.52%
Business model innovation	3	7.14%
Digital technologies	3	7.14%
Information technology	3	7.14%
Leadership	3	7.14%

Source Authors

Fig. 1 Word cloud of keywords plus. *Source* Authors

Fig. 2 Word cloud of authors' assigned keywords. *Source* Authors

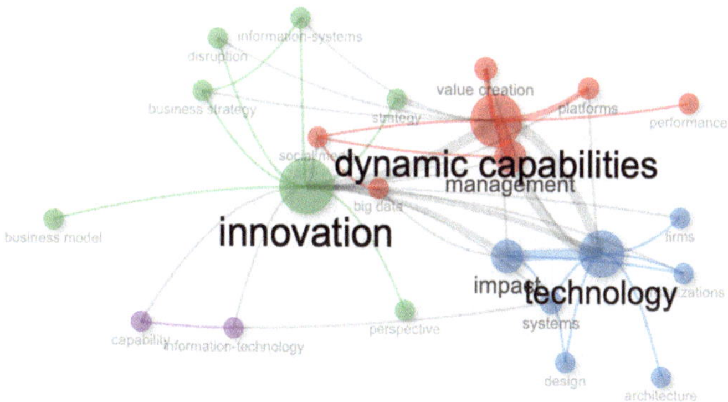

Fig. 3 Co-occurrence network. *Source* Authors

emphasizing disruption, data, and its impact on strategy and business models [18, 24]. The dynamic capabilities cluster focuses on management, value creation, and performance, highlighting their role in success [22, 23]. The technology cluster examines its impact on organizations, firms, and components like design, architecture, and systems [17].

Figure 4 highlights topic trends that capture the shifting focus of research and identifies any rising or declining areas of interest. The results show a shift from earlier research on technology to recent studies emphasizing innovation, dynamic

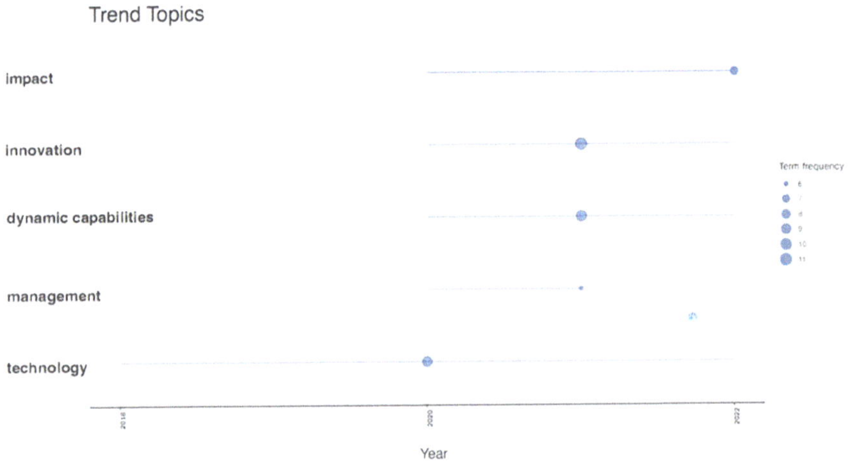

Fig. 4 Trend topics. *Source* Authors

capabilities, and the impact of digital transformation. Management-related topics remain less explored.

4.2 Content Analysis

We conducted a content analysis of the articles to identify key processes for optimizing digital transformation of business models. The resulting framework outlines eight critical variables [25–27], as detailed in Table 2, related to the main identified themes: technology and digital transformation, human-centric aspects, managerial aspects that foster sustainability and organizational resilience.

The frequencies in which these variables appeared and their interconnections allowed us to propose a conceptual framework for the digital transformation of business models that contemplates "Digital Transformation Choice", "Human Capital", "Scaling", "Client", "Security and Risk Management" as the main dimensions for the digital transformation of business models (see Fig. 5).

The content analysis also provided managerial implications of digital transformation for various industries, focusing on how they can innovate their business models. Below is a summary of the key implications for different industries.

In the manufacturing sector, digital transformation is largely driven by the adoption of technologies such as the Internet of Things (IoT), artificial intelligence (AI), and cloud computing [1, 2]. These innovations enable smart manufacturing and significantly enhance operational efficiency. A notable trend is the shift toward data-driven decision-making and predictive maintenance, which helps optimize resource usage and minimize downtime. Additionally, new business models have emerged,

Table 2 Definition of the variables for digital transformation of business models

Variable	Concept
Competitive and technological intelligence	Monitors the technological environment to identify opportunities and threats, focusing on digital transformation choices, value generation, and clients
Technological strategic planning	Guides organizational strategy, emphasizing digital transformation choices and value creation
Digital transformation strategy	Outlines the strategy to deliver products/services that meet market needs and generate value through digital transformation
These three first variables integrate multiple dimensions: the managerial aspect, which focuses on diagnosing the technological landscape, guiding and implementing technological choices; the human-centric perspective, centered on customer focus; and broader economic objectives such as value creation, aligned with principles of economic sustainability	
Infrastructure	Establishes internal and external infrastructure for digital transformation operations
Scaling digital value	Focuses on enhancing the efficiency, accessibility, and profitability of digital products as they grow
These two variables are mainly focused on technological advances and digital transformation	
Human capital	Prepares personnel with the necessary awareness, coordination, and competencies for successful digital transformation
Variable related to human-centric aspects	
Change management	Develops and implements strategies for managing change, addressing roles, responsibilities, team dynamics, and impact evaluation
Legal framework	Ensures regulatory convergence, interoperability, and information exchange, with a focus on security and risk management
These two variables integrate several key aspects: managerial dimensions related to change management and risk management, combined with a focus on sustainability and resilience. Together, they ensure adaptability and security in an increasingly dynamic and uncertain environment	

Source Authors

including servitization—offering products as services—and circular economy practices that promote sustainability. For instance, the integration of digital twins and real-time analytics has allowed manufacturers to simulate, monitor, and improve production processes with unprecedented precision.

The financial services industry is undergoing rapid digital transformation fueled by fintech innovations such as blockchain, digital wallets, and AI-driven customer service solutions. To stay competitive, financial institutions are increasingly forming ecosystem partnerships that provide seamless, end-to-end digital experiences for customers [3, 14].

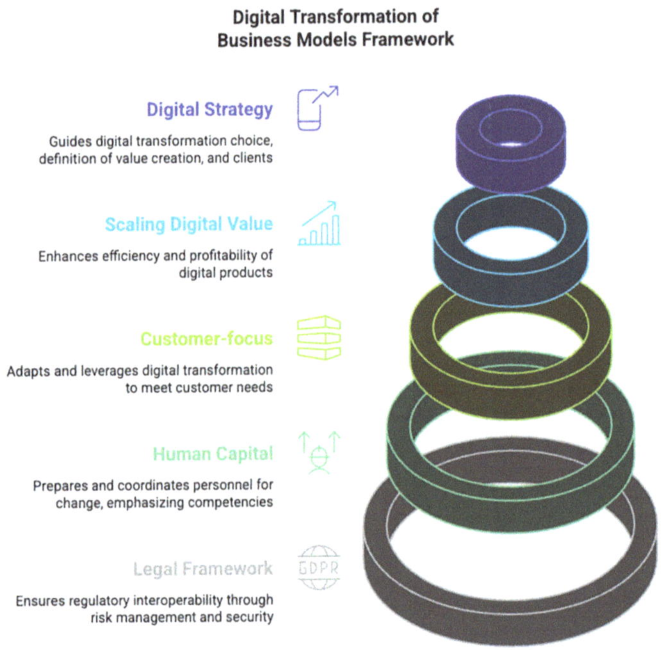

Fig. 5 Conceptual framework of the digital transformation of business models. *Source* Authors—Icons downloaded from napkin.ai

Digital transformation in healthcare is revolutionizing patient care through advancements such as telemedicine, AI-powered diagnostics, and the widespread adoption of electronic health records [28]. These technologies improve access and quality of care, but also bring challenges, particularly around data privacy and the integration of new tools with legacy systems. The industry is also witnessing a transition toward value-based care and personalized medicine, shifting business models to prioritize patient outcomes and tailored treatments.

In the retail and e-commerce sector, companies are leveraging big data and AI to create personalized shopping experiences and better understand consumer behavior [4, 13]. Omnichannel strategies have become essential, blending online and offline operations to deliver a seamless customer journey. Leading platforms exemplify the move toward ecosystem-based business models, where retailers create integrated digital environments that connect customers, suppliers, and service providers.

Even in more traditional sectors like textiles, digital transformation is helping address long-standing issues such as operational inefficiencies and rising labor costs. Firms are adopting automation and digital tools to enhance design, streamline production, and optimize supply chain management [29]. However, the success of digital initiatives in these industries often depends on cultivating an organizational culture that supports innovation and embraces change.

In the education sector, digital transformation is prompting universities and other institutions to adopt digital tools that enhance both learning and research [25]. There is a growing emphasis on lifelong learning and digital literacy to equip students for economies shaped by digital technologies. Collaborations between academia and industry are increasingly important, as they help align educational programs with the skills and competencies required in a digitally-driven job market.

Consulting firms are also evolving, shifting away from traditional advisory roles toward delivering comprehensive digital solutions [5]. This transformation is being facilitated by acquiring digital capabilities through talent recruitment and mergers and acquisitions. Additionally, consulting companies are focusing on building digital ecosystems and offering platform-based services that address a wide range of client needs in an integrated manner.

Finally, digital transformation in the public sector is improving service delivery and citizen engagement through more efficient processes and accessible platforms [30]. However, governments face significant challenges, including outdated legacy systems and institutional resistance to change. To navigate this landscape, agile governance models and data-driven policymaking are becoming increasingly important, enabling more responsive and transparent public administration.

5 Conclusions

This paper examines how companies can strategically drive digital transformation through a literature review and the development of a conceptual framework for transforming business models. The value of this framework lies in its cross-sector applicability, emphasizing the key drivers that enable successful transformation across diverse economic activities. These drivers include a well-defined digital strategy, the development of human capital, scalability of solutions, a strong customer-centric approach, and effective risk management. The framework emphasizes the alignment of digital initiatives with organizational goals and the cultivation of a culture centered on continuous innovation. While digital transformation can significantly increase company value through innovation, it requires careful investment of time and resources, alongside attention to technological and human factors and associated risks. The paper argues that successful transformation extends beyond technological infrastructure, deeply influencing company culture and leadership. In the context of Industry 5.0, the human element—skills, agility, and the synergy between technology and people—becomes critical [20, 21].

Across industries, several common themes emerge in the context of digital transformation and its managerial implications. One key shift is in the value proposition, as organizations move from offering purely physical products to delivering digital services and enhanced experiences. To stay competitive, firms must operate within digital ecosystems, fostering collaboration and interdependence across sectors. Organizational agility becomes essential, requiring flexible structures and adaptive cultures capable of responding to rapid technological change. At the same time, digital transformation supports sustainability by enabling more efficient resource use and promoting circular economy practices. Key takeaways include the need for visionary leadership to guide digital initiatives, strategic investment in both technology and human capital, and a strong emphasis on customer-centricity to ensure that transformation efforts ultimately enhance customer value and experience. These insights highlight not only the transformative potential of digital technologies across sectors but also the importance of developing tailored strategies to address the unique challenges faced by each industry.

Given the rapid pace of digital change, the theoretical and practical contributions of this research focus on offering organizations a model to guide the essential processes for successful digital transformation. Although the research offers valuable conceptual insights, further empirical studies are needed to assess the applicability of the proposed framework in real-world companies undergoing the 5.0 revolution. This study does not validate the proposed conceptual framework and relies solely on secondary data, rather than primary data collection. Validation is proposed as an objective for future research, along with enhancing the literature review by incorporating additional databases to capture a broader range of relevant studies on the topic. A more in-depth analysis of the methodologies and theoretical contributions of previous studies would enhance the critical evaluation of existing research, helping to identify gaps and areas that warrant further investigation.

References

1. Vaska, S., Massaro, M., Bagarotto, E.M., Dal Mas, F.: The digital transformation of business model innovation: a structured literature review. Front. Psychol. **11** (2021). https://doi.org/10.3389/fpsyg.2020.539363
2. Liu, Q.-R., Liu, J.-M., He, Z.-P.: Digital transformation ambidexterity and business performance. J. Enterp. Inf. Manag. **36**(5), 1402–1420 (2023). https://doi.org/10.1108/JEIM-08-2022-0280
3. Niu, Y., Wen, W., Wang, S., Li, S.: Breaking barriers to innovation: the power of digital transformation. Finance Res. Lett. **51** (2023). https://doi.org/10.1016/j.frl.2022.103457
4. Afzal, B., Li, X., Hernández-Lara, A.B.: The innovation journey and crossroads of sustainability, resilience and human-centeredness: a systematic literature review. Transform. Gov. People Process Policy **18**(3), 368–383 (2024). https://doi.org/10.1108/TG-10-2023-0151
5. Tavoletti, E., Kazemargi, N., Cerruti, C., Grieco, C., Appolloni, A.: Business model innovation and digital transformation in global management consulting firms. Eur. J. Innov. Manag. **25**(6), 612–636 (2021). https://doi.org/10.1108/EJIM-11-2020-0443
6. Vial, G.: Understanding digital transformation: a review and a research agenda. J. Strateg. Inf. Syst. **28**(2), 118–144 (2019). https://doi.org/10.1016/j.jsis.2019.01.003
7. Yoo, Y., Henfridsson, O., Lyytinen, K.: The new organizing logic of digital innovation: an agenda for information systems research. Inf. Syst. Res. **21**(4), 724–735 (2010). https://doi.org/10.1287/isre.1100.0322
8. Matt, C., Hess, T., Benlian, A.: Digital transformation strategies. Bus. Inf. Syst. Eng. **57**(5), 339–343 (2015). https://doi.org/10.1007/s12599-015-0401-5
9. Di Prima, C., Ferraris, A.: Rethinking the HR role: how digital transformation is changing HR departments. In: Visvizi, A., Troisi, O., Corvello, V. (eds.) Research and Innovation Forum

2023. Springer Proceedings in Complexity, pp. 641–648. Springer International Publishing, Cham (2024). https://doi.org/10.1007/978-3-031-44721-1_48
10. Steiber, A., Alvarez, D.: How culture and technology go hand-in-hand in digital transformations into an ecosystem business. In: Visvizi, A., Troisi, O., Corvello, V. (eds.) Research and Innovation Forum 2023. Springer Proceedings in Complexity, pp. 3–14. Springer International Publishing, Cham (2024). https://doi.org/10.1007/978-3-031-44721-1_1
11. Tekic, Z., Koroteev, D.: From disruptively digital to proudly analog: a holistic typology of digital transformation strategies. Bus. Horiz. **62**(6), 683–693 (2019). https://doi.org/10.1016/j.bushor.2019.07.002
12. Ravarini, A., Locoro, A., Martinez, M.: Digital transformation projects maturity and managerial competences: a model and its preliminary assessment. In: Exploring Digital Ecosystems: Organizational and Human Challenges. Lecture Notes in Information Systems and Organization, vol. 33, pp. 261–272 (2020). https://doi.org/10.1007/978-3-030-23665-6_19
13. Rodriguez-Lluesma, C., Garcia-Ruiz, P., Pinto-Garay, J.: The digital transformation of work: a relational view. Bus. Ethics Environ. Responsib. **30**(1), 157–167 (2021). https://doi.org/10.1111/beer.12323
14. Sia, S.K., Weill, P., Zhang, N.: Designing a future-ready enterprise: the digital transformation of DBS bank. Calif. Manage. Rev. **63**(3), 35–57 (2021). https://doi.org/10.1177/0008125621992583
15. Burchardt, C., Maisch, B.: Digitalization needs a cultural change—examples of applying agility and open innovation to drive the digital transformation. In: 29th CIRP Design Conference 2019, vol. 84. Procedia CIRP **84**, 112–117 (2019). https://doi.org/10.1016/j.procir.2019.05.009
16. Teece, D.J.: Business models, business strategy and innovation. Long Range Plan. **43**(2–3), 172–194 (2010). https://doi.org/10.1016/j.lrp.2009.07.003
17. Economic Commission for Latin America and the Caribbean (ECLAC): A Digital Path for Sustainable Development in Latin America and the Caribbean, LC/CMSI.8/3. Santiago de Chile (2022)
18. Caputo, A., Pizzi, S., Pellegrini, M.M., Dabić, M.: Digitalization and business models: where are we going? A science map of the field. J. Bus. Res. **123**, 489–501 (2021). https://doi.org/10.1016/j.jbusres.2020.09.053
19. Dabrowska, J., et al.: Digital transformation, for better or worse: a critical multi-level research agenda. R&D Manag. **52**(5), 930–954 (2022). https://doi.org/10.1111/radm.12531
20. Visvizi, A., Malik, R., Guazzo, G.M., Çekani, V.: The Industry 5.0 (I50) paradigm, blockchain-based applications and the smart city. Eur. J. Innov. Manag. **28**(1), 5–26 (2025). https://doi.org/10.1108/EJIM-09-2023-0826
21. Troisi, O., Visvizi, A., Grimaldi, M.: Rethinking innovation through industry and society 5.0 paradigms: a multileveled approach for management and policy-making. Eur. J. Innov. Manag. **27**(9), 22–51 (2024). https://doi.org/10.1108/EJIM-08-2023-0659
22. Kallinikos, J., Aaltonen, A., Marton, A.: The ambivalent ontology of digital artifacts. MIS Q. **37**(2), 357–370 (2013). https://doi.org/10.25300/MISQ/2013/37.2.02
23. Lanzolla, G., Lorenz, A., Miron-Spektor, E., Schilling, M., Solinas, G., Tucci, C.L.: Digital transformation: what is new if anything? Emerging patterns and management research. Acad. Manag. Discov. **6**(3), 341–350 (2020). https://doi.org/10.5465/amd.2020.0144
24. Diaz-Rainey, I., Ibikunle, G., Mention, A.-L.: The technological transformation of capital markets. Technol. Forecast. Soc. Change **99**, 277–284 (2015). https://doi.org/10.1016/j.techfore.2015.08.006
25. Teran-Bustamante, A., Martinez-Velasco, A., Lopez-Fernandez, A.M.: University-industry collaboration: a sustainable technology transfer model. Adm. Sci. **11**(4) (2021). https://doi.org/10.3390/admsci11040142
26. Verma, A., Dandgawhal, P.S., Giri, A.K.: Impact of ICT diffusion and financial development on economic growth in developing countries. J. Econ. Finance Adm. Sci. **28**(55), 27–43 (2023). https://doi.org/10.1108/JEFAS-09-2021-0185
27. Close, K., Franke, M.R., Grebe, M., Hrishikesh, H., Rogers, K.: The Keys to Scaling Digital Value. Boston Consulting Group, Boston (2022)

28. Ashby, S.E., Snodgrass, S.H., Rivett, D.A., Russell, T.: Factors shaping e-feedback utilization following electronic objective structured clinical examinations. Nurs. Health Sci. **18**(3), 362–369 (2016). https://doi.org/10.1111/nhs.12279
29. Naiwen, L., Rehman, M.Z.U., Wenju, Z., Naseem, S., Mohsin, M., Afzal, A.: The role of financial literacy and risk tolerance: an analysis of gender differences in the textile sector of Pakistan. Ind. Text. **72**(3), 300–308 (2021). https://doi.org/10.35530/IT.072.03.202023
30. Clarke, M.: Building employability through graduate development programmes: a case study in an Australian public sector organisation. Pers. Rev. **46**(4), 792–808 (2017). https://doi.org/10.1108/PR-12-2015-0321

Harnessing Local Potential Through Microprojects in Cross-Border Cooperation

Nicole Dujany and Alberto Ferraris

Abstract The Interreg program for European transnational cooperation aims to be a tool to encourage collaboration among the Union's neighbouring countries and is oriented toward sustainable territorial development. However, cross-border cohesion policies often seem to favour large urban organizations, leaving SMEs and less structured and peripheral local actors in the background. In response to this issue, the Interreg Alcotra program has introduced an experimental call for microprojects, aimed at fostering the inclusion of new actors by reducing administrative and financial barriers to participation. This study therefore explores the first Alcotra Microprojects 2021–2027 call, investigating the design of the instrument, access mechanisms, and the profile of beneficiaries. Through document analysis, comparison with other funding instruments, and interactions with stakeholders, the research highlights that, although procedural simplification has indeed encouraged the participation of new actors and expanded the cooperation network, some challenges remain. The results suggest that microprojects can constitute an effective model for making cooperation policies more inclusive and oriented toward local development, but policymakers should improve the instrument's operational aspects by further simplifying financial management and paying greater attention to the geographical distribution of beneficiaries. The analysis offers insights for the future improvement of funding mechanisms and the integration of innovative approaches into territorial cohesion policies.

Keywords Cross-border cooperation · Microprojects · SDG9 · Interreg · European Union (EU)

N. Dujany (✉) · A. Ferraris
University of Turin, Turin, Italy
e-mail: nicole.dujany@unito.it

A. Ferraris
University of Nicosia, Nicosia, Cyprus

A. Visvizi et al. (eds.), *Research and Innovation Forum 2025 Vol2*, Springer Proceedings in Complexity, https://doi.org/10.1007/978-3-032-09062-1_26

1 Introduction

The Interreg program, promoted by the European Commission, is one of the main tools through which activities of cooperation, networking, and trust-building are encouraged among European cross-border territories [1]. Border regions are not always economically thriving or attractive for innovation and entrepreneurship, and cross-border cooperation can often offer important productive opportunities and stimuli [2].

The European Union supports the concept that regional development is fostered by the creation of new and diverse networks of organizations [3] and is committed to supporting this through dedicated funding for cooperation between regions and border territories, mainly via various Interreg programs. However, the effectiveness of cross-border cohesion policies depends on several factors [4], and recently a critique of the program has been gaining traction.

Specifically, for positive effects to occur in the target territories of these policies, the role of partners and stakeholders embedded in the network as actors in funded projects is fundamental [5]. Interreg programs, however, seem to intrinsically favour larger, stronger organizations located in developed urban areas. This is because such entities have the managerial and financial capacity to participate in the funded projects and are already embedded within cross-border networks [6, 7].

The participation of SMEs, and in particular of entities located in more marginal areas, remains very limited, even though including these actors in the network could generate spillover effects of development and innovation for the peripheral territories targeted by Interreg and the Commission [8].

In this context, the Interreg Alcotra program's Joint Secretariat has committed to launching tailored calls aimed at facilitating the inclusion of new partners within the system. Alcotra is indeed one of the European cross-border programs that has introduced specific calls for microprojects, with the goal of directly involving citizens and supporting new types of beneficiaries. The program targets the Franco-Italian border region, encompassing Liguria, Piemonte, Valle d'Aosta in Italy, and Auvergne-Rhône-Alpes, Provence-Alpes-Côte d'Azur in France. This area, characterized by diverse geographical and socio-economic conditions, includes both densely populated urban centers and rural, mountainous territories, where cross-border cooperation plays a crucial role in fostering regional cohesion and sustainable development. The Interreg programme is not merely a tool to foster cross-border cooperation, but also a strategic component of the EU's broader approach to sustainable development. By promoting territorial cohesion, social inclusion, and transnational collaboration, Interreg directly contributes to several Sustainable Development Goals (SDGs) [9]. In particular, the specific objectives promoted by the call under analysis—namely digital innovation and environmental sustainability—are directly linked to SDG 9 (Industry, Innovation and Infrastructure) and SDG 13 (Climate Action). From this perspective, the Alcotra microprojects initiative can be seen as an experimental mechanism aligned with the sustainability agenda, aiming to strengthen local capacities, engage peripheral actors, and promote more inclusive and equitable cooperation practices.

Microprojects, therefore, represent an innovative and experimental tool aimed at promoting cross-border cooperation for sustainable regional development. Despite their potential, there is limited understanding of the mechanisms through which microprojects facilitate local engagement and address territorial disparities. The literature on cross-border cooperation tends to focus on large-scale initiatives, overlooking the role and impact of small-scale projects in bridging socio-economic gaps in marginal areas. This research addresses a critical gap in the literature by investigating how microprojects can function as entry points into transnational cooperation networks for underrepresented actors—such as small municipalities, SMEs, and grassroots organizations. Specifically, it asks: to what extent do Alcotra microprojects succeed in reducing territorial inequalities and promoting greater inclusion and cooperation across marginal areas?

This paper aims to delve into the motivations and organizational dynamics underlying microprojects, both from the perspective of policymakers and that of entities participating in funded activities, using the first Alcotra pilot call as a case study.

Through a combination of document analysis, comparative evaluation with other EU funding mechanisms, and stakeholder informal discussions, the study aims to provide a comprehensive understanding of how these initiatives function [10]. By examining their simplified procedures and inclusive design, the paper explores the potential of microprojects to address territorial challenges and foster collaboration across diverse sectors and regions.

This preliminary analysis establishes a foundation for future research on the broader effects and impacts of microprojects, contributing to the ongoing discourse on how tailored funding mechanisms can mitigate territorial inequalities and support sustainable development in peripheral regions, offering valuable insights for both scholars and practitioners. The paper contributes to the Rii Forum's discussion on how regulatory and policy frameworks can foster inclusive growth and well-being, offering insights into the potential of small-scale funding instruments to democratize access to cross-border cooperation. In the following sections, the relevant literature on cross-border cooperation and integration is reviewed. This is followed by a presentation of the methodology, the case study and its analysis. Finally, the discussion and conclusions examine the results in light of the literature and the identified research gap.

2 Cross-Border Cooperation in the Interreg Programme

Cross-border cooperation (CBC) is a key instrument in European territorial development policies. It refers to the set of interactions between public and private actors located in border regions belonging to different states [11], with the aim of improving connectivity, reducing economic and social disparities and promoting harmonised territorial growth [12].

However, CBC is often limited to collaboration on individual projects, without real structural integration between the territories involved. This is where the concept of Cross-Border Integration (CBI), which represents a more advanced stage of cooperation, comes into play. CBI is not limited to the sharing of resources or the resolution of common problems, but aims at the creation of an integrated economic and social space through the convergence of policies, infrastructures and local governance [13].

Three fundamental dimensions of CBC can serve as a foundation for an evolution towards true CBI [14]. The first dimension is multi-level governance, an approach that implies the simultaneous involvement of supranational, national and local institutions in shared decision-making processes. This model fosters greater administrative cohesion between border regions, reducing the risk of policy fragmentation.

The second dimension concerns economic and infrastructural integration, which aims to facilitate trade flows, promote investment and improve connectivity between adjacent territories. This process often results in the creation of economic corridors, the sharing of strategic infrastructure and cooperation for the sustainable development of the regions involved.

Finally, a crucial aspect is social and cultural cohesion, which aims to break down linguistic, administrative and cultural barriers between cross-border populations. Fostering mutual understanding and a sense of belonging to a common space is essential for strengthening long-term relations and consolidating the socio-economic fabric of border areas.

Regions with a high level of CBI tend to benefit from accelerated economic growth and convergence with national urban centres [15]. However, achieving CBI requires strong political alignment and a long-term commitment to building shared infrastructure, common development strategies and integrated governance mechanisms.

One of the main instruments through which the EU has promoted CBC is the Interreg programme, launched in 1989. European funds dedicated to territorial cooperation have surely played an important role in the exponential growth of cross-border processes, stimulating the creation of networks and partnerships between public and private entities [16]. It has also contributed to the decentralisation of governance, enabling regional and local actors to actively participate in supranational policies and influence the development of their areas [17].

However, although the CBC facilitated by Interreg has generated many benefits, the transition to full CBI remains complex. One structural limitation is the predominance of project-based financing, which, while offering flexibility to participants, may favour large companies at the expense of small and medium enterprises (SMEs) and less structured local entities [10, 18]. At the same time, these funding mechanisms allow cooperation networks a certain freedom in defining their strategies, while remaining aligned to the overall programme objectives [19]. The effectiveness of CBC largely depends on the composition and nature of partnerships. Differences in the motivations and capacities of the actors involved can significantly affect project results [20]. One of the main barriers to participation is the lack of adequate tools for entering cross-border networks, a problem particularly felt by local authorities and SMEs [21].

However, cross-border cooperation also represents a great opportunity for the organisations involved, as it enables them to share knowledge and resources that would be difficult to access individually [22]. This is particularly relevant for SMEs, which face even greater difficulties than large companies in participating in Interreg projects and in internationalising their activities [8].

To overcome these difficulties, the programme has progressively introduced procedural simplifications and more flexible access rules in order to facilitate the participation of less experienced actors in the management of European projects [23]. In addition, Interreg has responded to the needs of SMEs with targeted interventions, such as support for international market penetration, with particularly positive results in rural areas, where the need for support tools is greatest [24].

Despite this progress, some structural barriers to the transition from CBC to CBI remain, including the lack of financial resources, the fragmentation of development strategies, and the difficulty of implementing truly shared governance between the various institutional levels involved [25]. The first step toward overcoming these barriers could be to increase local participation in cross-border cooperation, by broadening the participatory base of EU-funded programs. Greater involvement of smaller entities may enhance the penetration of social and cultural cohesion within peripheral areas, bringing marginal territories closer to the benefits of cross-border integration. Indeed, while extensive research has addressed the role of large-scale Interreg initiatives in fostering regional development and cross-border integration, there is a significant gap in the literature regarding the role of small-scale, experimental tools such as microprojects. Most studies focus on established actors and formalized partnerships, often overlooking the challenges and opportunities involved in including new, underrepresented entities in cross-border frameworks. This paper contributes to filling this gap by providing an empirical analysis of the Alcotra microprojects initiative, assessing its capacity to lower entry barriers for peripheral actors and to promote more inclusive forms of territorial cooperation. By analyzing how simplified governance mechanisms and targeted support measures can alter participation dynamics, the case study offers new insights into how CBC programs can evolve toward more equitable and effective configurations.

3 Methods and Materials

In order to understand which mechanisms initiated and managed the Interreg programme's attempts to approach citizens and smaller local entities, it was decided to analyse the first Interreg Alcotra Microproject call for the 2021–2027 programming period. The case study is a useful tool for understanding phenomena related to cross-border cooperation policies [26]. The selected programme was singled out for the experimental nature of its proposal, which, although it is not the only example of approaching citizens through simplification of procedures and budget limitation, is an innovative tool in that it provides sufficient funding for the implementation of complex activities and at the same time includes objectives in line with those proposed by the programme as a whole. Other examples of Interreg microprojects are in fact developed with smaller budgets (around a €30,000 ceiling) and with objectives devoted only to cultural cooperation or exchange between organisations and institutions. The transversal nature of the call therefore makes the research interesting above all for its general implications on the extent and accessibility of cooperation to entities different and outside the usual networks promoted by traditionally sized projects. The research was therefore based on the collection of data from the review of the call and programme documents [27], participation in three project preparation and kick-off meetings, informal discussions in person, via telephone or email with seven organisations that submitted projects funded by the call, and analysis of results from data published on the programme website and official databases [28, 29]. The results seek to contribute to understanding how procedural innovations and simplified governance tools may affect participation dynamics and territorial inclusion within EU-funded programmes.

4 Findings

The first Interreg Alcotra 2021–2027 Microprojects call stems from the Joint Secretariat's appropriation of the European Commission's directive to structure territorial cooperation calls based on the needs and strategies of the target territories. This requirement has led to the creation of funding opportunities for projects with a limited financial scale, simplified application procedures, and more flexible reporting mechanisms. These modifications were introduced to enhance inclusivity within the program, encouraging the participation of private actors with smaller organizational structures compared to the entities traditionally involved in well-established transnational networks. The objective of broadening the base of beneficiaries participating in the funded cooperation between Italy and France is to increase the effectiveness of interventions in strategic areas that have historically been less impacted by the program [30].

The design and rationale behind this call are directly reflected in its operational structure. The call, issued in spring 2023, introduces four key innovations compared to conventional Interreg single-project calls.

First, the eligible budget for each project ranges between €25,000 and €75,000, with a total allocated budget of €1,160,000 and a co-financing rate of 80% of the eligible costs. These amounts reflect the program's intent to facilitate access for entities with limited financial and managerial capacity while still enabling the implementation of moderately complex activities involving multiple stakeholders with diverse objectives and approaches. In this context, for the first time, the program offers beneficiaries cash flow management support, providing an advance payment of 30% of the total grant—a significant departure from the traditional procedure, which limits pre-financing to 10% of the declared budget. This mechanism has been explicitly designed to support smaller entities, which often struggle to cover the initial costs of project implementation.

Second, given the cross-border nature of the cooperation, eligible beneficiaries must reflect the collaborative needs of the target territories. Partnerships must consist of a minimum of two and a maximum of four legal entities, with at least one from each of the two participating countries. However, this specific call introduces an evaluation criterion that prioritizes private entities (particularly SMEs) and those organizations that have never previously participated in Interreg Alcotra-funded projects. This explicit provision aims to expand the pool of participating entities in line with the objectives outlined above and represents a unique feature within the program. The measure appears to pursue a dual strategy: on the one hand, it encourages new entities to form partnerships and integrate into existing networks; on the other, it incentivizes traditional stakeholders, who benefit from institutionalized collaboration networks and accumulated experience in project proposal submission, to open their networks to new actors, fostering greater integration within their operational and social value chains.

Third, in addition to optimizing budget allocation and encouraging the participation of SMEs and new actors, the Secretariat has introduced a simplified proposal preparation process in the call, structured with a single work package and less demanding application requirements. Moreover, the specific objectives identified for the microprojects are limited to the digital sector, with the promotion of digitalization for citizens, businesses, and public administrations, and the environmental sector, with the promotion of renewable energy, climate change adaptation, and biodiversity protection. Finally, in line with the goal of incentivizing the participation of new and smaller entities, the Secretariat has committed to offering extensive guidance, support, and assistance throughout the proposal preparation process. This includes helping applicants identify potential partners, facilitating project activation, and providing training on project management and financial reporting requirements. Through a series of meetings and the direct availability of program staff, the Secretariat has sought to accompany interested stakeholders throughout the funding process, encouraging the broadest possible participation from the earliest stages. This initiative aims to bridge the skills gap that typically separates local actors from engagement in international cooperation.

From this analysis, it becomes evident that the program seeks to align with the European Commission's expectations by reducing the gap between Interreg and peripheral territories, as well as local enterprises and institutions. In doing so, it is trying to enhance inclusivity, streamlines participation processes, and maximizes the potential spillover effects of transnational collaboration.

However, the effectiveness in implementing these attempts is evaluated based on the projects actually funded by the call. The analysis of the approved micro-projects (Table 1) provides useful elements to understand the concrete performance of the measure in achieving the stated goals in terms of inclusivity and broadening the base of participation.

Overall, 17 projects were approved for funding, of which 7 were in the Digital Specific Objective and 10 in the Environmental Specific Objective.

The examination of the legal nature of the entities involved shows a heterogeneous distribution, with 23.7% being public bodies, 31.6% private entities (all of

Table 1 Profile of funded project partners

ID partner	Specific objective	Legal entity	Track record	NUTs
1	Digital	Private	No	Valle d'Aosta
2	Digital	Private	No	Auvergne-Rhone-Alpes
3	Digital	NPO	No	Provence-Alpes-Cote d'Azur
4	Digital	Public	Yes	Piemonte
5	Digital	NPO	No	Piemonte
6	Digital	NPO	No	Piemonte
7	Digital	NPO	No	Provence-Alpes-Cote d'Azur
8	Digital	Public	Yes	Piemonte
9	Digital	Private	No	Piemonte
10	Digital	Public	No	Provence-Alpes-Cote d'Azur
11	Digital	Public	Yes	Provence-Alpes-Cote d'Azur
12	Digital	NPO	No	Piemonte
13	Digital	Private	No	Provence-Alpes-Cote d'Azur
14	Digital	Private	No	Provence-Alpes-Cote d'Azur
15	Digital	Public	No	Auvergne-Rhone-Alpes
16	Digital	Public	Yes	Valle d'Aosta
17	Digital	NPO	No	Valle d'Aosta
18	Environment	NPO	No	Auvergne-Rhone-Alpes
19	Environment	Public	No	Piemonte
20	Environment	NPO	No	Provence-Alpes-Cote d'Azur
21	Environment	Private	No	Piemonte
22	Environment	Public	Yes	Piemonte
23	Environment	Private	Yes	Piemonte
24	Environment	NPO	Yes	Auvergne-Rhone-Alpes
25	Environment	Public	Yes	Liguria
26	Environment	NPO	No	Provence-Alpes-Cote d'Azur
27	Environment	Private	No	Piemonte
28	Environment	NPO	No	Provence-Alpes-Cote d'Azur
29	Environment	NPO	No	Piemonte
30	Environment	NPO	No	Auvergne-Rhone-Alpes
31	Environment	NPO	No	Piemonte
32	Environment	NPO	No	Provence-Alpes-Cote d'Azur
33	Environment	NPO	No	Piemonte
34	Environment	NPO	No	Provence-Alpes-Cote d'Azur
35	Environment	Private	No	Valle d'Aosta
36	Environment	Private	No	Valle d'Aosta

(continued)

Table 1 (continued)

ID partner	Specific objective	Legal entity	Track record	NUTs
37	Environment	Private	No	Auvergne-Rhone-Alpes
38	Environment	Private	No	Provence-Alpes-Cote d'Azur

Author's elaboration

which are SMEs, as the call did not allow large enterprises) and 44.7% non-profit organizations (NPOs) (Fig. 1). From a geographical point of view, considering the eligibility of the program's territories within the Italian and French cross-border area, the beneficiary entities have a multiple origin at the regional level (NUTs2). However, there is a predominance of actors located in Piedmont and Provence, with an intermediate presence of entities from Aosta Valley and Auvergne, and a minimal representation from Liguria (Fig. 2). This distribution reflects, at least in part, the consolidated participation of certain territories in Interreg projects, as well as the need to strengthen the involvement of some areas historically less active in the program [6]. A significant fact also concerns the previous experience of the entities involved in the funded projects. According to data from the Keep.eu database [29, 31], an analysis of participation records in Alcotra from 2003 to today reveals that approximately 79% of the entities had never taken part in Interreg Alcotra calls before, while only 21% had prior experience (Fig. 3). This data confirms that the Microprojects call has effectively contributed to expanding the audience of beneficiaries, favouring the entry of new actors in the cross-border cooperation landscape. In fact, out of 17 funded projects, 11 projects (64.7%) were composed entirely of entities with no prior participation in Interreg Alcotra, while 6 projects (35.3%) included at least one partner with previous experience. While the majority of projects were led by newcomers, the presence of experienced entities in a smaller but still relevant proportion of cases suggests that they may have facilitated the integration of new actors. This dynamic, although not predominant, indicates that collaboration and assistance networks between consolidated and new partners could still play a role in strengthening cross-border cooperation.

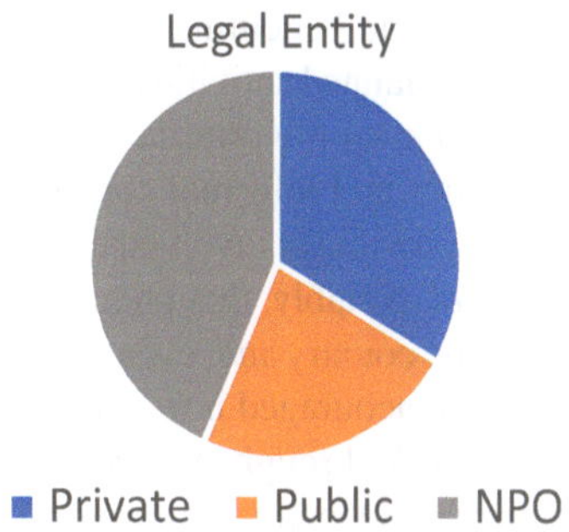

Fig. 1 Distribution by legal entity of partners. Authors' elaboration

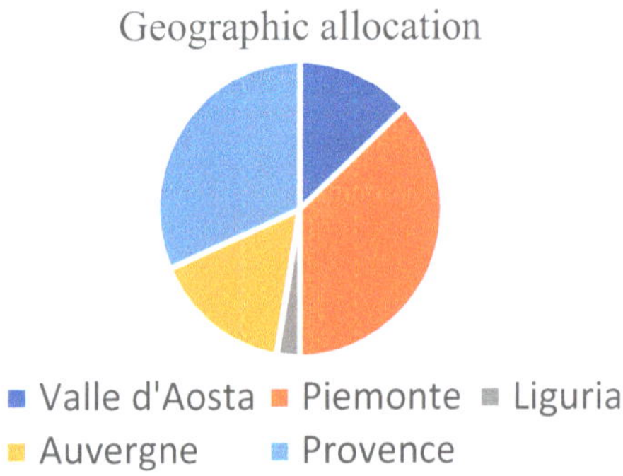

Fig. 2 Geographic allocation of partners. Authors' elaboration

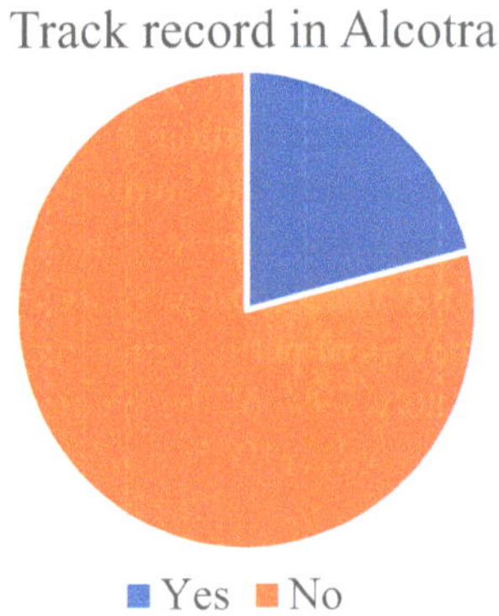

Fig. 3 Previous participation to the Interreg Alcotra program of partners. Authors' elaboration

In addition to the examination of the call's data, the opinions collected from various stakeholders involved in the funded projects offer an overview of the dynamics of participation and the challenges encountered during the application and implementation process.

Firstly, the analysis of the experiences reveals two distinct approaches in the construction of the submitted projects. Some participants have exploited the call as an opportunity to formalize projects already underway, consolidating existing collaborations and benefiting from the resources made available to expand or strengthen already planned activities. In other cases, instead, the projects were specifically structured to meet the criteria of the call, with partnerships formed through networking dynamics and informal contacts. This process often followed a snowballing effect, where previously involved entities played a key role in identifying and engaging new partners. Notably, this phenomenon was particularly prevalent among entities from the same country and, even more frequently, within the same region. As a result, while the call encouraged new actors to participate, many of them entered the cooperation framework through pre-existing local networks rather than through transnational outreach. Secondly, opinions on the ease of access to the call are conflicting. On the one hand, the support provided by the accompanying structures has been a facilitating element for many participants, especially for those with less experience in European project design. The role of information tools and guidance activities has been positively evaluated, as it has helped to clarify procedural aspects and simplify the application process. On the other hand, some critical issues remain related to administrative documentation and reporting mechanisms, which have been perceived as elements of complexity, especially for less structured entities. In some experiences, the requests for verification of financial capacity and budget management procedures have been considered challenging, despite the simplification compared to traditional calls.

These elements confirm that the micro-projects model represents a concrete opportunity to expand cross-border cooperation, but also highlight some operational challenges that could be addressed in future editions of the call. Possible improvement actions include further simplification of administrative procedures, a strengthening of animation activities in less involved areas, and greater technical assistance in the funding management phase.

This section of the analysis contributes to understanding the experimental value of the Microprojects call, not only as a funding instrument but also as an innovation laboratory for territorial cooperation, with potential implications for the evolution of inclusion policies in future Interreg programs.

5 Discussion

The insights of the first call for microprojects of the Interreg Alcotra programme highlighted how this instrument can be a significant lever for broadening the beneficiary base of cross-border cooperation and promoting greater territorial inclusiveness. The theoretical framework discussed showcased how cross-border cooperation policies are often characterised by privileged access for large organisations, leaving SMEs and less structured local realities on the margins. Microprojects are therefore an innovative response to this critical issue, facilitating the entry of new actors into the cooperation landscape and encouraging the creation of more diversified networks.

The implementation of the Alcotra call has confirmed the potential of this instrument, demonstrating that the reduction of administrative and financial barriers can encourage the participation of actors traditionally excluded from Interreg calls. The study of the projects funded revealed a heterogeneity in the types of beneficiaries involved, with a strong representation of entities without previous experience in Interreg projects. This finding supports the hypothesis that microprojects can help break down barriers to entry and make cross-border cooperation more accessible. However, some geographical limitations and operational challenges persist. The geographical distribution of beneficiaries within eligible regions remains concentrated in a few more developed territories, leaving peripheral realities behind. Furthermore, although the simplification of procedures has facilitated access to funds, reporting processes and financial audits continue to be an obstacle for less structured organisations. Although the partnership model required has favoured the involvement of new actors, interaction with more experienced partners has proved to be, still, an important factor in the success of applications. This suggests that, while reducing administrative barriers, the full inclusion of new actors in cross-border networks requires ongoing accompaniment and targeted capacity building tools.

6 Conclusions

The experience of Alcotra demonstrates that the inclusion of local actors and the promotion of innovative partnerships are not only desirable but achievable goals, provided that appropriate support tools, streamlined procedures, and continuous monitoring are put in place to fine-tune implementation strategies.

From a broader perspective, the Interreg programme emerges not only as a mechanism for territorial integration but also as a strategic pillar of the EU's commitment to sustainable development. The specific thematic focus of the call (digital innovation and environmental sustainability) aligns directly with Sustainable Development Goals 9 (Industry, Innovation and Infrastructure) and 13 (Climate Action), as well as with Goal 10 (Reduced Inequalities) and Goal 17 (Partnerships for the Goals). Against this backdrop, the Alcotra microprojects initiative can be seen as an experimental policy tool that translates high-level sustainability objectives into locally grounded cooperation practices.

In terms of academic contribution, this study helps fill a gap in the literature on cross-border cooperation by shedding light on small-scale funding instruments and their potential to democratize access to transnational governance. Whereas much of the existing research focuses on large-scale projects and established institutional actors, the findings presented here suggest that micro-level interventions can play a transformative role in reshaping participation dynamics and fostering social and territorial inclusion. Future research could usefully explore the long-term impact of such initiatives, particularly in terms of whether participation in microprojects facilitates sustained involvement in larger cooperation frameworks. Comparative studies across different Interreg contexts would also help refine our understanding of how design features and support structures influence the inclusiveness and effectiveness of CBC instruments.

References

1. Opilowska, E.: Euroregion. Od partnerstwa do sieci współpracy transgranicznej [Euroregion. From Partnership to a Cross-Border Cooperation Network]. Elipsa, Warsaw (2019) (review). Pogranicze. Polish Border. Stud. **8**, 39–43 (2020). https://doi.org/10.25167/ppbs1900
2. van den Broek, J., Benneworth, P., Rutten, R.: Border blocking effects in collaborative firm innovation. Eur. Plan. Stud. **26**(7), 1330–1346 (2018). https://doi.org/10.1080/09654313.2018.1476470

3. Glückler, J.: Economic geography and the evolution of networks. J. Econ. Geogr. **7**(5), 619–634 (2007). https://doi.org/10.1093/jeg/lbm023
4. Chamusca, P.: Public policies for territorial cohesion and sustainability in Europe: an overview. Sustainability **15**, 6890 (2023). https://doi.org/10.3390/su15086890
5. Nienaber, B., Wille, C.: Cross-border cooperation in Europe: a relational perspective. Eur. Plan. Stud. **28**(1), 1–7 (2019). https://doi.org/10.1080/09654313.2019.1623971
6. Martín-Uceda, J., Vicente Rufí, J.: Territorial development and cross-border cooperation: a review of the consequences of European INTERREG policies on the Spanish-French border (2007–2020). Sustainability **13**, 12017 (2021). https://doi.org/10.3390/su132112017
7. Esparza Masana, R., Montemayor Cejas, P.: Repeating partners and partnerships in European Territorial Cooperation: a quantitative analysis. Reg. Stud. Reg. Sci. **10**(1), 369–393 (2023). https://doi.org/10.1080/21681376.2023.2186803
8. van den Broek, J., Rutten, R., Benneworth, P.: Innovation and SMEs in Interreg policy: too early to move beyond bike lanes? Policy Stud. **41**(1), 1–22 (2018). https://doi.org/10.1080/01442872.2018.1539225
9. United Nations: https://sdgs.un.org/goals. Accessed Apr 2025
10. Shepherd, J., Ioannides, D.: Useful funds, disappointing framework: tourism stakeholder experiences of INTERREG. Scand. J. Hosp. Tour. **20**(5), 485–502 (2020). https://doi.org/10.1080/15022250.2020.1792339
11. Medeiros, E.: Territorial impact assessment and cross-border cooperation. Reg. Stud. Reg. Sci. **2**(1), 95–115 (2015)
12. Perkmann, M.: Cross-border regions in Europe: significance and drivers of regional cross-border co-operation. Eur. Urban Reg. Stud. **10**(2), 153–171 (2003). https://doi.org/10.1177/0969776403010002004
13. Wong Villanueva, J.L., Kidokoro, T., Seta, F.: Cross-border integration, cooperation and governance: a systems approach for evaluating "good" governance in cross-border regions. J. Border. Stud. **37**(5), 1047–1070 (2020). https://doi.org/10.1080/08865655.2020.1855227
14. Scott, J.W.: Transboundary cooperation on Germany's borders: strategic regionalism through multilevel governance. J. Border. Stud. **15**, 143–167 (2000). https://doi.org/10.1080/08865655.2000.9695545
15. Niebuhr, A.: Market access and regional disparities. Ann. Reg. Sci. **40**, 313–334 (2006). https://doi.org/10.1007/s00168-005-0047-1
16. Medeiros, E.: Cross-border cooperation in Inner Scandinavia: a territorial impact assessment. Environ. Impact Assess. Rev. **62**, 147–157 (2017). https://doi.org/10.1016/j.eiar.2016.09.003
17. Harguindéguy, J.-B., Bray, Z.: Does cross-border cooperation empower European regions? The case of Interreg III-A France–Spain. Eviron. Plann. C Gov. Policy **27**(4), 747–760 (2009). https://doi.org/10.1068/c08119
18. Nigohosyan, D., Vutsova, A., Vassileva, I.: Effectiveness and efficiency of the EU-supported energy efficiency measures for SMEs in Bulgaria in the period 2014–2020: programme design implications. Energy Effic. **14**, 24 (2021)
19. Tulppo, P., Aarrevaara, T., Harisalo, R.: 4: Interdependence, knowledge and cooperation in cross-border cooperation in the north. In: Future Public Governance and Global Crises. Edward Elgar Publishing, Cheltenham (2025). https://doi.org/10.4337/9781035318674.00011
20. Fricke, C.: Spatial governance across borders revisited: organizational forms and spatial planning in metropolitan cross-border regions. Eur. Plan. Stud. **23**(5), 849–870 (2014). https://doi.org/10.1080/09654313.2014.887661
21. Kurowska-Pysz, J., Szczepańska-Woszczyna, K.: The analysis of the determinants of sustainable cross-border cooperation and recommendations on its harmonization. Sustainability **9**, 2226 (2017). https://doi.org/10.3390/su9122226
22. Schermerhorn, J.R.: Determinants of interorganizational cooperation. Acad. Manag. J. **18**, 846–856 (1975)
23. Delecosse, É.: Cohesion policy: simplification put to the test by the complexity of Interreg. Reflets Perspect. Vie Econ. **59**(1), 65–78 (2020). https://doi.org/10.3917/rpve.591.0065

24. Pámer, Z., Buttinger, P., Márton, G.: The evolution of SME development in the EU's cross-border cooperation: the example of direct funding in the Hungary-Croatia Interreg programme. DETUROPE Cent. Eur. J. Reg. Dev. Tour. **16**(3), 82–98 (2024). https://doi.org/10.32725/det.2024.012
25. Medve-Bálint, G.: Incentives and obstacles to cross-border cooperation in post-communist central Europe. In: Bellini, N., Hilpert, U. (eds.) Europe's Changing Geography, pp. 145–170 (2013)
26. Castanho, R.A., Loures, L., Cabezas, J., Fernández-Pozo, L.: Cross-border cooperation (CBC) in southern Europe—an Iberian case study. The Eurocity Elvas-Badajoz. Sustainability **9**(3), 360 (2017). https://doi.org/10.3390/su9030360
27. Interreg Alcotra: https://www.interreg-alcotra.eu/it/bando-la-selezione-di-microprogetti. Accessed Feb 2025
28. Interreg Alcotra: https://www.interreg-alcotra.eu/it/i-progetti-finanziati-2021-2027. Accessed Feb 2025
29. Keep.eu: https://keep.eu. Accessed Feb 2025
30. Interreg Alcotra: Bando—La selezione di microprogetti. https://www.interreg-alcotra.eu/it/bando-la-selezione-di-microprogetti. Accessed 20 Feb 2025
31. Chilla, T., Lambracht, M.: Institutional mapping of cross-border cooperation. INTERREG programme analyses with KEEP data. Eur. Plan. Stud. **31**(4), 700–718 (2022). https://doi.org/10.1080/09654313.2022.2058321

Blockchain-Based Tool for the Management of Renewable Energy Generation and Consumption in El Salvador

Erick Varela-Guzmán, Carlos Juárez-Ramos, Mauricio Pohl-Alfaro, and Néstor Aldana-Rodríguez

Abstract The government of El Salvador recognizes the need to develop an energy system that is efficient, resilient, sustainable, and environmentally responsible. A major goal has been the diversification of the national energy matrix to maximize the use of available renewable energy sources, but there are some challenges such as the lack of policies for regulation and a structured framework for energy data registration. This work explores alternatives for the second issue through the development of a tool based on Blockchain technology that enables decentralized and transparent management of renewable energy generation and consumption data. A pilot test was conducted in a controlled environment using grid-connected photovoltaic systems on a university campus where smart contracts were used to ensure the integrity of the records as well as autonomy in the execution of conditions previously established. The findings from this controlled environment suggest the technical feasibility of the approach and provide a conceptual foundation for further exploration of blockchain applications in energy data governance in El Salvador.

Keywords Blockchain · Renewable energy · Sustainable development · Participative governance · El Salvador · SDG7

1 Introduction

In current literature, there is an ongoing debate on how some of the effects of climate change are becoming increasingly evident, such as temperature increase, sea level rise and the intensification of natural disasters [1, 2].

E. Varela-Guzmán (✉)
Kriete Institute of Engineering and Science, La Libertad, El Salvador
e-mail: erick.varela@keyinstitute.edu.sv

C. Juárez-Ramos · M. Pohl-Alfaro · N. Aldana-Rodríguez
Universidad Centroamericana José Simeón Cañas, La Libertad, El Salvador

A. Visvizi et al. (eds.), *Research and Innovation Forum 2025 Vol2*, Springer Proceedings in Complexity, https://doi.org/10.1007/978-3-032-09062-1_27

The demographic increase and the general advancement of society has led to a higher energy demand worldwide. To satisfy this demand, non-renewable electric energy generation methods are still used [3]. However, the literature discusses how dependence on these sources is not sustainable, since they promote an increase in greenhouse gas emissions, which generates serious environmental deterioration [1, 3]. There is also evidence on how the use of non-renewable energy can impact the economy of some countries [4–6]. In the case of countries such as El Salvador, located in Central America, the use of fossil fuels depends exclusively on imports from producing countries, so situations like natural disasters, conflicts or difficulties faced by these nations have an impact on the Salvadoran economy caused by pronounced upward variations in fuel prices.

To move towards a sustainable model of society, governments, industry, and the international community at large are promoting the development of strategies and policies to address the challenge of growing global energy demand efficiently and sustainably [3]. This work seeks to advance the digitization of the energy sector as an approach that leverages technology to enhance the efficiency of electricity generation, distribution, transmission, operation, and consumption [7].

The Salvadoran authorities recognize the importance of innovating in the electricity sector to promote the economic and social development of the country. A key objective established in the Salvadoran agenda has been the diversification of the national energy matrix to optimize the use of available renewable energy sources. Given the lack of tools that facilitate the recording of renewable energy generation and consumption data in El Salvador, and the growing participation of small-scale renewable energy producers in the energy matrix, this study investigates the following research question: How can a blockchain-based architecture be designed to enable transparent and decentralized registration of renewable energy data, while aligning with local needs and participatory governance principles in El Salvador?

To frame this work, we adopt a socio-technical perspective grounded in digital governance and participatory models, where decentralization, data sovereignty, and inclusive participation are considered pillars of sustainable and resilient energy systems [8, 9]. The proposed blockchain-based tool is conceived as an alternative that can shift responsibility for data management from a centralized model to a network of producers, distributors, and observers, a perspective aligned with the principles of energy democratization [10, 11].

In response to this inquiry, we designed, implemented, and evaluated a blockchain-based platform in a controlled university pilot environment, drawing on the insights and challenges encountered during the development process and incorporating complementary technologies such as IoT and web-based visualization.

2 Background and Technological Landscape

The following subsections review key contextual elements for this study, including the electricity management model in El Salvador and advances in the exploration of Blockchain applications in the energy sector.

2.1 The Electricity Generation Model in El Salvador

The electricity sector in El Salvador has operated under a conventional model that is regulated by several institutions: the General Superintendency of Electricity and Telecommunications (SIGET), which oversees tariffs and stakeholder interactions to ensure service quality; the National Energy Council (CNE), responsible for energy policies; and the Transactions Unit (UT), which manages the wholesale electricity market. Distribution companies are private institutions that operate under state regulations, acting as intermediaries between producers and consumers [12].

By 2022, energy produced from renewable sources accounted for 77.40% (see Table 1), with the main sources being hydropower (31.3%), geothermal energy (20.5%), and photovoltaic energy (15.7%) [12]. However, external factors such as climatic events can significantly reduce annual energy production. In 2023, for example, there was a reduction in hydropower production, a situation that was attributed to the El Niño phenomenon, a climatic event that occurred between June and December 2023 [13]. In such cases, the demand has traditionally been met through electricity generation from non-renewable sources such as liquefied natural gas and other fossil fuels. As a result, their share of use increased from 22.60% in 2022 to 39.10% in 2023 [12]. This situation has occurred in previous years with different climatic events and dry seasons, which indicates that the current power generation model in El Salvador has difficulties to adapt to unfavorable climatic situations.

Building on this goal, the National Energy Council (CNE) is working on the implementation of the National Energy Policy 2020–2050, which aims to develop a sustainable and competitive production model that emphasizes the diversification of renewable energy sources to meet the growing energy demand and protect natural resources [14].

In this regard, the authorities have promoted a strategy focused on the development of a distributed energy generation model. Within this model, a new key figure has been introduced: renewable self-producers [15]. Currently, these small producers do not interact directly with the wholesale market, nor are they regulated by UT. However, the legislation allows renewable self-producers to sell any surplus they do not consume directly to distribution companies within the network. With the implementation of this model, important progress has been made, for example, the participation of renewable sources in the national energy matrix has increased, along

Table 1 Annual electricity generation in El Salvador by energy source (2022–2023) [12]

Energy source	Type	Annual generation				Annual variation (%)
		2022		2023		
		GWh	(%)	GWh	(%)	
Hydraulics	Renewable	2234.13	31.30	1568.30	19.20	– 29.80
Geothermal	Renewable	1468.00	20.50	1482.71	18.20	1.00
Other fossil fuels	Nonrenewable	586.31	8.20	712.46	8.70	21.52
Liquefied natural gas	Nonrenewable	1027.12	14.40	2480.60	30.40	141.51
Biomass	Renewable	552.45	7.70	526.99	6.50	– 4.61
Photovoltaic	Renewable	1119.92	15.70	1185.24	14.50	5.83
Wind	Renewable	134.9	1.90	166.81	2.00	23.66
Biogas	Renewable	22.56	0.30	28.74	0.40	27.36
Total		7145.38	100	8151.84	100	14.09

with energy exports, positioning El Salvador as an important supplier in the Central American market [16, 17].

Despite the advances described above, key structural challenges remain before the full potential of the model can be realized. These challenges include: (i) the lack of policies for the regulation and control of renewable self-producers, and (ii) the absence of a standardized and structured framework for recording data on their operations. Given that data collection and its proper storage is a crucial step in the digitalization process of the electricity sector, this work aims to address the second identified challenge, by designing, developing, and implementing a digital tool that allows the registration and auditing of transactions between retail market participants and electricity distributors.

2.2 Distributed Ledger Technology (DLT) in the Energy Sector

Blockchain has quickly caught the attention of the international community due to the significant development of this technology in recent years [18, 19]. Its technical features have enabled applications across multiple sectors, including industry [19], the public sector [20, 21], and society in general [21–23].

While blockchain shares some features with traditional databases, it operates under a distributed ledger model (DLT), meaning that information is recorded chronologically, in a decentralized way, and without the mandatory need for a governing entity [24, 25]. Its decentralized nature allows any minor discrepancies in the data to be easily and automatically detected. This is achieved by storing in each new block a unique code that is generated from the information in the previous block, allowing the information to be protected against unintentional and malicious alterations [18, 26]. The high level of security, traceability, and reliability in the data provided by Blockchain allows it to be applicable in scenarios where multiple stakeholders with different interests interact, and, therefore, a single, trustworthy source of information is required, eliminating the need for a mediator, and reducing the potential for bad practice [26].

In the energy sector, blockchain is providing innovative solutions that empower citizens and communities with greater control. Cryptocurrencies and tokens are playing an important role. SolarCoin and NRGcoin are tokens that reward the activity of small producers to generate and sell energy [27, 28]. KWHCoin is a citizen-focused proposal, allowing individuals to use this token to purchase energy generated from renewable sources within their communities [29]. The idea of using a P2P network based on Blockchain and smart contracts to facilitate the transparent buying and selling of energy among community members has been explored [30]. In addition, Blockchain has been integrated with IoT to develop a microgrid that enables the decentralized management of energy production and consumption within a community [31–33]. Likewise, Blockchain has been used to develop an initiative based on renewable energy systems under the perspective of the circular economy [34]. Recent studies emphasize the potential of blockchain in decentralized energy governance scenarios, especially in scenarios that require reliable and tamper-proof data transactions [27, 34, 35]. Finally, the role of blockchain and its potential contributions to various aspects of energy management has been explored, including electric mobility, energy democratization, P2P-based energy trading, transparent metering and transaction recording, automation through smart contracts for the issuance of green certificates, and carbon trading [36, 37].

Ongoing research on the integration of Blockchain in the development of transparent and sustainable energy management systems suggests that this technology, characterized by its traceability, transparency, and decentralization, presents significant opportunities for further exploration and implementation. These developments illustrate the capacity of Blockchain to enhance trust, decentralize control, and improve the traceability and automation of energy-related transactions [35].

3 Blockchain-Based Tool for Recording Renewable Energy Generation and Consumption in a Test Microgrid

Considering the potential of blockchain in decentralized energy models and the challenges identified in the energy transition process of El Salvador, this study presents a digital tool designed to facilitate the transparent and traceable registration of renewable energy transactions. The proposed solution leverages blockchain technology and smart contracts to address the lack of appropriate tools for recording generation and consumption data, particularly in scenarios involving the integration of renewable self-producers into the energy distribution network. This scenario gives rise to several key challenges: (i) The lack of tools to facilitate the interaction of renewable self-producers with the wholesale market, which could make it difficult to collect accurate data on consumed energy and generated surpluses; (ii) Difficulty in coordinating between distribution companies and renewable self-producers; (iii) Difficulty in measuring the impact of renewable self-producers on the diversification process of the energy matrix in El Salvador; (iv) Potential data fragmentation, which may create obstacles for informed decision-making regarding incentives, changes in legislation, or the expansion of the distribution network; and ultimately (v) Lack of reliable data, as it is difficult to measure the progress of the objectives of the National Energy Policy 2020–2050 proposed by the government of El Salvador.

The literature reviewed above suggests that blockchain offers features that can contribute to overcoming the difficulties related to the lack of reliable and standardized energy data. To leverage this potential, the proposed tool implements smart contracts to collect key data related to the production of energy from renewable sources. This includes quality parameters, quantity of energy generated, date and time of production, as well as information about the producer and the consumer.

3.1 Proof of Concept

The pilot test was implemented at the Central American University (UCA) in El Salvador, which has existing grid-connected photovoltaic systems on some academic buildings. The selection of this site was based on the availability of renewable energy generation and consumption infrastructure, which allows for controlled experimentation with real-world installations. Access to these photovoltaic systems was facilitated by strong institutional support, which helped advance the development process.

The goal of the pilot test is twofold. First, the pilot aims to simulate the operation of a virtual power plant that integrates both decentralized energy generation and consumption points. Second, we seek to evaluate the technical feasibility and operational dynamics of the developed tool in a real-world but controlled environment. Although the scenario is academic, a node structure was defined for the generation and consumption units, along with the data flow and participant roles, to simulate the possible interactions that would occur in a larger, real-world context. The testbed included several buildings (see Fig. 1) designated as energy generation units (e.g., CEDITEC, Martín Baró, Andrés Escobar, ICAS, and NZEB) and others as consumption hubs (ICAS, CEDITEC, NZEB, and Jon de Cortina).

The first phase of the proof of concept involved installing IoT devices at the generation and consumption units on the university campus. In this phase, data was collected on energy production (kWh), energy consumption (kWh), timestamp, and associated node identifiers. Data transmission was performed at 5-min intervals using LoRa (long-range) connectivity. This communication protocol was chosen due to low Wi-Fi coverage in the areas where the devices were installed.

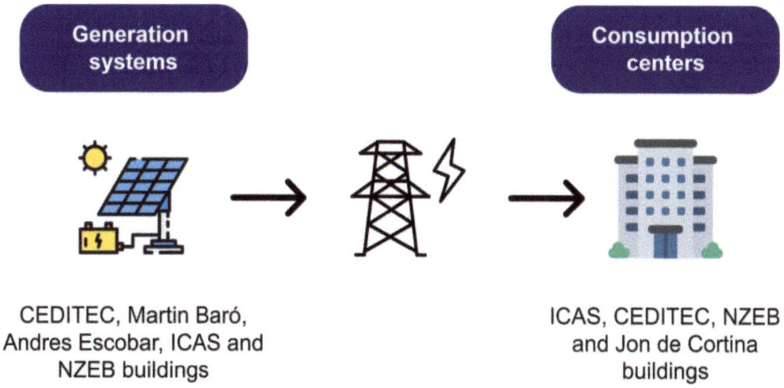

Fig. 1 Controlled testing environment for the blockchain-based energy management tool

Data was routed through an API and stored in a blockchain network via a decentralized application (DApp). The DApp was developed using a private chain perspective compatible with the Ethereum protocol, using the Proof of Authority (PoA) consensus mechanism to balance security and processing efficiency in a controlled environment. Smart contracts written in Solidity and tested with the Truffle framework, were used to validate data integrity and enforce the execution of the pre-established rules. These contracts were designed to retrieve data via an RPC (Remote Procedure Call), and run autonomously, ensuring that no actor can manipulate the data or modify the conditions once deployed.

As a complementary module to the blockchain, a relational database was integrated to store metadata that is not suitable for decentralized storage. This includes error logs, additional variables provided by the meters but not relevant to the current context, and user session information. A web-based interface was developed to provide intuitive, real-time visualization of energy generation and consumption data, enhancing user accessibility and engagement.

The validation procedures for the operation of the tool included periodic checks for data validation through periodic comparisons with physical electricity meter readings; the correct operation of the web service; and the correct storage of the data in the Blockchain network. The pilot was led entirely by the academic research team, with support from the university's technical staff. While no external market players were directly involved, the pilot was designed with scalability and real-world integration in mind. Future testing could include collaboration with energy distributors and regulatory authorities. Figure 2 shows the general architecture used for the development of the tool. The system integrates IoT devices to collect energy generation and consumption data, a blockchain layer for secure and immutable registration of transactions, a smart contract engine for automatic validation and execution, a relational database for storing additional metadata, and a web interface that allows users to visualize and manage energy records.

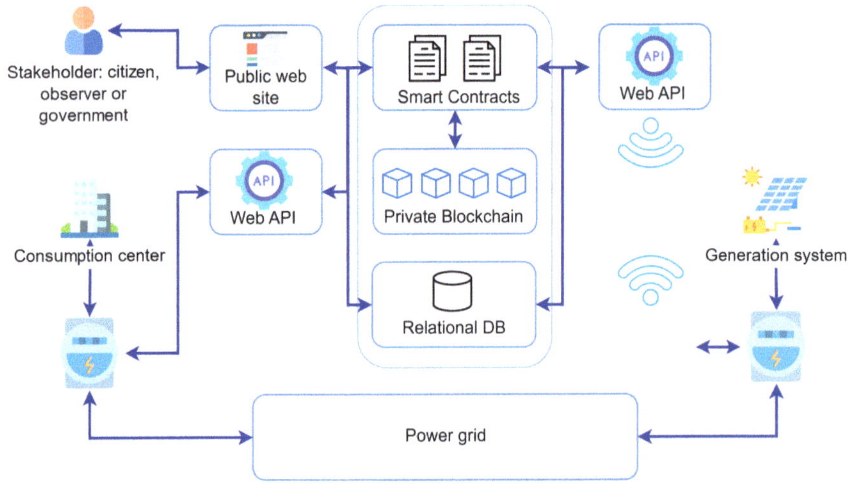

Fig. 2 General architecture of the proposed blockchain-based tool

The implementation of smart contracts allows the tool to operate autonomously according to predefined rules, minimizing the risk of data manipulation, either intentionally or unintentionally. The proposed tool serves as a functional prototype of this distributed model. In future large-scale implementations, the service could be maintained by nodes hosted in public, private and social institutions to ensure data transparency and reliability. This level of participation is possible because the implementation requirements suggest high scalability at a relatively low cost, allowing the integration of different stakeholders. Additionally, it enables the generation of basic reports on energy consumption and production through a user-friendly interface.

Although this pilot implementation was carried out in a controlled environment, the proposed tool was designed to support the interaction of multiple types of stakeholders. In a real deployment scenario, key actors would include renewable self-producers (responsible for registering and verifying their energy contributions), distribution companies (in charge of validating and aggregating energy data), and regulatory observers (who access the records for monitoring compliance and evaluating energy policies). Each stakeholder would interact with the system through specific user profiles, ensuring role-based access and the autonomous execution of conditions via smart contracts. While stakeholder engagement was not tested during this pilot, its inclusion is considered a key goal for future stages of implementation.

4 Discussion and Implications

The results obtained during the development of the tool described in this work, which was implemented as a pilot on a college campus, suggest relevant technical and functional insights. As a result, it is possible to say that the pilot met the initial technical and exploratory objectives established for this phase of the research. However, test results indicate potential bottlenecks under high concurrency conditions. This highlights the need to further evaluate alternative communication and concurrency mechanisms to ensure the scalability of the developed tool.

Although this prototype focuses primarily on demonstrating technical feasibility, it also explores the conceptual foundations of an innovative energy governance model in the context of El Salvador, through mechanisms that allow the transparent, traceable, and decentralized recording of renewable energy generation and consumption data. This approach outlines the potential to shift control of energy data from centralized institutions to a shared governance framework involving producers, distributors, regulatory observers, and potentially citizens themselves.

The implementation of smart contracts offers autonomous operation based on pre-established agreements guarantees transparency and reliability for all stakeholders. This capability may be relevant in the context of shared governance models, such as the one conceptually explored in this study. The approach has the potential to promote stakeholder participation and public trust in energy-related strategies and policies.

Deploying this tool in a real-world scenario requires careful consideration about the risks of relying on private blockchain hosted solely by institutions directly involved in the business logic. Such a setup may compromise the ideal of decentralization and transparency that underpins this kind of initiative. As an alternative, a hybrid blockchain architecture could be considered, allowing the participation of public, private, and social institutions. Under this scenario, where transparency and good practices are promoted, confidence in the model on the part of the general population could improve, and as a result, interest in participating could also increase [21, 23].

However, the successful deployment of such emerging technologies will likely require regulatory updates, capacity building among stakeholders, and effective public communication strategies. Without these efforts, the benefits of blockchain for energy democratization may remain limited to conceptual models and small-scale proofs of concept.

In theoretical terms, this work contributes to the discussion related to the adoption of emerging technologies in the context of decentralized energy management, with a particular focus on the case of El Salvador. The implementation of Blockchain as a medium for recording renewable energy generation and consumption data contributes to the understanding of decentralized energy governance and operational transparency frameworks. This work offers preliminary evidence of how a digital solution that promotes a decentralized work model can enhance the design of collaborative frameworks for energy data management, an area that remains underexplored in current literature.

From a practical perspective, the results offer valuable information for different stakeholders. First, tools like the one proposed in this study can support the digital transformation of current processes and the design of informed policies. In this regard, policymakers could develop regulatory frameworks that integrate emerging technologies to promote best practices such as transparency, real-time auditing, traceability, and efficiency in resource management. Second, related industries may explore the adoption of digital tools to reduce operational costs, promote best practices and enhance consumer trust in renewable energy systems and support more informed consumption decisions. Third, funding agencies may consider the approach explored in this work to support resource management solutions in contexts characterized by limited infrastructure but significant renewable energy potential.

Finally, although the tool offers an innovative alternative aligned with the principles of sustainable energy production, its real-world implementation would require careful consideration of the costs associated with adapting existing infrastructure. Challenges related to blockchain adoption—such as a shortage of specialists, limited public awareness, and the complexity of understanding its advantages—remain important barriers that must be considered. In addition, issues related to scalability, legal compatibility, and regulatory gaps must be addressed to support long-term success [26]. Overall, the findings suggest the technical and functional feasibility of the proposed solution. However, it is important to note that the pilot test was conducted in a simulated environment. Future stages of implementation are planned to involve directly involved stakeholders, such as renewable self-producers and distribution companies. The findings provide a basis for further exploration of decentralized digital tools for energy governance, especially in emerging contexts, such as El Salvador, where infrastructure and regulatory frameworks are still evolving.

5 Conclusions and Research Outlook

The development of practices based on a resilient and sustainable energy model requires strategies that allow the integration of alternative sources into national energy systems. The tool presented in this work proposes an alternative digital approach which contributes to current efforts in El Salvador to adopt a more efficient and sustainable energy model. The implementation of blockchain in this context enables more transparent and traceable data management, which could help address challenges related to data quality and coordination among various stakeholders, factors that are essential for the efficient and sustainable management of growing energy demand.

The test scenario provided preliminary evidence supporting the technical and functional viability of the proposed technology. The hybrid design used in the development of the tool, combining a Blockchain network with a relational database to store both key and supplementary data, was intended to address scalability challenges in real-world deployment. In addition, the development of a web interface contributed to improving the usability and accessibility of the tool.

The proposed tool has the potential to be adapted and expanded to other sectors where decentralized data control may be essential for the management of other shared resources, such as water systems. More broadly, this approach is expected to support the development of alternative solutions using emerging technologies, particularly in Latin America, where technological innovation remains a growing opportunity.

Future developments should involve collaboration with energy companies, policymakers, and civil society organizations to test the tool in environments that more closely resemble real-world conditions. Additionally, the collected data may be useful as a foundation for developing predictive analytics models, such as energy demand forecasting, identification of infrastructure bottlenecks, and assessment of the socioeconomic impacts of distributed energy systems. Future work may also include a socioeconomic impact analysis to evaluate whether the deployment of this tool promotes more equitable and transparent access to the energy market, particularly for small-scale renewable energy producers.

In conclusion, this work contributes to the ongoing debate on how emerging technologies can contribute to advancing sustainable development goals, especially in Latin America, where fragmented energy governance and limited digital infrastructure remain key challenges.

Acknowledgements This work was funded by Universidad Centroamericana José Simeón Cañas (UCA), El Salvador, and partially supported by the Ibero-American Program of Science and Technology for Development (CYTED), through the Ibero-American Network of Blockchain and Cybersecurity (RIBCI).

References

1. Abbass, K., Qasim, M.Z., Song, H., Murshed, M., Mahmood, H., Younis, I.: A review of the global climate change impacts, adaptation, and sustainable mitigation measures. Environ. Sci. Pollut. Res. **29**(28), 42539–42559 (2022). https://doi.org/10.1007/S11356-022-19718-6
2. Calvin, K.: IPCC, 2023: Climate Change 2023: Synthesis Report. In: Core Writing Team, Lee, H., Romero, J. (eds.) Contribution of Working Groups I, II and III to the Sixth Assessment Report of the Intergovernmental Panel on Climate Change. IPCC, Geneva, Switzerland (2023). https://doi.org/10.59327/IPCC/AR6-9789291691647
3. Olabi, A.G., Abdelkareem, M.A.: Renewable energy and climate change. Renew. Sustain. Energy Rev. **158**, 112111 (2022). https://doi.org/10.1016/J.RSER.2022.112111
4. da Silva, P.P., Cerqueira, P.A., Ogbe, W.: Determinants of renewable energy growth in Sub-Saharan Africa: evidence from panel ARDL. Energy **156**, 45–54 (2018). https://doi.org/10.1016/J.ENERGY.2018.05.068
5. Ivanovski, K., Hailemariam, A., Smyth, R.: The effect of renewable and non-renewable energy consumption on economic growth: non-parametric evidence. J. Clean. Prod. **286**, 124956 (2021). https://doi.org/10.1016/J.JCLEPRO.2020.124956
6. Jeon, H.: Renewable versus non-renewable: the role of electricity generation to economic growth. Electr. J. **35**, 107140 (2022). https://doi.org/10.1016/J.TEJ.2022.107140

7. Singh, R., Akram, S.V., Gehlot, A., Buddhi, D., Priyadarshi, N., Twala, B.: Energy system 4.0: digitalization of the energy sector with inclination towards sustainability. Sensors **22**, 6619 (2022). https://doi.org/10.3390/S22176619
8. von Scherenberg, F., Hellmeier, M., Otto, B.: Data sovereignty in information systems. Electron. Mark. **34**, 1–11 (2024). https://doi.org/10.1007/S12525-024-00693-4/TABLES/4
9. Fratini, S., Hine, E., Novelli, C., Roberts, H., Floridi, L.: Digital sovereignty: a descriptive analysis and a critical evaluation of existing models. Digit. Soc. **3**(3), 1–27 (2024). https://doi.org/10.1007/S44206-024-00146-7
10. Xuan, C.: The impact of public participation on energy policy: a comparative study. Int. J. Front. Sociol. **5**, 1–7 (2023). https://doi.org/10.25236/IJFS.2023.050501
11. Judson, E., Fitch-Roy, O., Soutar, I.: Energy democracy: a digital future? Energy Res. Soc. Sci. **91**, 102732 (2022). https://doi.org/10.1016/J.ERSS.2022.102732
12. SIGET: Bulletin of Electrical Statistics 2023. San Salvador (2024)
13. Varotsos, C., Sarlis, N.V., Mazei, Y., Saldaev, D., Efstathiou, M.: A composite tool for forecasting El Niño: the case of the 2023–2024 event. Forecasting **6**, 187–203 (2024). https://doi.org/10.3390/FORECAST6010011
14. National Energy Council: National Energy Policy 2020–2050. San Salvador (2016)
15. SIGET: Bulletin of Electrical Statistics No 22. San Salvador (2022)
16. Alemán, U.: The country increased energy exports by 86%, the main regional supplier. https://diario.elmundo.sv/ampArticle/el-pais-aumento-en-86-las-exportaciones-de-energia-el-principal-proveedor-regional (2024)
17. Rosales, D.: El Salvador exported more than 160 gigawatts of energy to the region. https://revistasumma.com/el-salvador-exporto-mas-de-160-gigavatios-de-energia-a-la-region/ (2024)
18. Mora, H., Mendoza-Tello, J.C., Varela-Guzmán, E., Szymanski, J.: Blockchain technologies to address smart city and society challenges. Comput. Hum. Behav. **122**, 106854 (2021). https://doi.org/10.1016/j.chb.2021.106854
19. Varela-Guzmán, E., Aguilar-Navarro, M., Reynosa-Pérez, G.: The role of blockchain in moving towards Industry 5.0 [in Spanish]. Real. Empres. 33–42 (2024). https://doi.org/10.51378/reuca.vi19.8712
20. Mora, H., Varela-Guzmán, E., Visvizi, A., Mollá-Sirvent, R.: Introduction to a Support Model Based on Cryptocurrencies for Social Inclusion Projects. Springer Proceedings in Complexity, pp. 3–12 (2021). https://doi.org/10.1007/978-3-030-62066-0_1
21. Varela-Guzmán, E., Mora, H., Visvizi, A.: Characterization of the Citizen Profile in the Context of the Smart Society. Springer Proceedings in Complexity, pp. 175–184 (2024). https://doi.org/10.1007/978-3-031-44721-1_14
22. Varela-Guzmán, E., Melara-Gálvez, C.: Token bienestar: a blockchain-based tool to address some financial inclusion challenges. In: 7th International Congress on Ambient Intelligence, Software Engineering, and e-Health and Mobile Health, AmITIC 2024, vol. 1, pp. 1–6 (2024). https://doi.org/10.1109/AMITIC62658.2024.10747598
23. Varela-Guzmán, E.: Technology-driven management models for the development of smart societies [in Spanish] (2023)
24. Rajasekaran, A.S., Azees, M., Al-Turjman, F.: A comprehensive survey on blockchain technology. Sustain. Energy Technol. Assess. **52**, 102039 (2022). https://doi.org/10.1016/J.SETA.2022.102039
25. Treiblmaier, H.: Toward more rigorous blockchain research: recommendations for writing blockchain case studies, pp. 1–31 (2020). https://doi.org/10.1007/978-3-030-44337-5_1
26. Visvizi, A., Mora, H., Varela-Guzmán, E.: The case of rWallet: a blockchain-based tool to navigate some challenges related to irregular migration. Comput. Hum. Behav. **139**, 107548 (2023). https://doi.org/10.1016/J.CHB.2022.107548
27. Almunawar, M.N., Ordóñez de Pablos, P., Anshari, M.: Solar coin and green revolution in the oil and gas industry. In: Sustainable Development and the Digital Economy: Human-centricity, Sustainability and Resilience in Asia, pp. 140–140. Routledge, London (2023). https://doi.org/10.4324/9781003388753

28. Mihaylov, M., Razo-Zapata, I., Nowé, A.: NRGcoin—a blockchain-based reward mechanism for both production and consumption of renewable energy. In: Transforming Climate Finance and Green Investment with Blockchains, pp. 111–131 (2018). https://doi.org/10.1016/B978-0-12-814447-3.00009-4
29. Johanning, S., Bruckner, T.: Blockchain-based peer-to-peer energy trade: a critical review of disruptive potential. In: International Conference on the European Energy Market, EEM, Sept 2019. https://doi.org/10.1109/EEM.2019.8916268
30. Lee, B.S., Song, J.G., Moon, S.J., Park, I.H., Jang, J.W.: Blockchain architectures for P2P energy trading between neighbors. In: ICTC 2019—10th International Conference on ICT Convergence: ICT Convergence Leading the Autonomous Future, pp. 1013–1017 (2019). https://doi.org/10.1109/ICTC46691.2019.8939856
31. Afzal, M., Huang, Q., Amin, W., Umer, K., Raza, A., Naeem, M.: Blockchain enabled distributed demand side management in community energy system with smart homes. IEEE Access **8**, 37428–37439 (2020). https://doi.org/10.1109/ACCESS.2020.2975233
32. Khalid, R., Javaid, N., Almogren, A., Javed, M.U., Javaid, S., Zuair, M.: A blockchain-based load balancing in decentralized hybrid P2P energy trading market in smart grid. IEEE Access **8**, 47047–47062 (2020). https://doi.org/10.1109/ACCESS.2020.2979051
33. Orsini, L., Kessler, S., Wei, J., Field, H.: How the Brooklyn microgrid and transactive grid are paving the way to next-gen energy markets. In: The Energy Internet: An Open Energy Platform to Transform Legacy Power Systems into Open Innovation and Global Economic Engines, pp. 223–239 (2019). https://doi.org/10.1016/B978-0-08-102207-8.00010-2
34. Yildizbasi, A.: Blockchain and renewable energy: integration challenges in circular economy era. Renew. Energy **176**, 183–197 (2021). https://doi.org/10.1016/J.RENENE.2021.05.053
35. Juszczyk, O., Shahzad, K.: Blockchain technology for renewable energy: principles, applications and prospects. Energies **15**, 4603 (2022). https://doi.org/10.3390/EN15134603
36. Andoni, M., Robu, V., Flynn, D., Abram, S., Geach, D., Jenkins, D., McCallum, P., Peacock, A.: Blockchain technology in the energy sector: a systematic review of challenges and opportunities. Renew. Sustain. Energy Rev. **100**, 143–174 (2019). https://doi.org/10.1016/J.RSER.2018.10.014
37. Teufel, B., Sentic, A., Barmet, M.: Blockchain energy: blockchain in future energy systems. J. Electron. Sci. Technol. **17**, 100011 (2019). https://doi.org/10.1016/J.JNLEST.2020.100011

Urban Transportation, Identity and Resilience: Indigenous Perspectives in the City of Quito

Aldrin Espín-León, Edison Torres-Parra, and Jaime Almeida-Cepeda

Abstract Cities have implemented mass and inclusive transportation solutions, with underground subways being one of the most effective. However, in Quito, Ecuador, the metro only began operating in late 2023, addressing historical gaps in infrastructure but also highlighting deep social and cultural inequalities, particularly among indigenous migrants from rural communities. This study analyzes how the metro not only transformed public transportation but also became a space of convergence of diverse social realities. It explores the resilience of indigenous people and their role in promoting inclusive urban development while influencing their identity and culture. Indigenous populations that have been historically marginalized in modernization processes. A survey was conducted among indigenous and non-indigenous metro users, focusing on adaptability, discrimination, and cultural impacts. The instrument consists of 20 items across 5 categories: (1) Urban Mobility, (2) Experiences of Discrimination, (3) Resilience and Adaptation, (4) Impact on Interculturality, and (5) Participation and Empowerment. Results showed negative differences in all categories for indigenous users, with Experiences of Discrimination showing the greatest disparity. The study highlights how indigenous people use resilience strategies to maintain their identity and challenge discrimination, emphasizing the need for inclusive policies that foster respect for diversity and transform the metro into an equitable space for coexistence.

Keywords Urban transportation · Indigenous resilience · Inclusion · Cultural identity

A. Espín-León (✉) · J. Almeida-Cepeda
Career of Social Work, Universidad Central del Ecuador, Quito, Ecuador
e-mail: amespin@uce.edu.ec

J. Almeida-Cepeda
e-mail: jsalmeidac@uce.edu.ec

A. Espín-León · E. Torres-Parra
Research Department, Instituto Tecnológico Superior Sudamericano Quito, Quito, Ecuador
e-mail: edison.torres@sudamericanoquito.edu.ec

A. Visvizi et al. (eds.), *Research and Innovation Forum 2025 Vol2*, Springer Proceedings in Complexity, https://doi.org/10.1007/978-3-032-09062-1_28

1 Introduction

In modern cities, mass transit systems have been tools to promote connectivity, economic development and urban sustainability. Since the implementation of the first subway systems, as a public transportation system in the nineteenth century, it has allowed advances in infrastructure and modernity [1]. However, in cities such as Quito, Ecuador, where the underground subway, just started operations at the end of 2023, not only represents a technical improvement in mobility but also a challenge in terms of social inclusion and respect for cultural diversity, adapting through a process of persistence, transition and transformation that does not lose its essence in the face of changes [2, 3]. Quito, a city marked by deep social inequalities, especially towards indigenous people, who have faced exclusion in the processes of urbanization and modernization, because of a marked colonialist perception of society, which stigmatizes and creates the conception that indigenous peoples are not part of urban areas [4].

The underground subway emerges as a space where diverse social realities converge, becoming a scenario where diverse cultures and identities of indigenous users who have migrated to the city from their rural communities converge. In this sense, the research aims to analyze how indigenous people experience cultural inclusion, discrimination and empowerment in the Quito Metro transportation system, and how these factors affect their resilience in the face of urban transformations.

This research explores how indigenous resilience manifests and evolves in response to a modern transportation system, through an analysis that incorporates factors such as urban mobility, experiences of discrimination, cultural adaptation and participation in urban processes. Likewise, how subway transportation can be a facilitator of inclusive processes or, on the contrary, a space that perpetuates attitudes of exclusion and marginalization.

This article is structured in the following parts: after this introduction, we present a context of the research, which addresses the indigenous issue and the Metro in the city of Quito. Section 2 presents a literature review with the most relevant contributions of the scientific community in relation to the research work. Section 3 describes the methodology used, the participants of the study and the instrument applied. After the methodology, Sect. 4 shows the main results of the study. Finally, the paper concludes with Sect. 5, which contains the conclusions and recommendations of the study.

1.1 Context of the Investigation

Ecuador is a plurinational and intercultural country, and its constitution mentions guaranteeing harmonious coexistence and respect for cultural diversity [5], however, public transport does not always guarantee the recognition and representation of all cultures, classism, harassment and antipathy among users have been accentuated [6].

The Quito Metro, inaugurated in 2023, has the potential to reflect indigenous and Afro-Ecuadorian cultures, but discriminatory attitudes and the lack of cultural elements, such as signage in native languages, perpetuate social exclusion, mainly affecting indigenous communities [7]. However, discrimination in public transportation takes verbal, physical and symbolic forms. A study carried out by Puma and Redroban [8], showed that people from rural communities frequently face derogatory comments related, among other things, to their traditional dress, due to the existence, since pre-Columbian times, of traditional classism, which has left its mark on present generations.

At the structural level, the digitization of the service and the lack of specific subsidies limit access for indigenous people, reflecting economic and technological exclusion [9]. This marginalization according to Dietz [10], can be interpreted as a double discrimination, that is, the intersection between their ethnic identity and other conditions of social exclusion, discrimination from non-indigenous users as well as from the indigenous community itself, hindering access and integration to urban environments.

Public transportation brings together people from different social classes and cultural backgrounds, creating new spaces for citizen coexistence, generating interactions that, although in some cases positive, can also reveal tensions or attitudes of exclusion in society [11]. In Quito, the lack of consultation with indigenous communities and popular sectors generated tensions during the planning of the subway [12]. Public transport can foster social cohesion if it is designed with an inclusive approach. Models such as TransMilenio in the city of Bogotá have integrated cultural activities to strengthen social cohesion [13].

Early subway systems, such as those in London (1863) and New York (1904), not only improved transportation efficiency, but also transformed urban dynamics by fostering social interactions and economic growth [14]. In many Latin American cities, their implementation was delayed due to various problems such as limited economies, poor urban planning and political decisions. Projects such as the Bogotá Metro have faced multiple postponements since the 1990s due to financial problems and corruption scandals [15], likewise, the Quito Metro, after almost a decade of facing problems, did not begin operations until the end of 2023.

With an extension of 22 km and 15 stations, the Quito Metro carries 140,000 passengers a day, from the north to the south of the city (Fig. 1). Mobility project considered the second largest in the history of Ecuador—after the first railroad. Its implementation and operation pose significant challenges in terms of the inclusion of indigenous ethnic groups that have historically been marginalized. An analysis by Ortiz Castañeda [16], examines how the design of the first metro line may contribute to the social exclusion and immobility of certain sectors of the population, including indigenous communities.

Indigenous communities in Ecuador, according to the Population and Housing Census [17] represent 7% of the national population, most of them live in rural areas of the Andes and the Amazon, however, its capital, with a 4% indigenous population, continues to be seen as a point of employment opportunities, promoting migration from rural communities to the popular neighborhoods of the city.

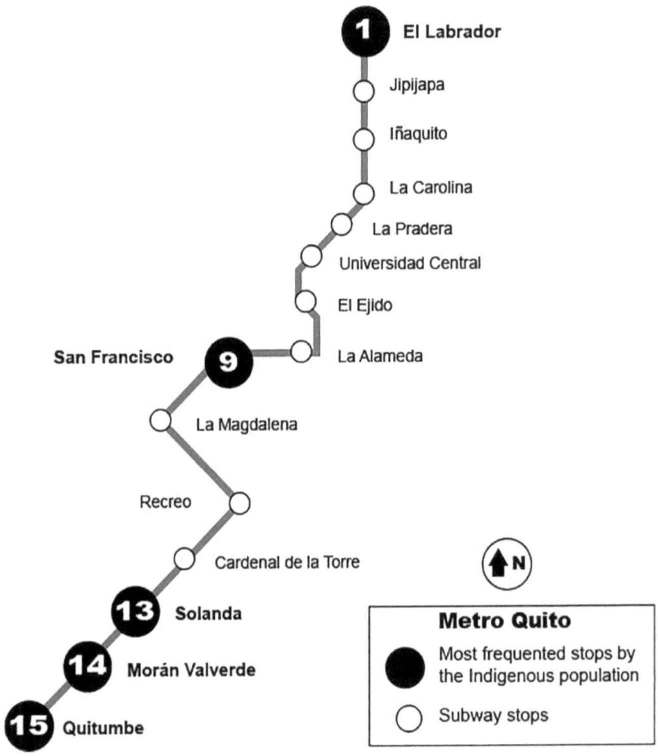

Fig. 1 Metro Quito, stations with the highest frequency of indigenous passengers. *Source* https://www.urbanrail.net/am/quito/quito.htm (public graph). Adapted by: Jaime Almeida

The objective of this study is to describe how the Quito Metro, not only transforms public transport but, through factors such as mobility, discrimination and exclusion, resilience, impact and participation, whether or not its space converges in the inclusion of indigenous people.

2 Literature Review

Urban transportation, considered a fundamental right, facilitates access to services and socioeconomic participation. The development of mass transit systems has been associated mainly with technical progress, urban connectivity and sustainability. However, several authors have warned that these advances alone do not guarantee the inclusion of historically excluded groups, such as indigenous peoples, whose cultural, symbolic and identity needs are often left out of the design of urban policies. For Littman [18], mass transit systems promote inclusive mobility, reducing urban inequalities. However, these infrastructures tend to reflect power dynamics that exclude historically marginalized groups, limiting their access to urban space. Rayner and Mérida [19], indicate that adversities reinforce the autonomy and cultural identity of the people, while Bhugra and Becker [20], emphasize that adaptation to new realities impacts the sense of belonging and individual expression.

To improve mobility and quality of life, many cities have transformed their urban transportation systems. However, these advances have had an uneven impact. According to Avellaneda [21], accessibility difficulties intensify poverty and social exclusion, showing that mobility cannot be reduced to a simple physical displacement. Rather, it is about effective access to resources, opportunities and services, where factors such as cultural identity and user resilience also play a role. Mobility and urbanization should strengthen economic, social and cultural development [22].

Social progress should not imply unequal sacrifices. For equitable transformation, participation and user-centered methodologies are key. These approaches articulate diverse communities, driving the creation of digital services [23]. In a study by Kashef et al. [24], criticize the top-down approach to technology adoption for its implications for civic rights. Rigid structures in urban models facilitate control but limit local adaptability. Low participation of the private sector, academia and civil society hinders efficient solutions for security in smart cities [24]. According to Velásquez [25], argues that accessibility must balance safety and environmental sustainability, but many planning strategies prioritize efficiency over equity. In the case of the metro, while it is seen as a modern mobility solution, fundamental social dimensions such as cultural diversity, appropriation of public space and effective participation of indigenous communities are often excluded. Despite its importance in transportation planning, the metro excludes social aspects, such as diversity and appropriation of public space [25].

Indigenous communities have historically been excluded from urban modernization. Bernal [26], points out that this exclusion is both physical and symbolic, since urban environments reproduce hegemonic discourses that make indigenous identities invisible, perpetuating discrimination and hindering their integration.

Cultural resilience which, according to Ungar [27], is the capacity to adapt to adverse contexts without renouncing one's own values and practices, in urban indigenous peoples, is expressed through strategies of social adaptation to modernism. In Quito, initiatives such as "Quito Resiliente" [28], address the structural tensions of the city, although they tend to prioritize infrastructure over inclusion and interculturalism. For Walsh [29], interculturality is fundamental to overcoming the historical exclusion that indigenous peoples face in the processes of modernization and urbanization. In the transportation sector, this should translate into inclusive policies that recognize indigenous practices, as mentioned by Pacari [30], interculturality recognizes differences in the construction of local worlds and values marginalized knowledge. Nevertheless, mass transportation systems produce unequal impacts, benefiting some sectors while others face economic and cultural barriers [31].

In urban contexts, daily commuting represents a considerable burden of time and resources, especially affecting the most vulnerable sectors residing in peripheral areas. Lizárraga Mollinedo [32], highlights that urban mobility imposes physical and emotional burdens, limiting access to employment and opportunities, inequality in transportation restricts the improvement of quality of life.

In Ecuador, recent studies have examined how indigenous migration to cities such as Quito and Guayaquil has transformed social and cultural dynamics. Indigenous migrants develop strategies to adapt to urban life, although this entails high symbolic costs, including the loss of cultural elements and exposure to new forms of discrimination, often accentuated by regionalism [33]. At the same time, the concentration of the indigenous population in certain urban areas is evidence of the need for an efficient mobility system to sustain their productive and community activities [34]. Quito's mobility policies consider four dimensions and components, as follows: (1) Leadership and Strategy, (2) Health and Wellness, (3) Economy and Society, and (4) Infrastructure and Environment towards a smart and inclusive city [35]. Important dimensions for urban development-oriented activities. However, according to the same secretariat, user resilience faces challenges due to the lack of monitoring and evaluation in its implementation. For Troisi et al. [36], a smart city must integrate surveillance technologies and wireless networks to improve security and urban sustainability.

3 Methods and Materials

3.1 Participants

The study focuses on users of the Quito Metro, taking into account the total population, which, according to data from the National Institute of Statistics and Census [17], there are 2,679,722 inhabitants of which 192,607 self-identify as indigenous, however, the vast majority of the indigenous population are not natives of the city, they are migrants from rural communities.

By the year 2024, the Quito Metro mobilized an average of 140,000 people daily; however, for the research, the study population was established as men and women over 18 years of age, differentiating those persons who self-identify as indigenous from non-indigenous persons. A total of 284 people from four main areas of the city participated in the study (Table 1).

Table 1 Participants in the study

Population		Sex		Origin (zone)			
		Males	Females	North	South	Central	Valleys
Non-indigenous	99	41	58	25	38	18	18
Indigenous	185	108	77	44	61	55	25
Total	284	149	135	69	99	73	43

Source The authors

3.2 Instrument

For this study, an instrument was developed based on five categories selected from a literature review on urban mobility, cultural resilience, interculturality and discrimination towards self-identified indigenous people and, on the basis of a general review of the reproduction of inequalities and direct and symbolic exclusions [37]. The instrument was also based on user experience, perception of urban space, and factors such as access, safety, and user well-being [38]. The structure of the instrument follows the Likert scale model, which allows the measurement of perceptions and experiences through linguistic quantifiers of frequency, frequency, and frequency [39].

The first category, "perception of urban mobility," groups items on citizens' experience with the metro and its impact on the city's connectivity. It evaluates whether the metro facilitates access to various urban areas, whether it is seen as a convenient means of transport, and whether it improves mobility options in Quito. According to Chará Ordóñez et al. [40], the perception of mobility is influenced by dissatisfaction with road infrastructure, increased travel times and lack of integration. Thus, the perceptions of indigenous and non-indigenous people will be indicators of the real impact on urban mobility and the quality of life of users. The second category, "experiences of discrimination and exclusion" analyzes how the Quito metro can reproduce inequalities. It assesses whether users experience differential treatment due to their dress or cultural appearance. According to Oviedo [41], racial hierarchies persist in urban spaces, and although there are public policies against discrimination, no significant structural changes have been achieved, allowing prejudice and exclusion to remain present in everyday life. The third category, "resilience and cultural adaptation", examines how the subway system impacts the cultural preservation and adaptive capacity of indigenous users. For Romero-Castro [42], indigenous peoples organize themselves in urban processes that have offered opportunities and difficulties in preserving their identity. The fourth category, "impact of interculturality on resilience," examines users' perceptions of cultural diversity in the Quito metro. It evaluates whether the metro promotes intercultural respect, whether it integrates cultural elements and whether it strengthens the identity of indigenous communities through coexistence. Limón Aguirre and López Bracamonte [43], highlight that shared spaces can reinforce identity and social cohesion. Finally, the fifth category "participation and empowerment" evaluates citizens' perception of their ability to influence public transport, specifically to promote the empowerment of indigenous people. For Rojas [44], empowerment seeks to transform power relations and requires full awareness of their implications, which fosters citizens' willingness to participate in community projects.

Validity of content. The instrument was validated through Lawshe's Content Validity Index—CVI, according García-Cueto et al. [45], consists of a group of experts evaluating each item, giving responses of "acceptable," "observed" or "not acceptable." Determined by the following equation:

$$CVI = \frac{n - N/2}{N/2}$$

where N is the number of experts who evaluated the item and n is the number of experts who evaluated the item as acceptable. The instrument—5 categories and 25 items—was sent for evaluation to 14 experts, including social workers, sociologists and transportation technicians of the city municipality, but only 10 of them responded. Considering the number of 10 experts, to consider an item as accepted, a CVI value of at least 0.80 is required for each item. After applying the CVI, 5 items were discarded, leaving the instrument with 5 categories and 20 items (Table 2).

Reliability of the instrument. To determine the reliability of the data and its internal consistency, the Cronbach's alpha coefficient analysis was applied. Alfa, evaluates whether the questions, elaborated with a Likert-type scale, are correlated with each other [46]. The statistics place the value between 0 and 1; the closer it is to unity, the greater the internal consistency of the items to be measured and the higher the correlation [47]. Acceptable alpha values, according to George and Mallery [48], must be greater than 0.6 whereas, for Nunnally [49], mentions a standard of at least 0.7.

The reliability of the construct, after applying Cronbach's Alpha statistic, obtained a value of 0.742 for the 20 items, a reliability value considered acceptable.

Prior to the application of the survey, user observation activities were carried out both in the streets surrounding the stations and in the commercial establishments that surrounded them to determine the stations where the greatest influx of indigenous people went. The surveys were applied in 5 of the 15 Metro stations—stations 1, 9, 13, 14, 15—and in the nearby places where there are commercial establishments and street vendors, places where their workers are migrants, many of them indigenous (Fig. 1). After reading the informed consent form and receiving it by means of a written signature, the survey was carried out by directly asking the individuals—indigenous and non-indigenous—by students of the Social Work Career of the Central University of Ecuador, while their responses were recorded in digital applications—Google Forms and KoboToolbox.

3.3 Data Analysis

Once the surveys were completed, the data were tabulated on a spreadsheet and analyzed using SPSS statistical software. For this purpose, the responses to each item were standardized with values on a scale of 0–3, with 3 being the maximum value from the positive point of view of each item, while for the categories, once the values of their corresponding items were added, values were established in a range of 0–100. During the tabulation process, it was identified that 13 of the 20 items were negatively formulated, so the scale was inverted to a range from 3 to 0.

Table 2 Instrument, categories and items

Cod	Category and items
Cat 1	**Perception of urban mobility**
PM1	Do you consider that the metro system facilitates access to different parts of the city?
PM2	Do you feel comfortable using the metro as a means of transportation in your daily life?
PM3	Do you think that the metro system significantly improves transportation options in the city?
Cat 2	**Experiences of discrimination and exclusion**
ED 4	Have you felt that other people treat you differently because of your dress or cultural appearance when using the subway?
ED 5	Do you feel that the metro does NOT sufficiently respect the cultural diversity of Quito?
ED 6	In your experience, is the metro a space where discriminatory attitudes are reproduced?
ED 7	Have you felt excluded or uncomfortable when using the metro due to the attitudes of other users?
Cat 3	**Resilience and cultural adaptation**
RA 8	Do you feel that, despite cultural differences, you can adapt to using the subway without problems?
RA 9	Do you feel that your cultural identity is not affected by using the urban transport system?
RA 10	Does using the metro allow you to find new ways to maintain your connection to your culture?
RA 11	Do you feel that you can face the challenges of the modern city without losing your cultural identity?
Cat 4	**Impact of interculturality on resilience**
IMR 12	Do you consider that the metro should be a space where Quito's cultural diversity is valued and respected?
IMR 13	Have you observed efforts to include intercultural elements in the urban transportation system?
IMR 14	Do you think that the coexistence of different cultures in the metro can strengthen the resilience of indigenous communities?
IMR 15	Do you feel empowered by sharing your culture with other users in the subway?
IMR 16	Do you think that the metro could include more elements that reflect Quito's cultural diversity?
Cat 5	**Participation and empowerment**
PYE 17	Do you feel that your voice is heard on issues related to public transportation in Quito?
PYE 18	Do you think that the metro can be a space to promote the empowerment of indigenous communities?
PYE 19	Do you feel that you have the capacity to contribute to the improvement of the urban transportation system in your city?
PYE 20	Would you actively participate in projects to improve cultural inclusion in the public transportation system?

Source The authors

4 Results

The results show that indigenous people give a higher value to Category 1—perception of urban mobility—with a value of 85.40 while they give a lower value to Category 2—experiences of discrimination and exclusion—with a value of 74.05. However, in all five categories, indigenous people have negative differences with respect to non-indigenous people, with Category 2—experiences of discrimination and exclusion—having the greatest difference with – 18.003 points. 003 points, likewise, category 3—resilience and cultural adaptation—has an important difference with a value of – 17.681 points, while Category 1—perception of urban mobility—is the category with the smallest difference, with – 1.80 points, see Fig. 2 and Table 3.

The results show that indigenous people live different and, in general, unfavorable experiences in Metro Quito, especially in terms of discrimination, exclusion and empowerment. These perceptions affect their sense of belonging and resilience. However, respect for cultural diversity showed a positive difference, suggesting that indigenous users see the system as more respectful of diversity than non-indigenous users.

When analyzing the differences of the two populations by items—values between 0 and 3—the results show that 15 of the 20 items have negative differences – 75%— while 5 have positive differences – 25%. The item with the greatest negative difference is ED 6—In your experience, the subway is a space where discriminatory

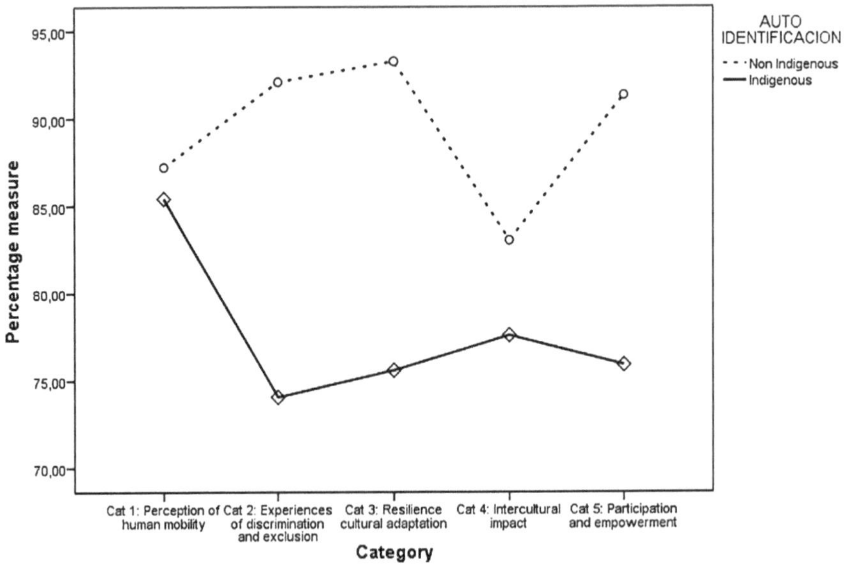

Fig. 2 Average population of transport users by category. ↔: Maximum negative difference between non-indigenous and indigenous Cat 2. O: Minimal negative difference between non-indigenous and indigenous Cat 1. *Source* Results taken from SPSS, own adaptation

Table 3 Differences in the 5 categories between non-indigenous and indigenous users

Category	Non indigenous		Indigenous		Difference
	Mean	Standard deviation	Mean	Standard deviation	(Mean)
Cat 1: Perception of urban mobility	87.207	12.331	85.407	12.631	– 1.800**
Cat 2: Experiences of discrimination and exclusion	92.088	13.330	74.055	15.452	– 18.033*
Cat 3: Resilience and cultural adaptation	93.266	11.885	75.586	14.383	– 17.681
Cat 4: Impact of interculturality	83.030	15.157	77.586	11.555	– 5.444
Cat 5: Participation and empowerment	91.330	15.609	75.900	14.194	– 15.430

* Maximum difference (Cat 2: Experiences of discrimination and exclusion)
** Minimum difference (Cat 1: Perception of urban mobility)
Source The authors

attitudes are reproduced?—with a value of – 1.6792/3 which belongs to the Category Experiences of discrimination and exclusion—Cat 2, while the item with the greatest positive difference is ED 5—Do you consider that the subway does not sufficiently respect the cultural diversity of Quito?—which belongs to the same Category Cat 2 with a value of 0.2294/3 see Fig. 3. Although the results are mostly negative, the positive value of item ED 5—inverse tabulated item—indicates that indigenous people consider that cultural diversity is respected in the metro even more than non-indigenous users.

To determine whether the differences between indigenous and non-indigenous people are statistically significant, t-tests were used to determine the differences for the 5 categories and the 20 items (Table 4).

Although all the differences in the categories are negative, the t-tests show significant differences in 4 of the 5 categories ($p < 0.05$), with Category 1—Perception of urban mobility—having a non-significant difference ($p = 0.250$, $p > 0.05$), while, in the case of the items, 13 of the 20 have significant differences, of which only 2 items ED5 and item IMP16 have positive differences with values favorable to indigenous people, items that refer to respect for cultural diversity and the incorporation of cultural elements in the Metro.

The results of the t-tests corroborate the difference between indigenous and non-indigenous users, showing that the Quito Metro transportation system is used in a differentiated manner and, for the most part, with negative manifestations. The 13 significant differences found in the items support the perception of discrimination and cultural exclusion faced by indigenous people, as well as their limited inclusion and empowerment in the urban space, validating the theoretical discussion of this

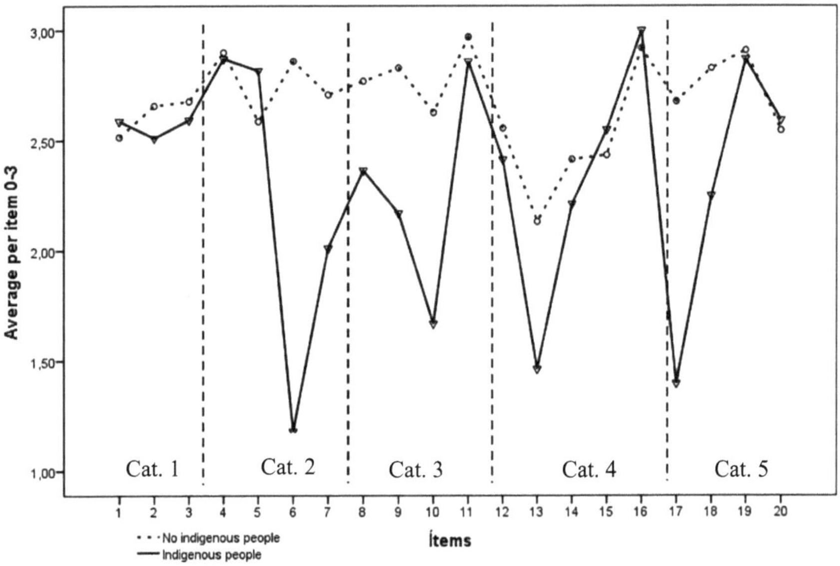

Fig. 3 Population means, for the 20 items for indigenous and non-indigenous transport users. *Source* Results taken from SPSS, own adaptation

study, which identifies these phenomena as recurrent challenges in transportation and mobility systems.

5 Discussion and Conclusions

From the literature reviewed, there is no evidence of studies focused specifically on transportation systems such as the subway and their relationship with cultural inclusion, marginalization and empowerment of indigenous users, as well as their resilience in the face of urban transformations. Cultural inclusion in public transportation is not a challenge exclusive to the city of Quito, Latin American cities that also have indigenous users such as Mexico City and Santiago de Chile face similar problems; in contrast, cities such as Toronto and Vancouver have incorporated concrete measures such as indigenous languages in signage or physical spaces for cultural expression, showing progress towards a more inclusive and diversity-sensitive urban transportation.

In the study of the Quito City Metro, In the metro, transportation users, in all five categories analyzed, reflect a negative perception of the indigenous population compared to the non-indigenous population. The indigenous perception was significantly lower in four of the five categories. Particularly noteworthy is category 2—"Experiences of discrimination and exclusion"—with a difference of − 18.03, followed by category 3—"Resilience and cultural adaptation"—with − 17.69. These

Table 4 t-tests for independent samples: by category and by item

By category (5)		t-test for equality of means				
		t	Sig. (bilateral)	Mean difference	95% confidence interval for the difference	
					Lower	Upper
Cat 1: Perception of urban mobility		1.154	0.250*	– 1.800	– 1.271	4.870
Cat 2: Experiences of discrimination and exclusion		9.819	0.000	– 18.033	14.418	21.648
Cat 3: Resilience and cultural adaptation		10.465	0.000	– 17.681	14.355	21.006
Cat 4: Impact of interculturality		3.384	0.001	– 5.444	2.277	8.611
Cat 5: Participation and empowerment		8.428	0.000	– 15.430	11.826	19.033
By ítems (20)	Cod					
Cat 1	PM1	– 1.109	0.268**	0.0686	– 0.190	0.053
	PM2	2.418	0.016	– 0.1485	0.028	0.269
	PM3	1.361	0.174**	– 0.0822	– 0.037	0.201
Cat 2	ED 4	0.445	0.657**	– 0.0287	– 0.098	0.156
	ED 5	– 3.208	0.001Δ	0.2304	– 0.372	– 0.089
	ED 6	12.750	0.000	– 1.6748	1.416	1.933
	ED 7	6.698	0.000	– 0.6909	0.488	0.894
Cat 3	RA 8	4.794	0.000	– 0.4001	0.236	0.564
	RA 9	7.063	0.000	– 0.6553	0.473	0.838
	RA 10	10.004	0.000	– 0.9560	0.768	1.144
	RA 11	2.436	0.015	– 0.1102	0.021	0.199
Cat 4	IMR 12	1.461	0.145**	– 0.1393	– 0.048	0.327
	IMR 13	5.334	0.000	– 0.6773	0.427	0.927
	IMR 14	2.245	0.026	– 0.1979	0.024	0.371
	IMR 15	– 1.532	0.127**	0.1170	– 0.267	0.033
	IMR 16	– 3.234	0.001Δ	0.0808	– 0.130	– 0.032
Cat 5	PYE 17	10.970	0.000	– 1.2714	1.043	1.499
	PYE 18	8.773	0.000	– 0.5783	0.449	0.708

(continued)

Table 4 (continued)

By category (5)		t-test for equality of means				
		t	Sig. (bilateral)	Mean difference	95% confidence interval for the difference	
					Lower	Upper
	PYE 19	0.763	0.446**	− 0.0388	− 0.061	0.139
	PYE 20	− 0.425	0.671**	0.0491	− 0.277	0.178

Calculated with alpha = 0.05, significance values less than or equal to 0.05
* Non-significant difference in category
** Non-significant difference in items
Δ Positive significant difference, favors indigenous people in cultural diversity and in-corporate cultural elements in metro
Source SPSS software report, authors' adaptation

results show that, despite existing inclusion policies, indigenous people continue to experience exclusion, difficulty of adaptation and perceptions of loss of identity in this new transportation system.

Of the 20 items evaluated, 15 show negative results and only 5 positives; of the positive ones, only 2 have significant differences that favor indigenous people over non-indigenous people. They feel that there is respect for their cultures and that the subway could include more elements that identify their identities. These perceptions could be due to the fact that, in the stations, there are visual elements that include culture, history and tradition. The indigenous people could feel represented in these visual elements.

The findings indicate that indigenous people face disadvantages in the Quito Metro in terms of integration processes, experiencing exclusion, discrimination and difficulties to adapt culturally. These experiences affect their resilience to urban transformations. The literature review contextualizes these dynamics as part of structural exclusion and highlights how transport systems can reproduce or challenge inequalities. The study highlights the need for inclusive approaches that value indigenous identities in urban policies and transportation systems.

Although there are policies on coexistence and avoiding exclusion in the Quito Metro, most of these policies were not incorporated by the indigenous cultures themselves. Cultural activities typical of the Andean indigenous cultures, such as carrying prepared food in baskets or baskets or carrying bundles of clothing on their backs, are prohibited on the metro. Policies that, with the participation of minority groups, should be reviewed in their entirety.

The study faced some limitations, such as data collection in densely populated areas and the difficulty in obtaining in-depth testimonies from indigenous people, due to factors such as distrust or historical resentments towards external actors. These difficulties were partly overcome thanks to the direct involvement of community leaders and the use of participatory methodologies that strengthened trust during the research process.

From the results obtained, it is evident the need to redesign public transportation policies from an intercultural approach that actively integrates the visions, practices and needs of indigenous peoples. This implies overcoming superficial symbolic approaches and moving towards effective mechanisms of inclusion, participation and representation. The data show that exclusion and perception of discrimination persist, so measures must be aimed at correcting these structural gaps. Recommending the implementation of strategies such as multilingual signage, cultural accessibility measures in public transportation, spaces for music, dance and indigenous folklore, as well as the inclusion of indigenous personnel as workers in administrative, operational and control areas. In addition, it is suggested to enable permanent spaces for cultural visibility within the transportation system, such as areas for music, art, visual narratives or identity symbols. Regarding the category of participation and empowerment, the creation of consultative tables with indigenous representation is proposed for the evaluation and continuous improvement of the system, as well as the promotion of labor inclusion of indigenous people in the different levels of operation of the subway, encouraging participation in the construction of mobility works. These actions would contribute to transforming the subway into a space of mutual recognition, effective participation and cultural empowerment.

6 Support

The Universidad Central del Ecuador, the Instituto Tecnológico Superior Sudamericano, the team of students of the Social Work Career of the Universidad Central and the indigenous people of different cultures from all over Ecuador who have migrated to the capital of Ecuador and who use the urban transportation system Metro de Quito.

References

1. Rose, M.H., Seely, B.E., Barret, P.F.: The Best Transportation System in the World: Railroads, Trucks, Airlines, and American Public Policy in the Twentieth Century. Ohio State University Press, Ohio, Columbus (2006)
2. Cretney, R.: Resilience for Whom? Emerging Critical Geographies of Socio-Ecological Resilience (2014). [En línea]. Available: https://doi.org/10.1111/gec3.12154
3. Meerow, S., Newell, J., Stults, M.: Defining urban resilience: a review (2016). [En línea]. Available: https://doi.org/10.1016/j.landurbplan.2015.11.011
4. Peters, E.: Emerging Themes in Academic Research in Urban Aboriginal Identities in Canada, 1996–2010 (2011). [En línea]. Available: https://journals.library.ualberta.ca/aps/index.php/aps/article/view/9242/7944
5. Asamblea Constituyente: Constitución de la República del Ecuador (2008)
6. Nazareth, N.: Arte y Movilidad: Metro de Estocolmo, 2020 Julio 2025. [En línea]. Available: https://www.duabitad.com/explorando-1/arte-y-movilidad-metro-de-estocolmo
7. Huayhua, M.: Interacción social y racismo en el transporte público peruano. In: Racismo y lenguaje, pp. 83–114 (2017)

8. Puma, J., Redroban, W.: Comunidades indígenas y su combate diario ante la discriminación 593 (2023). [En línea]. Available: https://www.593dp.com/index.php/593_Digital_Publisher/article/view/1736/1492
9. Bernal, D., Murillo, A.: The access of indigenous peoples to information and communication technologies in Colombia: inclusión or social and political exclusión? (2012). [En línea]. Available: https://revistas.uptc.edu.co/index.php/derecho_realidad/article/download/4862/3954
10. Dietz, G.: Entre Estado-Nación y Comunidad indígena: la identidad del magisterio bilingüe purhépecha en México, junio 2000. [En línea]. Available: https://www.researchgate.net/publication/330369549_Entre_Estado-Nacion_y_Comunidad_indigena_la_identidad_del_magisterio_bilingue_purhepecha_en_Mexico
11. Quito, M.: MetroCultura (2025). [En línea]. Available: https://metrodequito.gob.ec/metrocultura/
12. Rivas, N., Ramón, P.: El Metro está pasando y Solanda se está hundiendo (2015). [En línea]. Available: https://www.labarraespaciadora.com/ddhh/el-metro-de-quito/
13. Castiblanco, C.: Colectivos culturales enseñan 10 reglas para viajar seguros en TransMilenio, 27 Ago 2022. [En línea]. Available: https://bogota.gov.co/mi-ciudad/movilidad/intervenciones-de-cultura-ciudadana-en-sistema-transmilenio-en-bogota
14. Civitatis: The Oldest Subways in the World, 23 Ago 2023. [En línea]. Available: https://www.civitatis.com/blog/en/10-oldest-subways-world/?srsltid=AfmBOoqSFbi_-pPflXSeJt16_H_6L6womcU40M_23yFo4KY2wxFWOci1
15. Poole-Fuller, E.: Letras Verdes (2017). [En línea]. Available: https://doi.org/10.17141/letrasverdes.21.2017.2445
16. Ortiz Castañeda, C.A.: Exclusión social e Inmovilidad. Un análisis al diseño de la Primera Línea del Metro de Quito. Tesis de maestría, Flacso Ecuador, Quito (2022)
17. INEC: Censo Ecuador (2022). [En línea]. Available: https://censoecuador.ecudatanalytics.com/
18. Littman, T.: Transportation Planning Principles, Dstortions and Reforms, 30 Dic 2024. [En línea]. Available: https://vtpi.org/distort.pdf
19. Rayner, J., Mérida, J.: Las comunas del Ecuador: autonomía, terriotorio y la consitución del Estado plurinacional (2019). [En línea]. Available: https://editorial.iaen.edu.ec/wp-content/uploads/sites/12/2020/01/Las-comunas-del-Ecuador.pdf
20. Bhugra, D., Becker, M.: Migration, cultural bereavement and cultural identity. [En línea]. Available: https://pmc.ncbi.nlm.nih.gov/articles/PMC1414713/
21. Avellaneda, P.G.: 27 Feb 2008. [En línea]. Available: file:///C:/Users/User/Downloads/ecob,+AGUC0808220009A.PDF.pdf
22. Ryder, R., Brown, L.: Evolución de los sistemas urbanos en la frontera de la Amazonía ecuatoriana (2000). [En línea]. Available: https://doi.org/10.1111/j.1931-0846.2000.tb00352.x
23. Mora, L., Deakin, M., Reid, A.: Strategic Principles for Smart City Development: A Multiple Case Study Analysis of European Best Practices (2019). [En línea]. Available: https://napier-repository.worktribe.com/preview/2311245/Strategic%20Principles%20for%20Smart%20City%20Development.pdf
24. Kashef, M., Visvizi, A., Troisi, O.: Smart city as a smart service system: human-computer interaction and smart city surveillance systems (2021). [En línea]. Available: https://doi.org/10.1016/j.chb.2021.106923
25. Velásquez, C.: Espacio público y movilidad urbana (2015). [En línea]. Available: https://hdl.handle.net/2445/67821
26. Bernal, A.: De la exclusión a la participación. Pueblos indígenas y sus derechos colectivos en Ecuador, 30 Ene 2025. [En línea]. Available: https://digitalrepository.unm.edu/cgi/viewcontent.cgi?article=1075&context=abya_yala
27. Ungar, M.: El Social Ecología de la Resliciencia: Direcccionamiento Contextual y Cultural Ambigüedad de a Naciente Contrsuir (2011). [En línea]. Available: https://www.researchgate.net/publication/49740885_The_Social_Ecology_of_Resilience_Addressing_Contextual_and_Cultural_Ambiguity_of_a_Nascent_Construct
28. Alcaldía del Distrito Metropolitano de Quito: Resilent Quito (2017). [En línea]. Available: https://resilientcitiesnetwork.org/downloadable_resources/Network/Quito-Resilience-Strategy-English.pdf

29. Walsh, C.: Interculturalidad y (de)colonialidad: Perspectivas críticas y políticas. In: XII Congreso AIC, Florianópolis, Brasil (2009)
30. Pacari, N.: Reflexiones sobre territorios indígenas e interculutales en Estados plurinacionales (2019). [En línea]
31. Miralles-Guasch, C.: Ciudad y transporte: el binomio impoerfecto (2002). [En línea]. Available: https://books.google.com.cu/books?id=qR8wFfHovsYC&printsec=copyright#v=onepage&q&f=false
32. Lizárraga Mollinedo, C.: Movilidad urbana sostenible: un reto para las ciudades del siglo XXI (2006). [En línea]. Available: https://est.cmq.edu.mx/index.php/est/article/view/260/265
33. Espín-León, A., Jimeno-Morenilla, A., Pertegal-Felices, M., Azorín-López, J.: Effect of migration flows on the ethnic identity of the indigenous peoples of the Amazonian communities. In: Research and Innovation Forum 2020: Disruptive Technologies in Times of Change, pp. 643–659
34. Valdés, C.: Dimensiones culturales de la movilidad urbana (2011). [En línea]. Available: https://doi.org/10.4067/S0718-83582011000100004
35. Secretaría General de Planificación del Distrito Metropolitano de Quito: Informe de proyecto y de cooperación que mantiene la Dirección de Resiliencia de la Secretaría General de Planificación (2022). [En línea]. Available: https://www7.quito.gob.ec/mdmq_ordenanzas/Administraci%C3%B3n%202019-2023/Comisiones%20del%20Concejo%20Metropolitano/Desarrollo%20Econ%C3%B3mico/2022/2022-04-18/Documentos%20para%20tratamiento/Punto%202/informe_-_comisio%CC%81n_de_desarrollo_econo%CC%81mi
36. Troisi, O., Visvizi, A., Kashef, M.: Smart city as a smart service system: interacción humano-ordenador y sistemas inteligentes de vigilancia de la ciudad, Nov 2021. [En línea]. Available: https://www.sciencedirect.com/science/article/abs/pii/S0747563221002466
37. Delgado, M.: El Espacio Público Como Ideología. Universidad de la Rioja, España (2007)
38. Reicher, C.: Reclaiming Public Space Through Intercultural Dialogue. Deutschen Nationalbibliothek/LIT VERLAG GmbH & Co. KG Wien, Vienna (2019)
39. Sánchez Bruno, A., Cañadas Osinski, I.: Categorías de respuesta en escalas tipo Likert. Universidad de La Laguna (1998)
40. Chará Ordóñez, W.D., Palta Limas, M.A., López Martínez, J.C.: Percepción ciudadana, gestión pública e intervención institucional de la movilidad urbana. Summa Iuris **5**(2), 334–356 (2017)
41. Oviedo, A.O.: Ecuador: racismo y discriminación. Mundos Plur. **9**(2), 111–133 (2022)
42. Romero-Castro, M.d.R.: Pueblos indígenas: resistencia y resiliencia. Ra Ximhai **18**(5), 75–93 (2022)
43. Limón Aguirre, F., López Bracamonte, F.: Componentes del proceso de resiliencia comunitaria: conocimientos culturales, capacidades sociales. Latino Am. Cienc. Psicol. **9**(3), 14 (2017)
44. Rojas, A.D.C.: La formación de los. Redalyc **16**(55), 385–394 (2012)
45. García-Cueto, E., Suárez-Álvarez, J., Pedrosa, I.: Evidencias sobre la validez de contenido: Avances. SciELO 16 (2013).
46. Bojórquez Molina, J.A., López Aranda, L., Hernández Flores, M.E., Jiménez López, E.: Utilización del alfa de Cronbach para validar la confiabilidad de un instrumento de medición de satisfacción del estudiante en el uso del software Minitab. Microsoft Word, p. 9 (2003)
47. Welch, S., Comer, J.: Quantitative Methods for Public Administration: Techniques and Applications. Houghton Mifflin Harcourt, Boston (1988)
48. George, D., Mallery, M.: Using SPSS for Windows Step by Step: A Simple Guide and Reference. Allyn and Bacon, Boston (2003)
49. Nunnally, J.: Psychometric Methods. McGraw-Hill, New York (1978)

Kinetic Energy Harvesting from Human Footsteps: An Insight into Modern Architecture's Response to Energy Conservation

Mady A. A. Mohamed

Abstract Kinetic energy, derived from motion, manifests through the movement of objects, its magnitude increasing with the object's mass and velocity. In today's era, where the integration of design and science is central to advancing architectural and energy-related practices, both scientists and designers are working collaboratively to develop more efficient built environments. Harnessing motion within architectural systems for energy conservation presents a growing challenge in the field of Renewable Energy (RE). This paper offers a comprehensive and systematic review of contemporary approaches and motivations for capturing kinetic energy within buildings. It also underscores the significance of harvesting energy from motion through a detailed analysis of global case studies involving kinetic architecture. Drawing upon these analyses, the study presents initial findings regarding the feasibility of applying footstep-based kinetic energy technologies in a local context. The results suggest that kinetic energy holds promise not only for environmental and economic gains but also for enhancing human well-being. In particular, kinetic tiles demonstrate dual utility, generating renewable energy while also serving as tools for data collection and modelling pedestrian movement.

Keywords Kinetic energy · Kinetic tiles · Architecture · Footsteps · Renewable energy · SDG 7 · SDG 11

1 Introduction

Energy is vital in modern-day life, and the current global resources supplying this demand are declining [1]. According to the International Energy Agency (IEA) forecasting data, the global energy demand growth will exceed 25% by 2040 [2]. All known worldwide oil and gas reserves will be consumed by 2066 [3]. Oil, coal

M. A. A. Mohamed (✉)
Architecture Department, Effat University, Jeddah, Saudi Arabia
e-mail: momohamed@effatuniversity.edu.sa

Architecture Department, Faculty of Engineering, Zagazig University, Zagazig, Egypt

A. Visvizi et al. (eds.), *Research and Innovation Forum 2025 Vol2*, Springer Proceedings in Complexity, https://doi.org/10.1007/978-3-032-09062-1_29

and gas reserves will be depleted by 35, 107 and 37 years, respectively [3]. By 2050, renewable and nuclear energy will reach 30% of the energy supply. In addition, approximately 40% and 30% of the total will be supplied by coal and oil/gas, respectively. To attain the 2050 European Green Deal, the European Union (EU) has launched a "zero-emission" economy program [4]. By 2030, CO_2 emissions reduction should reach at least 50% with respect to the 1990s value [4]. To cope with this objective, the wind and solar installed energy capacities in the EU-28 countries have increased by 104.1 GW and 100.4 GW, respectively, in the last decade [5]. Moreover, the recent European Commission calls and the Paris Agreement aim to promote and reach carbon-neutral goals in stationary and automotive systems by 2050 [6]. Despite the various carbon emission international agreements during COP21 in Paris in late 2015 and the diverse annual green climate funds assigned to assist developing countries in responding to climate change programs, increased fuel demand is still witnessed in developing countries [7]. Therefore, defining solutions for reducing energy consumption or saving has become a real challenge facing worldwide decision-makers, stakeholders and governments [8]. Furthermore, passive strategies and measures that respond to either macro- or micro-climate can achieve considerable energy savings in the building industry [9–17]. Additionally, proposing new approaches for energy system policies becomes necessary to further converge towards the desired sustainable energy transitions [18–20].

Moreover, designing efficient and straightforward rating systems for green building operations to cope with the new energy agendas is due [21]. In addition, identifying renewable energy (RE) sources and solutions is a straightforward means of compensating for the escalating energy demand. Different research studies have pointed to alternative energy sources to accommodate social, environmental, political and economic fossil fuel-related problems. About 8 and 92% of global energy consumption is covered by renewable and non-renewable sources [22]. As a result, new energy generation means diverse industrial development to decrease fossil fuel needs. The consumption of industrial developments' fossil fuels has been identified as the primary reason for climate change. Different human-powered technologies have been considered for generating electricity for appliances. The widely used industrial development has resulted in environmental damage and the depletion of natural resources. The global ecosystem relevant to climate change will lead to substantial economic losses [23]. Different sorts of RE have been recently used: hydroelectric systems, solar technologies, biomass/biofuel systems, geothermal technologies and wind energy turbines [20, 24, 25]. These CO_2 emission-free energy sources have no environmental impacts. Figure 1 depicts the worldwide RE-technology installed capacity and its share percentage of the overall installed RE capacity [22].

Taking this into account, against the backdrop of the existing body of research and empirical insights, the objective of this paper is to explore the prospects of capturing kinetic energy in buildings. The latter could serve as a vital, alternative source of energy in urban contexts. To this end, the argument in this paper addresses the following research questions:

RQ1: What is the importance of extracting energy from motion?

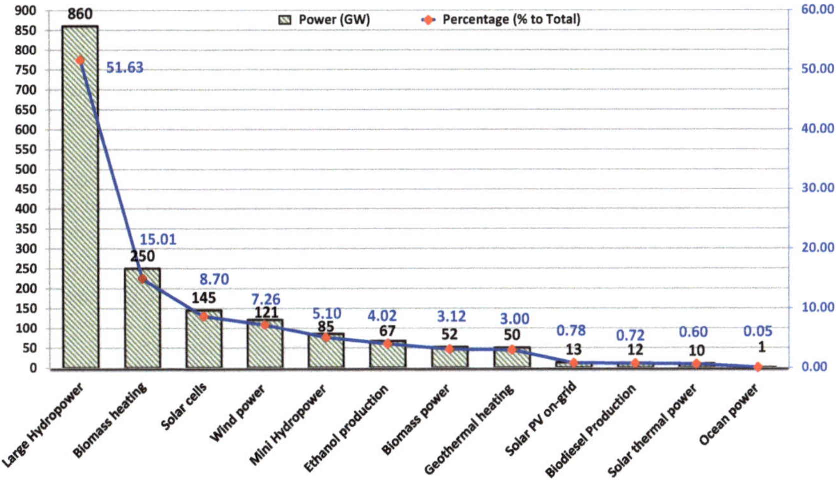

Fig. 1 The worldwide RE-technology installed capacity and its share percentage of the overall installed RE capacity. *Source* Author, based on [22]

RQ2: How is kinetic energy measured, and how is it done internationally?
RQ3: How to ensure that modern architecture adopts solutions aimed at capturing footsteps, a source of kinetic energy?

The argument in this paper will be structured as follows. Section 2 elaborates on the methods and materials. Section 3 elaborates on the kinetic energy field and the energy from footsteps. Section 4 elaborates on precedent analysis, and finally, Section 5 elaborates on a project that while implemented at the Effat University is founded on the precepts specific to the human kinetic energy paradigm.

2 Methods and Materials

The current research follows a multi-methodological approach, which includes a systematic literature review, case study analysis and field study. The Systematic Literature Review (SLR) conducted for this study employed a chronological and thematic content analysis approach covering research that evolved from material science (pre-2020) to architectural systems (post-2023). The methodology involved exploring all sourced peer-reviewed publications (books, book chapters, articles and conference papers) using academic databases such as ScienceDirect, PubMed, Springer, and MDPI. The case studies to analyze the kinetic energy measurements in several international case studies. The field study will be conducted by employing one of the spaces at Effat University and analyzing the potential extracted kinetic energy. The field study employed the survey and observation methods to collect the required data about students' movement in the space. The basketball space was observed to

Table 1 SLR versus case studies

Aspect	Systematic literature review	Case studies
Scope	Academic studies, prototypes, and experimental setups globally	Real-world implementations (e.g. Tokyo Subway, Simon Langton School, Effat University)
Objective	Analyse and categorize technologies (piezoelectric, triboelectric, electromagnetic, hybrid, etc.) and materials for footstep energy harvesting	Assess actual feasibility, user experience, and energy production in public/high-traffic environments
Methodology	Thematic and chronological content analysis across > 100 references	Observational data, measured outputs, user interaction, and spatial assessment
Application context	Primarily lab-based or prototype-focused studies	Implemented in public spaces: corridors, stations, sports facilities
Insights on efficiency	Power densities vary by technology; hybrid and kinetic tile systems offer the highest output	Energy yield varies by crowd density and usage frequency, with Pavegen tiles performing reliably underfoot

Source Author

determine the expected number of students who use the space daily. Also, a semi-structured interview was conducted to ensure the students' perception of such an idea. 30 students were involved in this survey, exercise time, sport type and spent hours.

Table 1 shows how the two methods of the current research (Systematic Literature Review "SLR" and the Case Studies) reflect the main aspects of the research.

3 Kinetic Energy

Electricity is no longer generated only in centralized power plants. Instead, it can be supplied at lower power rates at different points as distributed sources [26]. The popularity of both wearable and portable devices has mainly been witnessed. Devices such as smartwatches, emergency locators, radio frequency identifiers and health monitors consume many batteries. Replacing these batteries by harvesting wasted energy from the human body is a potential alternative, enabling the devices to be autonomous and fit-and-forget [27–29]. Kinetic energy is created in an object by motion. It can be defined as the energy needed to either slow it down or speed it up [30, 31]. The human body's movement generates kinetic energy. This movement results in generated heat, which in turn could be converted to another sort of energy, which is then used to power everything from electric lamps to radios … etc. Generally, the human body could be the ultimate alternative source of energy. Depending on the body, the latter may be the most convenient among the different energy sources [30]. Typically, bicycles can be adopted for generating energy and power hand-crank tools [23]. Yet, there are four ways of generating energy from the human body, from simple exercise to human waste, such as [32], the flow of the bloodstream, exercise, body heat, footsteps.

The flow of the bloodstream can operate a micro-turbine similar to the hydroelectric power plants [33, 34]. Pfenninger and his team tested the energy generated by bloodstream turbines and revealed that they can generate up to 800 μW of power [35]. The heart production is approximately 1 or 1.5 W of hydraulic power. The pacemaker requires around 10 μW only. Today, the generation of microturbines is limited to feeding neurostimulators, drug-delivery pumps, and blood-pressure sensors [32].

Nowadays, many people are doing exercises daily. Stationary bicycles, elliptical trainers and steppers are the main tools in any gym. Each one can generate electricity. This can give 'manpower' a whole new meaning. Different companies enable wiring the ellipticals with DC generators through inverters. Then, the power is transferred to A.C. to supply the building or for grid-connection purposes. Hooking up exercise bikes to batteries or building stationary bikes can generate energy. The 'hand cranks' and pedals can operate the flywheel linked to the generator. This latter can be connected to multiple bikes simultaneously. However, these machines have not yet produced a grid-saving amount of power. The average hourly power output can reach 150 W, while the corresponding top-level cyclist can deliver 400 W [32].

Recent research [36–40] has developed wearable fabric for generating electricity, e.g. Power Felt [33]. The flexible fabric can either/both conduct electricity or/and provide thermal insulation. Through Power Felts, body heat can be captured and reused for charging phones and other purposes. The human body can produce 100–120 W. At the same time, you might get one or 2 W out of that, which is enough for operating small electronic devices like cell phones [32].

Converting the human walk into energy can produce a lot of energy. Several companies have started to produce footstep-powered tiles for near-future energy generation. A single footstep on such tiles can produce 1–7 W of power based on how complex the step is. This power is considered enough electricity for street Light-Emitting Diode (LED) operation during the 30 s [32].

4 Energy from Footsteps

Extracting energy from foot strikes by converting the vertical linear motion of footsteps into a rotation will generate electricity from the airflow at high frequencies. The idea is achieved by converting the resulting force of the human body's weight, which is transferred to the ground by footsteps, to a proper mechanical torque. Hence, electricity can be generated by connecting this device to an electric generator. This idea suits crowded places such as airports, trains and underground stations, theatres, stadiums, universities, sports halls, churches and mosques [41].

4.1 The Technology of Energy Harvesting from Footsteps

Among the earliest piezoelectric material implementations in 1917, the SONAR system has been considered the quartz for transmitting and receiving ultrasonic waves [36]. Both Lead Zirconate Titanate (PZT) wafers and flexible multilayer polyvinylidene fluoride (PVDF) films can be used inside shoes for converting motion energy into electricity [37]. These systems, conveniently proposed for mobile computing, can provide 1.3 mW at 3 V while walking at a 0.8 Hz rate [38]. In other projects, piezoelectric films have been used for extracting electrical energy (< 5 μW) from machines' mechanical vibrations for Micro-Electro-Mechanical Systems (MEMS) devices [39]. In [36], the power extraction from machine and building vibrations has reached 70 μW. Siddappa and Ahmed developed the footstep electricity-generating device for producing 4.1 V. In [40], both commonly available piezoelectric materials, PZT and PVDF, have been impartially discussed to define the best productive material regarding the generated energy feature. According to the study, the PZT provided higher voltage than PVDF, which were about 2 V and 0.4 V, respectively. The tile has been fabricated from piezo material. The voltage was used to recharge the battery and supply the DC loads. Through an inverter, the A.C. loads can be provided.

The LCD is interfaced to the tile. The PIC microcontroller has been considered for displaying the voltage across the piezo-tile. About 40 V has been reached with a parallel/series combined connection [40]. In [42], the Foot Stop Electric Converter has enabled the production of 12 V. The lead acid battery can be used as energy storage through the AC-DC converter bridge. The authors have claimed that every footstep can produce 2.4 J (0.67 mWh). In addition, 1000 footsteps can produce around 2500 J (0.67 Wh). This energy can generate a lighting bulb with 100 W "230 V" for 25 s. The same footsteps can be appropriately used for lighting bulbs with 40 W "230 V" for the 60 s. Accordingly, the lighting of one average-size flat can include 10 lighting bulbs with 40 W "230 V" needs 600 thousand footsteps. Different researchers [27, 43–60] have investigated the application of kinetic energy in different daily functions such as staircases, dance floors, urban areas, street pavements, and sidewalks. Farouk and Abdel Gawad [41] have designed and tested a simple device that converts the kinetic energy of human footsteps to electrical energy. This device consists of a piston, Wrist Pin, Connecting Rod, Crank Shaft, Flywheel, Bearings, Springs, Top Plate, Main Frame, Support Rods, Central Support and Generator [61–65].

When someone moves and puts their foot on the top plate of the device, the LED operates effectively. After removing the foot, the top plate returns to its original position and prepares for the following footstep.

Xu and Tam [43] reviewed piezoelectric tiles' technologies and techniques to harvest energy from human footsteps. It delves into the design and mechanisms of these tiles and the electrical circuits needed to convert the harvested energy into usable electricity. The article also discusses these systems' economic feasibility and energy efficiency, along with the challenges and potential solutions for their implementation. Yue et al. [66] discussed the state-of-the-art achievements, stretchable piezoelectric nanowires, nanocomposite nanogenerators, and future possibilities of lead-free flexible energy harvesting devices. Xu and Tam presented innovative designs for piezoelectric energy harvesters that capture energy from human footsteps [43, 67]. Both designs employ a multi-stage force amplification mechanism to increase energy conversion efficiency. The first design prioritizes load-bearing capacity and protection of the piezoelectric stack, while the second design focuses on maximizing power output within a compact footprint. Both designs have been optimized through simulations and physical prototypes, demonstrating their potential for practical applications in energy harvesting from human movement.

Fu et al. in several research studies, tried to develop energy harvesting in different applications, including vibrating shoe-mounted piezoelectric cantilevers [27]. They invested in an energy harvester through a piezoelectric shoe. This device is straightforward in terms of structure but does not produce too much energy due to the low operating frequency and the small volume of piezoelectric material [27, 44]. Fu et al. proposed a piezoelectric shoe energy harvester with an air-pumped turbine system [27]. This solution can overcome the limitations of operating frequency and the limitations of vertically distorting harvesting energy from foot strike.

As a result, the airflow operates the turbine with a high frequency, which increases the produced foot-strike energy [27]. Fu et al. worked on a harvester prototype, fixing an air pump on a stainless-steel sheet of 120 mm × 60 mm × 13 mm. The air pump and a small turbine were connected by a plastic tube with a diameter of 4 mm. A Ø7 mm × 15 mm DC motor was utilized to achieve the required transduction.

The device generated 6 mW with 4.9 Ω when a 65 kg person conducts regular footsteps. The Sustainable Energy/Dance Floor (SEF/SDF) uses kinetic tiles directly applied in buildings to benefit from human footsteps. SDF uses the movement of dancers to generate electricity that can light up the floor. Dance Floor allows for different sizes, shapes and setup designs. It can produce up to 35 W, with 5–20 W for each person who drives a small generator and can light up the LED lights [45, 46].

Hwang and colleagues sought to develop a piezoelectric tile that efficiently captures energy from human footsteps [47]. They devised a system that indirectly transfers energy using springs and a weight to protect the fragile piezoelectric modules from direct impact. The goal was to align the tile's natural vibration frequency with the frequency of footsteps. Experiments showed that a 10 g weight on the module produced a resonance frequency of 22.5 Hz, matching the vibrations caused by dropping an 80 g ball from a meter high. By optimizing the electrical components, they achieved a power output of 770 μW (RMS) and 55 mW (peak).

A dynamic model was created considering friction, electrical resistance, and mechanical damping. Using this model, the researchers analyzed the pavement's Design to maximize energy harvesting from human footsteps. Experiments revealed that the paver can generate an average power of 3.6 W and a peak power of 12 W during normal walking. Each step produces an average of 1.8 J of energy. The study also explores how the flywheel and electrical load affect the amount of energy harvested. The impact of the flywheel on energy harvesting during walking, fast walking, and running is compared and analyzed. The energy-harvesting paver shows promise for use in high-traffic pedestrian areas like sports stadiums, airports, train stations, shopping malls, offices, and apartment complexes [48].

Chand et al. used a system that uses two square tiles with fluid-filled bags connected by valves and mini-hydro generators [68]. As people walk or run on these tiles, the movement of the fluid generates electricity. The researchers used computer simulations to optimize the Design, including fluid flow, power output, and bag shape. Experimental tests showed that the system can produce an average of 1.4 W of power per step during normal walking, enough to power a 390 Ω LED [49].

A new type of energy-harvesting floor tile that combines two technologies: triboelectric nanogenerators (TENG) and electromagnetic generators (EMG) was introduced Islam et al. [50]. The TENG component uses friction to generate electricity, while the EMG uses magnetic induction. The researchers tested the device under different loads and walking speeds and achieved impressive results, including a maximum open-circuit voltage of 1200 V and a short-circuit current of 5 mA. They confirmed that this hybrid nanogenerator produces 25 times more open-circuit voltage and 20% more power than a commercially sold energy-harvester floor tile, Pavegen. Wu and Xu [51] introduced a new type of underfloor energy harvester that can generate electricity from vertical and horizontal forces caused by human footsteps. Unlike traditional underfloor harvesters that only capture vertical forces, this dual-axial design utilizes a unique force amplification mechanism to increase energy output. The device efficiently converts two-dimensional forces into a single direction that the piezoelectric stack can harness. Experimental results demonstrate the effectiveness of the design, with the device capable of generating an average power of 1.25 mW and harvesting 7 mJ of energy per footstep. This innovative approach offers a sustainable and efficient way to generate electricity from human activity.

Yingyong et al. investigated how factors like weight, pace, and density of pedestrians affect the energy harvested by a piezoelectric floor tile [52]. Experiments were conducted to measure the energy output under different conditions. The study found that body weight and walking speed significantly impact energy generation, while pedestrian density has a less pronounced effect. The study found that a single footstep generated 35 mW of power. A group of 10 lighter individuals produced 11 mJ of energy per step, while a group of 10 heavier individuals generated 32 mJ. 30 people walking randomly generated 289 mJ, while 30 people running generated 736.9 mJ. 30 people walking in a line with different gaps produced similar energy outputs. The key factors influencing energy generation were pedestrian weight and pace, while pedestrian density had a minimal impact.

Alajlouni et al. [53] introduced a new method for tracking people's movement within a building using sensors embedded in the floor. The proposed ML localization method was evaluated using Monte Carlo simulations and single-occupant walking experiments for 2 different test subjects on a 16 m × 2 m instrumented floor section. Results show the superiority of the proposed method to previous RSS footstep localization methods. Asadi et al. focused on designing and testing a floor tile that can generate electricity from human footsteps [54]. The tile uses a unique design to amplify the force of footsteps and increase energy output. A detailed analysis, including simulations and experiments, was conducted to optimize the Design.

The resulting tile can generate power from various human activities, including walking, running, and jogging. A tile of 30 × 30 cm^2 is capable of generating energy while walking, running, jogging, slow running, and fast running, with power outputs of 0.57, 0.85, 1.11, and 1.42 W (0.57, 0.86, 1.11, and 1.58 mW/cm^2) respectively. It is estimated that a single footstep can produce 0.511 J of energy. The study also assesses the potential energy output and environmental benefits of using these tiles in specific locations like subway stations. In a following study by the same authors [55], they delved into the application of Halbach arrays (HAs) in electromagnetic energy harvesting systems. HAs are a specific arrangement of magnets that enhance magnetic field strength on one side. The paper provides a detailed overview of HA-based systems and their potential applications. Fu et al. introduced a new type of electromagnetic energy harvester designed to capture energy from human footsteps in high-speed rail environments [69]. The harvester uses a unique double-rocker mechanism to convert vertical motion into rotational motion, which is then used to generate electricity. A flywheel is incorporated to improve the system's stability and power output. Through simulations and experiments, the researchers demonstrated that the harvester can achieve a maximum power output of 466.6 mW under optimal conditions. The addition of the flywheel resulted in a 49.45% increase in average power output. A prototype with four harvesters was successfully tested in a real-world setting, generating a maximum voltage of 54.4 V and a power output of 1.034 W. This innovative technology has the potential to power various low-power devices and sensors in high-speed rail environments, contributing to energy-efficient and sustainable infrastructure.

Bahmanziari and Zamani [56] introduced a new design for piezoelectric energy-harvesting floor tiles, combining three mechanisms: magnetic plucking, mechanical impact, and mechanical vibration force. These mechanisms work together to maximize energy generation from human footsteps. The researchers designed and tested various tile configurations to optimize energy output, including combining all three mechanisms.

In the energy harvesting evaluations of the proposed structure, the energy and average power of 71.0 mJ and 13.6 mW were obtained at the optimal resistance load of 100 kΩ. Megdich et al. introduced a novel piezoelectric energy harvester (PEHs) design using metamaterials [57]. By employing a specific auxetic structure, a reentrant shape with carefully optimized dimensions was found to be the best choice for the development of an efficient Polyvinylidene fluoride/multi-walled carbon nanotubes (PVDF/MWCNTs) PEH with optimized composition. The unique structure of the PEH enables it to activate three distinct piezoelectric modes. When stretched, it activates d31 and d32 modes due to its negative Poisson ratio, generating a voltage of 14.8 V.

When compressed, it activates the d33 mode, producing a voltage of 28.2 V. This Design allows the harvester to charge a 1 μF capacitor to 22 V in 260 s. The feasibility of the structure is further validated by its successful 3D printing using FDM technology. PEH can efficiently convert mechanical energy into electricity through multiple modes of operation, including stretching and compression. This versatility enables the device to generate significant electrical energy, even from low-intensity forces like human footsteps. The researchers have successfully fabricated a prototype using 3D printing and demonstrated its ability to power a small alarm system. As a result, from the above, it can be concluded that all human activities, such as walking, exercising, dancing, and sports activities. These sorts of energy could be converted to electricity, stored, or used directly to charge mobile phones, lighting bulbs, electrical tools and wearable devices.

4.2 Systematic Literature Review

References were grouped into core thematic categories based on three classifications:

1. The research fields include Architectural Integration, Global Energy Trends, Piezoelectric Fundamentals, Footstep Harvesting, and Material Innovations.
2. The primary energy conversion mechanism include Piezoelectric, Triboelectric, Electromagnetic, Hybrid Systems, and Emerging Materials.
3. Overview of energy harvesting technologies, summarising power outputs and researchers associated with each technology.

Each group was analyzed for methodological trends, technological innovations, and application scenarios. Finally, the thematic categories were ordered chronologically based on the earliest appearance of the cited works, allowing for an examination of how the research focus has shifted over time. This methodology was selected to ensure comprehensive coverage of foundational and cutting-edge contributions while maintaining analytical clarity through thematic consistency.

Table 2 presents the classification of the research field, including: Architectural Integration, Global Energy Trends, Piezoelectric Fundamentals, Footstep Harvesting, and Material Innovations.

The thematic analysis of energy harvesting technologies reveals distinct patterns in performance capabilities across different technological approaches. When examining the power output comparison chart, clear performance tiers emerge that align with our thematic groupings. Technologies within the **Architectural Integration** category demonstrate remarkable variation, with Kinetic Tiles producing the highest output (35 W) while Dual-Axial Underfloor Harvesters generate merely 0.00125 W. This significant disparity highlights how implementation methodology dramatically impacts energy capture efficiency within built environments.

The **Footstep Harvesting** category shows promising results across multiple approaches, with Hybrid TENG and EMG systems (6 W) representing a particularly successful combination of harvesting mechanisms. This aligns with findings from the **Material Innovations** theme, where hybrid material systems consistently demonstrate synergistic effects. The electromagnetic-based systems, including the EM Harvester with Flywheel (0.47 W), occupy the middle tier of performance, suggesting that electromagnetic induction represents a reliable energy conversion approach for moderate power needs.

Table 2 Research field thematic groups, structured by key findings, citations, and methodology

Thematic group	References	Key findings	Methodology/ technique
Architectural integration	[51, 52, 54, 58, 59]	5–20 W/person in dance floors; crowd-flow data utility; kinetic tiles show highest output (35 W) among all technologies	Field measurements, implementation case studies, performance monitoring
Global energy trends	[3, 60, 70, 70–77]	Oil/coal/gas depletion in 35/107/ 37 years; growing importance of decentralized energy harvesting	Modelling, statistics analysis, trend forecasting
Policy/ regulation	[60, 78, 78–85]	Carbon neutrality targets by 2050; emerging standards for energy harvesting technologies in building codes	Policy analysis, regulatory framework assessment, implementation studies
Piezoelectric fundamentals	[86–88]	PZT (2 V/step) > PVDF (0.4 V/ step); piezoelectric tiles achieve 35–55 mW; spring-mass systems enhance output	Lab testing, material characterization, comparative analysis
Footstep harvesting	[27, 41, 69]	6 mW via air-pumped turbine system; hybrid systems showing output of 6 W; flywheel systems reaching 466.6 mW	Prototyping, real-world implementation, efficiency testing
Material innovations	[27, 63, 89–99]	Lead-free devices achieve 85% PZT efficiency; metamaterials enhance conversion rates; hybrid material systems show synergistic effects	Synthesis, materials characterisation, performance testing

Source Author

Technologies utilizing **Piezoelectric Fundamentals** generally produce lower outputs, ranging from 0.006 W (Piezoelectric Shoe with Air Pump) to 0.055 W (Piezoelectric Tiles with Spring-Mass). While these outputs appear modest compared to technologies like Kinetic Tiles, their implementation versatility makes them suitable for specialized applications where space constraints or material flexibility are prioritized over raw power generation. The logarithmic scale of the chart effectively illustrates the nearly five orders of magnitude difference between the highest and lowest-performing technologies, underscoring the importance of matching harvesting technology to specific application requirements as highlighted in the **Policy/Regulation** analysis. This performance data provides crucial empirical evidence that can guide implementation decisions for sustainable energy solutions aligned with the carbon neutrality targets identified in our **Global Energy Trends** analysis. Table 3 presents the classification of the literature based on the primary energy conversion mechanism, including Piezoelectric, Triboelectric, Electromagnetic, Hybrid Systems, and Emerging Materials.

The final synthesis of over 100 references offers a holistic view of the technological landscape and research trends in footstep-based energy harvesting systems. The review reveals a progressive shift from single-mechanism designs to hybrid and multifunctional systems, reflecting a broader trend in sustainable micro-energy harvesting towards maximizing efficiency and adaptability.

Piezoelectric systems emerged as the most studied and implemented technology in early research due to their ease of integration and material availability. Studies from 2015 through 2023 demonstrate an incremental improvement in energy output, material durability, and design optimization, often targeting pedestrian traffic zones and wearable applications. However, limitations in energy density and mechanical fatigue prompted diversification into alternative mechanisms.

Triboelectric nanogenerators (TENGs) have gained significant traction in the past five years, characterized by their lightweight nature, high voltage output, and compatibility with flexible substrates. This thematic area is still emerging but rapidly expanding, with strong applications in smart surfaces and ambient environment harvesting. Research also highlights the synergy between triboelectric effects and surface engineering, demonstrating how material choice can amplify output by over 200%.

Electromagnetic systems were prominently featured in both foundational and recent works, often used in hybrid configurations due to their suitability for lower-frequency, high-force movements like human steps. Halbach arrays, enhanced coil geometries, and ferrofluid-based designs have allowed these systems to compete with or exceed piezoelectric outputs under specific conditions.

Combining piezoelectric and electromagnetic mechanisms, hybrid systems have become the gold standard in experimental setups, aiming to harness complementary advantages such as voltage generation from piezo elements and current delivery from electromagnetic components. These systems promise to power off-grid sensors, public infrastructure, and low-power electronics in urban environments.

Finally, emerging materials such as MXenes, 2D nanomaterials, and layered ferroelectrics have opened new frontiers in energy harvesting. These materials offer superior mechanical compliance and tunable electrical properties, crucial for developing flexible and miniaturized harvesters. While still under experimental development, their potential to revolutionize the energy harvesting landscape is undeniable.

The thematic review also uncovered a methodological trend: early studies relied heavily on conceptual modelling and small-scale prototyping, while more recent works integrate multiphysics simulations, real-world field testing, and material science insights. This reflects a maturing field moving from proof-of-concept to deployment-ready technologies.

In conclusion, the systematic review demonstrates that footstep energy harvesting is a vibrant, interdisciplinary research area poised for real-world impact. The chronological-theoretical mapping not only tracks the evolution of ideas but also identifies underexplored directions—such as energy storage integration and long-term reliability studies—that warrant future investigation. This review lays a solid

Table 3 Primary energy conversion mechanism thematic groups, structured by key findings, citations, and methodology

Thematic grouping	References	Main finding	Methodology/techniques used
Footstep energy harvesting	[31, 35, 40, 65, 46, 50, 100–103]	General footstep energy harvesting approaches explore conceptual, experimental, and hybrid techniques for converting human motion into electricity	Experimental or conceptual analysis
General renewable energy	[32–34, 41, 37, 39, 63, 64, 64, 67, 43, 45, 49, 51, 53, 56, 104, 105]	Diverse renewable energy harvesting systems are analyzed, setting the context for kinetic harvesting relevance	Experimental or conceptual analysis
Electromagnetic systems	[48, 106, 107]	Electromagnetic harvesters use coils and magnets to convert mechanical movement into electrical energy, often in hybrid designs	Experimental or conceptual analysis
Piezoelectric systems	[41, 38, 54, 55, 69, 89, 92, 96, 108–110, 110, 111]	Piezoelectric materials are widely used in footstep tiles due to their high responsiveness and ease of integration	Experimental or conceptual analysis, lab experiments, analytical review

(continued)

Table 3 (continued)

Thematic grouping	References	Main finding	Methodology/techniques used
Triboelectric systems	[77, 97, 112, 112–114]	Triboelectric systems provide high voltage output using friction-based mechanisms, useful in wearable or floor-based harvesters	Material studies, conceptual review
Electromagnetic systems	[59, 59, 115–118]	Electromagnetic harvesters use coils and magnets to convert mechanical movement into electrical energy, often in hybrid designs	Simulation, experimental validation
Hybrid energy systems	[103, 114]	Hybrid systems combine mechanisms to improve efficiency and adapt to multiple input conditions	Device design, integration studies
Emerging materials and technologies	[107, 111, 112, 117]	Emerging materials like MXenes and 2D compounds offer novel energy harvesting capabilities and miniaturization potential	Material analysis, experimental setup

Source Author

foundation for advancing both academic inquiry and practical innovation in sustainable energy technologies. Table 4 presents an overview of power outputs energy from harvesting technologies.

To provide a fair and meaningful comparison, power outputs for all technologies must be normalized to watts per square meter (W/m^2). This approach accounts for each device's actual area and reveals the true energy harvesting potential in real-world installations. Power values reported in diverse units, such as joules (J), millijoules (mJ), microwatts (μW), volts (V), and step-based energy values were converted into a common unit of milliwatts (mW). Where outputs were expressed per step (e.g., 7 mJ/step), an average walking rate of 1 step/s (1 Hz) was assumed, resulting in the conversion formula Power (mW) = Energy (mJ/step) $\times$ 1 Hz. For voltage and current outputs (e.g., 1200 V and 5 mA), the power was calculated using the standard formula $P = V \times I$, with all units adjusted to SI equivalents. Where original power was reported in watts, the values were multiplied by 1000 to convert to mW. This harmonized approach allows for a direct, fair comparison across technologies in terms of their effective power generation potential in everyday use scenarios.

Table 5 and Fig. 2 present the calculated power output across energy harvesting technologies per square meter after normalization.

Table 4 Categorized overview of energy harvesting technologies, summarising power outputs and researchers associated with each technology

	Technology	References	Power output
1	Dual-axial underfloor harvester	[51]	1.25 mW, 7 mJ/step
2	EM harvester with flywheel	[59]	466.6 mW max
3	Floor-embedded sensors	[53]	Not quantified
4	Fluid-filled tiles + mini-hydro	[49]	1.4 W per step
5	Force amplification tiles	[54, 55]	0.57–1.42 W
6	Hybrid TENG and EMG	[50]	1200 V open-circuit, 5 mA short-circuit
7	Kinetic tiles	[104]	Up to 35 W
8	Mechanical device	[41]	Not quantified
9	Metamaterial-based PEHs	[57]	28.2 V
10	Multi-mechanism tiles	[56]	71.0 mJ, 13.6 mW
11	Multi-stage force amplification	[67, 43]	Not quantified
12	Piezoelectric floor tiles	[52]	35 mW per step
13	Piezoelectric shoe with air pump	[27, 44]	6 mW
14	Piezoelectric tiles	[100]	Not quantified
15	Piezoelectric tiles with spring-mass	[47, 87]	770 μW RMS, 55 mW peak
16	Stretchable piezoelectric nanowires	[89]	Not quantified

Source Author

Table 5 Power output across energy harvesting technologies per square meter

Technology	Power output (W)	Conversion	W/m^2
Kinetic tiles	35.0	• Power output: 2–8 W per step, up to 35 W when dancing/jumping 12456 • Tile size: 0.75 m × 0.75 m = 0.5625 m^2 • W/m^2 (walking): • 2 W/0.5625 m^2 ≈ 3.6 W/m^2 • 8 W/0.5625 m^2 ≈ 14.2 W/m^2 • W/m^2 (dancing/jumping): • 35 W/0.5625 m^2 ≈ 62.2 W/m^2	3–15 up to 60
Fluid-filled tiles + mini-hydro	1.4	• Power output: 1.4 W per step • Assumed tile size: ~ 0.5 m × 0.5 m = 0.25 m^2 (typical for floor tiles) • W/m^2: 1.4 W/0.25 m^2 = 5.6 W/m^2	5.6
Force amplification tiles	1.0	• Power output: 0.57–1.42 W • Assumed tile size: 0.25 m^2 • W/m^2: 1.42 W/0.25 m^2 = 5.68 W/m^2	5.68
Hybrid TENG and EMG	6.0	• Power output: 6 W (from chart) • Assumed device area: 0.25 m^2 • W/m^2: 6 W/0.25 m^2 = 24 W/m^2	24
EM harvester with flywheel	0.4666	• Power output: 0.47 W • Assumed device area: 0.25 m^2 • W/m^2: 0.47 W/0.25 m^2 = 1.88 W/m^2	1.88
Multi-mechanism tiles	0.0136	• Power output: 0.0136 W • Assumed tile size: 0.25 m^2 • W/m^2: 0.0136 W/0.25 m^2 = 0.054 W/m^2	0.054
Piezoelectric floor tiles	0.035	• Power output: 0.035 W per step • Assumed tile size: 0.25 m^2 • W/m^2: 0.035 W/0.25 m^2 = 0.14 W/m^2	0.14
Piezoelectric shoe w/air pump	0.006	• Power output: 0.055 W • Assumed tile size: 0.25 m^2 • W/m^2: 0.055 W/0.25 m^2 = 0.22 W/m^2	0.22
Piezoelectric tiles w/spring-mass	0.055	• Power output: 0.006 W • Shoe area: ~ 0.03 m^2 (sole of a shoe) • W/m^2: 0.006 W/0.03 m^2 = 0.2 W/m^2	0.2
Dual-axial underfloor harvester	0.00125	• Power output: 0.00125 W • Assumed device area: 0.25 m^2 • W/m^2: 0.00125 W/0.25 m^2 = 0.005 W/m^2	0.05

Source Author

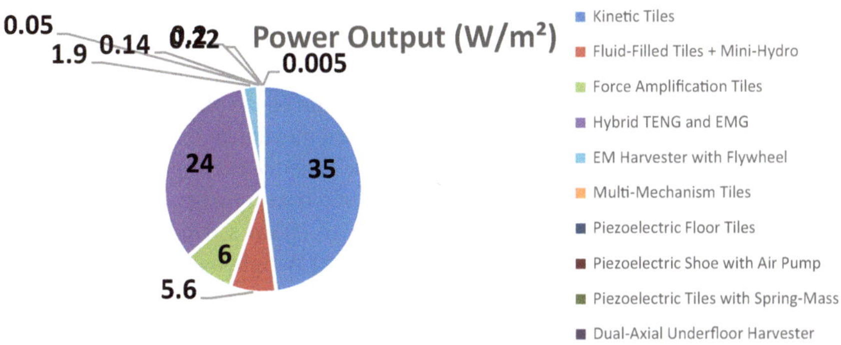

Fig. 2 Power output across energy harvesting technologies per square meter. *Source* Author

- Kinetic Tiles and Hybrid TENG/EMG systems deliver the highest power densities, especially under dynamic, high-traffic, or dancing conditions, reaching up to 60 W/m^2 12456.
- Fluid-Filled Tiles and Force Amplification Tiles provide moderate power densities (2–6 W/m^2), suitable for regular pedestrian areas.
- Piezoelectric and Multi-Mechanism Tiles generally yield lower power densities (< 0.25 W/m^2), indicating their role is best suited for low-power sensor applications or where only minimal energy harvesting is feasible.
- Dual-Axial Underfloor Harvesters have negligible output per area, highlighting a significant gap compared to newer, more efficient technologies.
- For large-scale, high-impact energy harvesting in public spaces, Kinetic Tiles and Hybrid TENG/EMG systems are the most promising, while piezoelectric-based solutions are best reserved for niche, low-power applications.

4.3 Precendent Analysis

Companies try to disseminate the knowledge and spread the technology of kinetic tiles all over the world. Pavegen is one of these companies that implemented the idea in many projects, such as Hillsborough North in San Francisco, BI Group Kazakhstan in Abu Dhabi, Uniqlo Heattech in London and the Schneider Paris marathon in Paris [46]. It managed to implement it in tens of case studies around the world. Two real case studies have been selected to study the estimated production of energy from the footstep's renewable energy source. These are the Tokyo Subway Station in Japan and the Simon Langton School in the USA.

4.4 Tokyo Subway Station

Tokyo Station is considered the busiest subway station on the planet. Hence, the East Japan Railway Company benefited from the crowd movement inside the station by generating energy using a piezoelectric floor. The generated energy is used to power display systems and ticket gates. The floor is covered with rubber and stone tiles to cover around 6 m^2. and produces daily electricity of 10 kW s. This produced energy is equivalent to the required electricity to light a 100 W light bulb for around 2 min.

4.5 Simon Langton School

Schools are one of the places that have daily foot traffic, especially in their corridors. There are many ideas to apply this idea to encourage the students and increase their interactions with the tiles. Such as making a game, providing valuable information or making the tile generate some sounds when they move above it. At Simon Langton School, located in Kent, they installed 24 kinetic tiles in its corridors. They encouraged the students to run and jump, not just walk on the tile, to generate energy and have fun simultaneously. The generated energy from the tiles is used in many applications, such as lighting the corridors.

The school guides the students and teachers by explaining the processes and the information in diagrams on the walls. The school is expected to generate 100 W of power during busy times. As a result, over a year, the 24 tiles generated power that can charge 853 cell phones or power a phone call for 2.5 years [104]. This energy could also charge an electric car to move for 7 miles.

Many building types and places worldwide can generate lots of energy from footsteps. These buildings include museums, stadiums, schools, airports, universities, underground stations, and train stations. Specific crowded places and projects worldwide can benefit particularly from generating energy from footsteps. Holy religious places such as Makkah and the Vatican are the most crowded places, where many people, who may reach millions, gather at times of the year. Also, some other tourist places, such as the Louvre Museum and Hong Kong Disneyland, are visited by millions of people regularly.

5 Application Proposals

Religious places such as the grand mosque in Makkah in Saudi Arabia, and St. Peter's Square can generate big quantities of power due to the high number of visitors to these places. In Makkah, people march long distances when performing their worship activities. One advantage of Makkah is that it is crowded for the entire year, 365 days. At the same time, the number of pilgrims can reach 3–4 million. Only

Al Kaaba "Tawaaf", where the people walk around the great house all the time, can accommodate 105,000 people [109].

There are several historical monuments in St. Peter's Square, such as the Swiss Guard, St. Peter's Basilica, and the statues of both Peter and Paul, among other vendors. It is worth mentioning here that religious people visit St. Peter's Square, and non-religious people travel to this place to enjoy the work of Raphael, Bernini, and Michelangelo, the most famous Renaissance artists. According to the Catholic Herald [108], about four million Catholics visited this place in 2016.

Squares are also considered hectic places. Times Square, New York City, is the most crowded square in the US. This square is always crowded during the day and year. According to the official website of Square, more than 300,000 pedestrians enter the heart of the square every day [102].

The Louvre in Paris, in France, has hundreds of the most well-known art pieces in the world. These art pieces include Eugène Delacroix's "Liberty Leading the People" and Leonardo Davinci's "The Mona Lisa". According to the museum's website, 9.6 million people visited in 2018 [105].

Disney Lands and entertainment places worldwide are also considered very busy places at all times. A famous example is Hong Kong Disneyland, considered one of the most crowded parks [108]. The Hong Kong Park has more than 7,400,000 annual visitors. The park increased its size by 25% to meet the demand over the last three years. More than 750,000 tourists and commuters pass through the Grand Central Terminal in the USA every day [108], where a train arrives every 58 s, 65 shops and many restaurants.

The number of steps is estimated based on the number of visitors, assuming each step takes one second, to calculate the generated energy in these proposed places. Based on analyzing the real case studies, it has been concluded that each step will generate 7 J. The number of steps is multiplied by 60 s and 60 min to find the produced energy in one hour. The total energy for the entire year is calculated based on the estimated duration of time for each visitor. After that, the total energy is compared to the required energy to operate the lighting for one public building with an average Gross Floor Area (GFA) of 33,000 m^2 that needs about 100 kWh for illumination [9]. Table 6 shows the estimated energy from the proposed energy generated from footsteps. Exact numbers of visitors were found for each hour in one case, while for each day for others, and for the whole year for the others. Therefore, each case's estimated generated energy was calculated based on the available information.

5.1 Power Output Comparison

Power Output Gap: Laboratory-tested or industrial tile systems (like Pavegen) under dynamic use (dancing, jumping) can produce up to 60 W/m^2, while real-life stationary-use settings like Tokyo's subway yield well below 1 W/m^2.

Table 6 The estimated generated energy for the proposed crowded places from footsteps

No.	Place name	Number of visitors	Produced energy per day	Produced energy per year	No. of operated the lighting in public buildings that need 100 kWh
1	"Al Kaaba" The Great Mosque	Al Kaaba is occupied by 105 thousand people all the time, and this means that the 105 thousand stay for 24 h	17,640 mWh	6438 mWh	64,000 buildings
2	St. Peter. Square	4 million visitors/year Assumption: each visitor stays 5 h in St. Peter's Square	–	140 mWh	1400 buildings
3	Times Square	300 thousand every day Assumption: each visitor stays 5 h in Times Square	10,500 kWh	3832.5 mWh	38,320 buildings
4	The Louvre	9.6 million/year Assumption: each visitor stays 5 h in the Louvre Museum	–	336 mWh	3360 buildings
5	Hong Kong Disneyland	7.4 million/year Assumption: each visitor stays 5 h in Times Square	–	259 mWh	2590 buildings
6	USA Grand Central Terminal	750 thousand/day Assumption: each visitor stays one hour on average in Grand Central Terminal	5250 kWh	1916 mWh	19,160 buildings

Source Author

Use Frequency and Behaviour Matter: Real-world cases highlight usage behaviour (walking vs. jumping) and density of foot traffic as critical to achieving high output. For instance, Simon Langton School's high output relied on active engagement (running, jumping), not just casual walking.

Technology Maturity: The literature highlights hybrid and triboelectric systems as future-focused with higher voltage/current output potential. However, these are often still experimental. In contrast, case studies use commercial or simplified piezoelectric systems for reliability and cost control.

Space Efficiency: The literature provides output in W/m^2, while case studies report total daily/annual Wh or J, making area-normalized performance comparisons essential. Literature-normalized values far exceed most case implementations.

While the literature review suggests significant energy harvesting potential, especially with hybrid and kinetic tile systems, the case studies reveal the limitations of real-world implementations due to user behaviour, scale, and installation conditions. While lab results report impressive figures, real case studies like Simon Langton and Effat University deliver more modest but still useful energy outputs, mainly suitable for low-power applications like LED lighting or phone charging. This contrast reinforces the importance of site-specific feasibility and user engagement strategies in maximizing the practical value of kinetic energy harvesting.

5.2 *Local Proposal: Human Kinetic Energy in Effat University*

Effat University, Jeddah, Saudi Arabia, has been chosen for the field study. Since the energy generated from the footsteps depends on the number of people and the frequency and strength of the steps, basketball in the sports facility zone was selected for the study. This hall is being used for several activities throughout the year, including exhibitions that will help disseminate the idea of kinetic energy and increase the electricity generated. The basketball space was observed to determine the expected number of students who use the space daily. Figure 3 shows Effat University Basketball.

Also, a semi-structured interview was conducted to ensure the students' perception of such an idea. 30 students were involved in this survey, exercise time, sport type and spent hours. The questionnaire was designed to capture students' demographic information (age, weight), exercise routines (timing and duration), types of sports played, and perceptions of energy expenditure. The aim was to analyze the relationship between exercise habits (especially time and type of sport) and the potential for human kinetic energy conversion to electricity. The survey specifically sought

Fig. 3 Effat University Basketball, captured by the author

to identify which days and times students exercised most, which sports were most energy-intensive, and the typical duration of exercise sessions across the week. The questionnaire includes 4 categories and 10 questions as follows:

Section 1: Demographics

What is your age?
What is your weight?

Section 2: Exercise Habits

On which days of the week do you exercise?
At what time do you usually exercise each day?
How many hours do you typically exercise per day?

Section 3: Sport Type

What type(s) of sport do you participate in? (Select all that apply)
Basketball, Badminton, Running, Other (please specify)

Section 4: Energy and Effort

On which days do you spend the most hours exercising?
Which sport do you feel requires the most energy?
How many hours do you spend exercising on Monday and Wednesday?
How many hours do you spend exercising on other days?

(a) Time

The more hours they exercise, the more electricity is generated. The students used to do basketball exercises from 11:00 a.m. to 2:00 p.m. on Mondays and Wednesdays. On the other days of the week, between 11:00 a.m. and 3:00 p.m.

(b) Sport type

The different sports activities that students do in basketball are basketball, badminton, and running. The sport that can generate more energy is basketball.

(c) Hours

The survey revealed that the average hours used were between 1 and 3 h, Mondays and Wednesdays, but between 1 and 2 h on the other days. This is because Effat University has two hours on Mondays and Wednesdays for activities (from 12:30 to 2:30 p.m.). Based on the precedents studied, the expected electricity generated has been calculated (Table 7).

Assumptions

1. Number of users

From the observation and the survey, it has been noted that the average number of students who use the space frequently is:

Table 7 The summary of the survey results

Category	Details	Notes
Students age	From 21 to 23 years	
Weight	From 40 to 70 kg	
Exercise time (Mon and Wed)	11:00 a.m.–2:00 p.m.	More hours of exercise correlate with more electricity energy work
Exercise time (other days)	11:00 a.m.–3:00 p.m.	More hours of exercise correlate with more electricity energy work
Sport type	Basketball, badminton, running (basketball highest energy)	Basketball spends more energy, best for electricity energy
Average hours (Mon and Wed)	1–3 h	Monday and Wednesday have more hours of exercise
Average hours (other days)	1–2 h	Other days have fewer

Source Author

6 students from 8:0 to 10:0 a.m.
15 students from 10:0 a.m. to 2:0 p.m.
4 students from 2:0 to 4:0 p.m.

To calculate the average number of students per hour, each student's number will be multiplied by the number of hours, and the total will be divided by the total number of hours. ((6 * 2) + (15 * 4) + (4 * 2))/8 = This gives an average of 10 students/h.

2. Pavegen tiles will cover the whole floor area.
3. Each student will have one step/s.
4. Each step will generate 7 J (based on the case study analysis).

Expected Generated Power in each hour = 10 Person * 60 s * 60 min * 7 J per step = 252 kJ, which equals 70 Wh. In "Simon Langton School", they mentioned that each 100 Wh, can charge 853 cell phones or power a phone call for 2.5 years. This means that the generated energy from the basketball, "70 Wh", can charge 597 cell phones or power a phone call for 1.75 years. It can also lighten two saving bulbs 35 W each. However, the cumulative energy production from the basketball could be calculated for an average of two hours daily on the three semesters (around 300 days) to be as follows: 70 W * 8 h * 300 days = 168 kWh. Suppose the university increases basketball use by renting it and using it at different times throughout the year. In that case, it can generate a large amount of electricity.

6 Conclusion and Recommendations

An object's kinetic energy is its energy due to its motion. It is characterized as the work needed to accelerate a body of a given mass from rest to its expressed speed. Having picked up energy during its acceleration, the body keeps up this active energy except if its speed changes. The body finishes a similar measure of work while decelerating from its current speed to a condition of rest. Moreover, kinetic energy can be transferred among items and changed into different sorts of energy. Kinetic energy is considered a new energy source that could benefit humanity's environment, economy, and well-being.

The review of literature indicates a growing interest in harvesting kinetic energy from human footsteps as a potential source of renewable energy. Various technologies and designs have been explored to capture and convert this energy into usable electricity. Early research focused on piezoelectric materials, such as PZT and PVDF, to convert mechanical stress into electrical energy. Studies have explored their application in shoes and for harvesting energy from vibrations. However, these systems often suffer from low energy output. More recent studies have explored innovative designs to improve energy conversion efficiency. These include:

- Multi-stage force amplification mechanisms to increase energy conversion.
- Air-pumped turbine systems to overcome limitations of operating frequency.
- Hybrid nanogenerators that combine triboelectric and electromagnetic effects.
- Dual-axial harvesters that capture both vertical and horizontal forces.
- Designs that incorporate springs, weights, and flywheels to optimise energy capture.
- Fluid-filled bags connected to mini-hydro generators.
- Metamaterial-based piezoelectric energy harvesters.

Researchers have also investigated the factors affecting energy harvesting performance, such as pedestrian weight, walking speed, and density. The findings suggest that these factors can significantly impact the amount of energy generated. The literature highlights the potential applications of this technology in various settings, including:

- High-traffic pedestrian areas (airports, train stations, shopping malls)
- Dance floors
- Urban areas, street pavements, and sidewalks
- Buildings.

Overall, the literature suggests that kinetic energy harvesting from human footsteps holds promise as a sustainable and renewable energy source. However, further research and development are needed to improve energy conversion efficiency, reduce costs, and address challenges related to practical implementation.

Reviewing the case studies that implemented the idea of kinetic energy using footsteps, it has been concluded that each step can give up to 7 J. A local study on Effat University's basketball showed that covering the floor with Pavegen tiles can generate "70 Wh" that can charge 597 cell phones or power a phone call for 1.75 years or light two, saving electrical bulbs. Kinetic tiles are valuable for generating energy and helping with data collection. This is useful for crowd-flow modelling, seeing how people move through cities, events, exhibitions, and touristic activities.

Analyzing the estimated energy generated from footsteps in crowded places proved that this renewable energy source can cover the required energy for lighting hundreds and thousands of buildings in each place. It is recommended that this renewable energy source be implemented in stadiums, trains and metro stations, malls, museums, squares, commercial streets, and religious and tourist places. The entrances of campuses, parks, commercial, service and public buildings could be a big energy source.

- A wide range of technologies is being explored for harvesting energy from human footsteps, including piezoelectric materials, mechanical devices, electromagnetic generators, and fluid-based systems.
- Significant progress has been made in improving energy conversion efficiency through design, materials, and mechanisms innovations.
- Practical applications are being explored in various settings, including sidewalks, dance floors, and transportation systems.
- Challenges remain in areas such as cost-effectiveness, scalability, and integration into existing infrastructure.

Gaps

- Only 8% of references address tropical/humid environments.
- No studies on vandalism resistance in public installations.
- Emerging Trend: Integration with IoT for real-time crowd analytics (2024–2025 references).

Acknowledgements The author would like to thank the students Asmaa Ghurab, Smaher Qari, Rozana Alkhzami, Magferah Alasmman, and Jawan Hassanan—Effat University, Jeddah—for their contribution in the first phase of this work during their Energy and Design Class at Effat University.

References

1. Khater, E.-S., et al.: Development of a bio-solar house model for Egyptian conditions. Energies **13**(4) (2020). https://doi.org/10.3390/en13040817
2. Yanine, F., et al.: Grid-tied distributed generation systems to sustain the smart grid transformation: tariff analysis and generation sharing. Energies **13**(5) (2020). https://doi.org/10.3390/en13051187
3. Shafiee, S., Topal, E.: When will fossil fuel reserves be diminished? Energy Policy **37**(1), 181–189 (2009)
4. Brodny, J., Tutak, M.: Analyzing similarities between the European union countries in terms of the structure and volume of energy production from renewable energy sources. Energies **13**(4) (2020). https://doi.org/10.3390/en13040913

5. Gailani, A., et al.: On the role of regulatory policy on the business case for energy storage in both EU and UK energy systems: barriers and enablers. Energies **13**(5) (2020). https://doi.org/10.3390/en13051080
6. Wulff, N., et al.: Comparing power-system and user-oriented battery electric vehicle charging representation and its implications on energy system modeling. Energies **13**(5) (2020). https://doi.org/10.3390/en13051093
7. Durrans, B., Whale, J., Calais, M.: Benchmarking a sustainable energy engineering undergraduate degree against curriculum frameworks and pedagogy standards from industry and academia. Energies **13**(4) (2020). https://doi.org/10.3390/en13040822
8. Mohamed, M.: Saving energy through using green rating system for building commissioning. Energy Procedia (Spec. Issue Emerg. Renew. Energy Gener. Autom.) **162**, 369–378 (2019). https://doi.org/10.1016/j.egypro.2019.04.038
9. Mohamed, M.: Green building rating systems as a design target for public building—case study analysis. Prog. Ind. Ecol. Int. J. (PIE) **11**(2), 118–134 (2017). https://doi.org/10.1504/PIE.2017.10004936
10. Mohamed, M.: Green building rating systems as sustainability assessment tools: case study analysis. In: Bastante-Ceca M.J. (ed.) Sustainability Assessment. IntechOpen (2019). https://doi.org/10.5772/intechopen.87135
11. Mohamed, M., et al.: Guidelines to utilize the wind catcher "Malqaf" technique in Jeddah's new residential buildings. In: 7th Annual International Conference on Architecture and Civil Engineering ACE 2019. Global Science and Technology Forum (GSTF), Singapore (2019). https://doi.org/10.5176/2301-394X_ACE19.506
12. Mohamed, M., Abdel-Megeed, S., Hussein, H.: Influence of smart technology on IEQ of office buildings. In: The 1st Memaryat International Conference MIC 2017; Architecture of the Future, Challenges and Visions. Universal Publisher Co. Publisher & Distributor, Beirut, Lebanon, Jeddah, KSA (2017)
13. Mohamed, M.: Investigating the environmental performance of government primary schools in Egypt: with particular concern to thermal comfort. In: Dundee School of Architecture, p. 442. Dundee University, Dundee (2009)
14. Mohamed, M., Gado, T., Osman, M.: Investigating the intelligence of the low-tech earth architecture of the Sahara: a feasibility study from the western desert of Egypt. Intell. Build. Int. (IBI) **2**(2), 179–197 (2010). https://doi.org/10.3763/inbi.2010.0038
15. Mohamed, M., et al.: Quantifying the effectiveness of mass proportions and the orientation for buildings on thermal performance in Tebessa, Algeria. IOP Conf. Ser. Earth Environ. Sci. **397**, 012008 (2019). https://doi.org/10.1088/1755-1315/385/1/012050
16. Luo, Y., et al.: Active building envelope systems toward renewable and sustainable energy. Renew. Sustain. Energy Rev. **104**, 470–491 (2019). https://doi.org/10.1016/j.rser.2019.01.005
17. Omrany, H., et al.: Application of passive wall systems for improving the energy efficiency in buildings: a comprehensive review. Renew. Sustain. Energy Rev. **62**, 1252–1269 (2016). https://doi.org/10.1016/j.rser.2016.04.010
18. Chandel, S.S., Sharma, A., Marwaha, B.M.: Review of energy efficiency initiatives and regulations for residential buildings in India. Renew. Sustain. Energy Rev. **54**, 1443–1458 (2016). https://doi.org/10.1016/j.rser.2015.10.060
19. Robertson Munro, F., Cairney, P.: A systematic review of energy systems: the role of policymaking in sustainable transitions. Renew. Sustain. Energy Rev. 109598 (2019). https://doi.org/10.1016/j.rser.2019.109598
20. Benedek, J., Sebestyén, T.-T., Bartók, B.: Evaluation of renewable energy sources in peripheral areas and renewable energy-based rural development. Renew. Sustain. Energy Rev. **90**, 516–535 (2018). https://doi.org/10.1016/j.rser.2018.03.020
21. Li, J., Wang, Q., Zhou, H.: Establishment of key performance indicators for green building operations monitoring—an application to China case study. Energies **13**(4) (2020). https://doi.org/10.3390/en13040976
22. Quaranta, E., Revelli, R.: Gravity water wheels as a micro hydropower energy source: a review based on historic data, design methods, efficiencies and modern optimizations. Renew. Sustain. Energy Rev. **97**, 414–427 (2018). https://doi.org/10.1016/j.rser.2018.08.033

23. AENews: Human Power: News and Information About Human Powered Kinetic Energy Technology (2019). Available from: http://www.alternative-energy-news.info/technology/human-powered/
24. Mohammed, Y.S., et al.: Existing and recommended renewable and sustainable energy development in Nigeria based on autonomous energy and microgrid technologies. Renew. Sustain. Energy Rev. **75**, 820–838 (2017). https://doi.org/10.1016/j.rser.2016.11.062
25. Behrouzi, F., et al.: Global renewable energy and its potential in Malaysia: a review of hydrokinetic turbine technology. Renew. Sustain. Energy Rev. **62**, 1270–1281 (2016). https://doi.org/10.1016/j.rser.2016.05.020
26. Taddeo, P., et al.: Management and activation of energy flexibility at building and market level: a residential case study. Energies **13**(5) (2020). https://doi.org/10.3390/en13051188
27. Fu, H., et al.: Energy harvesting from human motion using footstep-induced airflow. J. Phys. Conf. Ser. **660**(1), 012060 (2015). https://doi.org/10.1088/1742-6596/660/1/012060
28. Prakash, S., Agarwal, A.: Human power harvesting. Int. J. Eng. Res. Technol. (IJERT) **3**(2), 1950–1953 (2014)
29. Gokul, S.V., Lakshmi, M.S., Swetha, T.: Design of RF energy harvesting antenna using optimization techniques. Int. J. Eng. Res. Technol. (IJERT) **9**(3), 595–601 (2020)
30. Sorbothane: Kinetic Energy: What It Is and How Sorbothane Is Used, 15 May 2019. Available from: https://www.sorbothane.com/kinetic-energy.aspx
31. Wani, S.A.: Kinetic energy recovery system for vehicle suspension. Int. J. Eng. Res. Technol. (IJERT) **9**(5), 106–109 (2020)
32. Savenije, D.: 5 Ways You Can Use the Human Body to Generate Electricity (2014). Cited 2019
33. Poulter, R., et al.: Quantified dual energy computed tomography perfusion imaging using myocardial iodine concentration: validation using CT derived myocardial blood flow and invasive fractional flow reserve in a porcine model. J. Cardiovasc. Comput. Tomogr. **13**(2), 86–91 (2019). https://doi.org/10.1016/j.jcct.2019.01.020
34. Abdelmageed, M.G., Fath El-Bab, A.M.R., Abouelsoud, A.A.: Design and simulation of pulsatile blood flow energy harvester for powering medical devices. Microelectron. J. **86**, 105–113 (2019). https://doi.org/10.1016/j.mejo.2019.02.021
35. Pfenniger, A., et al.: Performance analysis of a miniature turbine generator for intracorporeal energy harvesting. Artif. Organs **38**(5) (2014). https://doi.org/10.1111/aor.12279
36. Siddapa, S., Ahmed, S.A.: Design and analysis of power generating tiles. Int. J. Mech. Eng. Robot. (IJMER) **4**(1) (2016)
37. Kymissis, J., et al.: Parasitic power harvesting in shoes. In: 2nd IEEE International Conference on Wearable Computing (1998)
38. Shenck, N.S., Paradiso, J.A.: Energy scavenging with shoe-mounted piezo electrics. IEEE Micro **21**(3), 30–42 (2001)
39. Glynne-Jones, P., Beeby, S.P., White, N.M.: Towards a piezoelectric vibration-powered micro generator. In: IEE Proceedings-Science, Measurement and Technology (2001)
40. Boby, K., et al.: The footstep power generation using piezo electric transducers. Int. J. Eng. Innov. Technol. (IJEIT) **3**(10) (2014)
41. Farouk, A., Abdel Gawad, A.: Utilization of human footsteps for power generation. In: The First Conference of Recent Trends in Energy Systems (RTES), Banha, Egypt (2015)
42. Taliyan, S.S., et al.: Electricity from Footsteps. In: BARC Newsletter (2010)
43. Xu, Q., Tam, L.M.: Design of a piezoelectric energy harvester based on two-stage force amplification frame, chap. 4. In: Xu, Q., Tam, L.M. (eds.) Mechanical Design of Piezoelectric Energy Harvesters, pp. 67–91. Academic Press (2022). https://doi.org/10.1016/B978-0-12-823364-1.00012-2
44. Zhao, B., et al.: A review of piezoelectric footwear energy harvesters: principles, methods, and applications. Sensors (Basel) **23**(13) (2023). https://doi.org/10.3390/s23135841
45. Abdalla, A., Eleshy, A., Eltantawy, A.: Life energy architecture: crowd farms as human power plants (main entrance for Mansoura University "El-Baron Gate"). Energy Procedia **115**, 272–289 (2017). https://doi.org/10.1016/j.egypro.2017.05.025

46. Pavegen: Pavegen creates high engagement with citizens by converting their footsteps into energy, data and reward (2020). Cited 02-01-2020. Available from: https://pavegen.com/author/admin/
47. Hwang, S.J., et al.: Designing and manufacturing a piezoelectric tile for harvesting energy from footsteps. Curr. Appl. Phys. **15**(6), 669–674 (2015). https://doi.org/10.1016/j.cap.2015.02.009
48. Liu, M., et al.: Design, simulation and experiment of a novel high efficiency energy harvesting paver. Appl. Energy **212**, 966–975 (2018). https://doi.org/10.1016/j.apenergy.2017.12.123
49. Chand, A.A., et al.: Design simulation of a novel fluid based footstep energy harvesting system. Sustain. Energy Technol. Assess. **39**, 100708 (2020). https://doi.org/10.1016/j.seta.2020.100708
50. Islam, E., et al.: Electromagnetic-triboelectric-hybrid energy tile for biomechanical green energy harvesting. Nano Energy **77**, 105250 (2020). https://doi.org/10.1016/j.nanoen.2020.105250
51. Wu, Z., Xu, Q.: Design and testing of a new dual-axial underfloor piezoelectric energy harvester. Sens. Actuators A Phys. **303**, 111858 (2020). https://doi.org/10.1016/j.sna.2020.111858
52. Yingyong, P., et al.: Evaluation of harvesting energy from pedestrians using piezoelectric floor tile energy harvester. Sens. Actuators A Phys. **331**, 113035 (2021). https://doi.org/10.1016/j.sna.2021.113035
53. Alajlouni, S.E., Baker, J., Tarazaga, P.: Maximum likelihood estimation for passive energy-based footstep localization. Mech. Syst. Signal Process. **163**, 108158 (2022). https://doi.org/10.1016/j.ymssp.2021.108158
54. Asadi, M., Ahmadi, R., Abazari, A.M.: Footstep-powered floor tile: design and evaluation of an electromagnetic frequency up-converted energy harvesting system enhanced by a cylindrical Halbach array. Sustain. Energy Technol. Assess. **60**, 103571 (2023). https://doi.org/10.1016/j.seta.2023.103571
55. Asadi, M., Ahmadi, R., Abazari, A.M.: Halbach magnet arrays in electromagnetic kinetic energy harvesters: a review. Energy Convers. Manag. X **22**, 100544 (2024). https://doi.org/10.1016/j.ecmx.2024.100544
56. Bahmanziari, S., Zamani, A.-A.: A new framework of piezoelectric smart tiles based on magnetic plucking, mechanical impact, and mechanical vibration force mechanisms for electrical energy harvesting. Energy Convers. Manag. **299**, 117902 (2024). https://doi.org/10.1016/j.enconman.2023.117902
57. Megdich, A., et al.: Enhanced piezoelectric performance of PVDF/MWCNTs energy harvester through a 3D-printed multimodal auxetic structure for smart security systems. Mater. Today Sustain. **27**, 100847 (2024). https://doi.org/10.1016/j.mtsust.2024.100847
58. Sekhar, A.C., Kishore, B.M., Raju, T.J.: Electromagnetic foot step power generation. Int. J. Sci. Res. Publ. **4**(6) (2014)
59. Cao, H., et al.: An electromagnetic energy harvester for applications in a high-speed rail pavement system. Int. J. Mech. Sci. **243**, 108018 (2023). https://doi.org/10.1016/j.ijmecsci.2022.108018
60. United Nations: Paris Agreement Under the United Nations Framework Convention on Climate Change. Conference of the Parties (COP21). United Nations, Paris (2015)
61. You, J., et al.: Constructal design and experimental validation of a non-uniform heat generating body with rectangular cross-section and parallel circular cooling channels. Int. J. Heat Mass Transfer **148**, 119028 (2020). https://doi.org/10.1016/j.ijheatmasstransfer.2019.119028
62. Wang, Y., et al.: Wearable thermoelectric generator to harvest body heat for powering a miniaturized accelerometer. Appl. Energy **215**, 690–698 (2018). https://doi.org/10.1016/j.apenergy.2018.02.062
63. Park, H., et al.: High power output based on watch-strap-shaped body heat harvester using bulk thermoelectric materials. Energy **187**, 115935 (2019). https://doi.org/10.1016/j.energy.2019.115935

64. Nozariasbmarz, A., et al.: Thermoelectric generators for wearable body heat harvesting: material and device concurrent optimization. Nano Energy **67**, 104265 (2020). https://doi.org/10.1016/j.nanoen.2019.104265
65. Hyland, M., et al.: Wearable thermoelectric generators for human body heat harvesting. Appl. Energy **182**, 518–524 (2016). https://doi.org/10.1016/j.apenergy.2016.08.150
66. European Commission: The European Green Deal: Delivering the EU's 2030 Climate Targets. Brussels (2021)
67. Xu, Q., Tam, L.M.: Design of a piezoelectric energy harvester based on multistage force amplification frame, chap. 6. In: Xu, Q., Tam, L.M. (eds.) Mechanical Design of Piezoelectric Energy Harvesters, pp. 115–140. Academic Press (2022). https://doi.org/10.1016/B978-0-12-823364-1.00003-1
68. U.S. Congress: Energy Independence Act of 2024. U.S. Congress, Washington, DC (2024)
69. Fu, Y., et al.: A novel multimodal piezoelectric energy harvester with rotating-DOF for low-frequency vibration. Energy **278**, 129878 (2023). https://doi.org/10.1016/j.energy.2023.129878
70. United Nations Framework Convention on Climate Change: National Inventory Submissions 2020. Bonn (2020)
71. International Energy Agency: World Energy Outlook 2022. OECD/IEA, Paris (2022)
72. European Commission: EU Energy Report 2023. Publications Office of the European Union, Brussels (2023)
73. International Renewable Energy: Renewable Capacity Statistics 2023. Abu Dhabi (2023)
74. BP: BP Statistical Review of World Energy 2024. London (2024)
75. IPCC: Climate Change 2024: Impacts, Adaptation, and Vulnerability. Cambridge University Press (2024)
76. U.S. Energy Information Administration: Annual Energy Outlook 2025. Washington, DC (2025)
77. Min, Z., et al.: The dynamic relationship among technological innovation, international trade, and energy production. Front. Environ. Sci. **10** (2023). https://doi.org/10.3389/fenvs.2022.967138
78. United Nations: Transforming Our World: The 2030 Agenda for Sustainable Development. United Nations, New York (2015)
79. European Commission: The European Green Deal. European Commission, Brussels (2021)
80. United Nations: Sustainable Energy Policy Framework. United Nations, New York (2022)
81. Energy Harvesting Standards Organization: International Standards for Energy Harvesting Devices. Energy Harvesting Standards Organization, Geneva (2023)
82. European Commission: Smart City Policy Initiative: Urban Innovation and Sustainability. European Commission, Brussels (2023)
83. European Commission: COP21 Follow-Up: Progress and Policy Updates. European Commission, Brussels (2024)
84. International Code Council: Building Code Integration Guidelines for Energy Harvesting. International Code Council, Washington, DC (2024)
85. International Energy Agency: Zero Energy Building Regulations and Guidelines. International Energy Agency, Paris (2024)
86. Zhu, D., et al.: Buckled MEMS beams for energy harvesting from low frequency vibrations. Research **2019**, 1087946 (2019). https://doi.org/10.34133/2019/1087946
87. Hwang, J.: Three-layered piezoelectric energy harvesting tile with spring-mass mechanism (2020)
88. Siddappa, L., Ahmed, S.: A review of piezoelectric energy harvesting tiles: available designs and performance. Energy **215**, 119153 (2020)
89. Yue, R., et al.: A review of flexible lead-free piezoelectric energy harvester. J. Alloys Compd. **918**, 165653 (2022). https://doi.org/10.1016/j.jallcom.2022.165653
90. Li, Z., Wang, Z.L.: Flexible nanogenerators for wearable electronic applications based on piezoelectric and triboelectric mechanisms. Nano Today **36**, 101016 (2021). https://doi.org/10.1016/j.nantod.2020.101016

91. Zhou, Y., et al.: Enhanced ferroelectric and piezoelectric properties in graphene/ $Pb(Zr_{0.52}Ti_{0.48})O_3/SrRuO_3$ capacitors. ACS Appl. Mater. Interfaces **14**(15), 17987–17994 (2022). https://doi.org/10.1021/acsami.2c02277
92. Kim, J., Zhang, Y.: Quadrant-electroded nanogenerators for decoupling piezoelectricity and flexoelectricity. Nano Energy **104**, 107987 (2023). https://doi.org/10.1016/j.nanoen.2022.107987
93. Kim, S., Zhang, Y.: Polyvinylidene fluoride-based nanofiber utilizing electrospinning for flexible nanogenerators. Molecules **29**(15), 3576 (2023). https://doi.org/10.3390/molecules29153576
94. Li, J., et al.: The versatility of piezoelectric composites. Annu. Rev. Mater. Res. **53**, 1–26 (2023). https://doi.org/10.1146/annurev-matsci-080921-092839
95. Zhang, X., et al.: Blue light emitting piezoelectric few-layered borophene nanosheets for flexible nanogenerators. Commun. Mater. **4**, 375 (2023). https://doi.org/10.1038/s43246-023-00375-2
96. Liu, H., et al.: Porous flexible molecular-based piezoelectric composite achieves near-milliwatt power density. Nat. Commun. **15**, 53031 (2024). https://doi.org/10.1038/s41467-024-53031-9
97. Liu, Y., et al.: Enhanced piezoelectric nanogenerator based on tridoped graphene and Ti_3CNT_x MXene. Ind. Eng. Chem. Res. **63**(36), 15853–15868 (2024). https://doi.org/10.1021/acs.iecr.4c00987
98. Wang, X., et al.: Biodegradable flexible triboelectric nanogenerator for winter sports. Front. Mater. **11**, 1400258 (2024). https://doi.org/10.3389/fmats.2024.1400258
99. Yue, Y., et al.: Flexible nanogenerators based on enhanced flexoelectricity in Mn_3O_4 membranes. Small **20**(12), 2307167 (2024). https://doi.org/10.1002/smll.202307167
100. Sharma, S., et al., A review of piezoelectric energy harvesting tiles: available designs and future perspective. Energy Convers. Manag. **254**, 115272 (2022). https://doi.org/10.1016/j.enconman.2022.115272
101. Jung, S., Yoon, S.: Changes in sunlight and outdoor thermal environment conditions based on the layout plan of flat type apartment houses. Energies **8**(9), 9155–9172 (2015). https://doi.org/10.3390/en8099155
102. Kazmierski, A.F.: Times Square—New York. INSIDER, New York (2020). Available from: https://www.insider.com/the-most-crowded-places-on-earth-2017-10#forbidden-city-beijing-china-8
103. Wang, T.: Hybrid piezoelectric-electromagnetic energy harvesters. Rev. Sci. Instrum. (2023)
104. Kemball-Cook: Kinetic Tiles—Case Study: Schools (2015). Cited 2019. Available from: https://kinetictiles.wordpress.com/category/uncategorized/
105. Louvre Museum: (2019). Available from: https://www.louvre.fr/
106. Qian, F., et al.: Equipment operation and maintenance management of Shanghai power distribution network after power system reform. Energy Procedia **152**, 1182–1187 (2018). https://doi.org/10.1016/j.egypro.2018.09.156
107. Zhang, J.: Advances in translational electromagnetic harvesters. Energies (2023)
108. DiNuzzo, E.: 20 places that are just way too crowded (2017). Available from: https://www.insider.com/the-most-crowded-places-on-earth-2017-10
109. Makkah AlMukarama, W.: The third stage of the Tawwaf extension, in Al-Riadh (2014)
110. Shabara, H.M.: Energy flow in piezoelectric energy harvesters. Sensors (2023)
111. Demiss, B.: 2D cobalt telluride as a piezo-tribo generator. arXiv (2021)
112. Wang, Z.L.: Advances in triboelectric nanogenerators. Tribol. Int. (2023)
113. Li, X.: Rotating triboelectric nanogenerators for motion energy harvesting. Nanogener. J. (2023)
114. Chen, Y.: Review of electromagnetic-triboelectric hybrid nanogenerators. Nano Energy (2021)
115. Yang, B.: Electromagnetic vibrational energy harvesting: a review. Sensors (2022)
116. Zhang, M.: MXene-based energy harvesters. Energy Mater. (2023)
117. Zhang, M.: MXene-enhanced energy harvesters. Energy Mater. (2023)

118. Liu, R.: Design of piezoelectric and electromagnetic harvesters for human motion. Sens. Actuators A (2024)

Author Index

A. Visvizi et al. (eds.), *Research and Innovation Forum 2025 Vol2*, Springer Proceedings in Complexity, https://doi.org/10.1007/978-3-032-09062-1